Self-Organization and the City

Springer
Berlin
Heidelberg
New York
Barcelona
Hong Kong
London
Milan
Paris
Singapore
Tokyo

Springer Series in Synergetics

An ever increasing number of scientific disciplines deal with complex systems. These are systems that are composed of many parts which interact with one another in a more or less complicated manner. One of the most striking features of many such systems is their ability to spontaneously form spatial or temporal structures. A great variety of these structures are found, in both the inanimate and the living world. In the inanimate world of physics and chemistry, examples include the growth of crystals, coherent oscillations of laser light, and the spiral structures formed in fluids and chemical reactions. In biology we encounter the growth of plants and animals (morphogenesis) and the evolution of species. In medicine we observe, for instance, the electromagnetic activity of the brain with its pronounced spatio-temporal structures. Psychology deals with characteristic features of human behavior ranging from simple pattern recognition tasks to complex patterns of social behavior. Examples from sociology include the formation of public opinion and cooperation or competition between social groups.

In recent decades, it has become increasingly evident that all these seemingly quite different kinds of structure formation have a number of important features in common. The task of studying analogies as well as differences between structure formation in these different fields has proved to be an ambitious but highly rewarding endeavor. The Springer Series in Synergetics provides a forum for interdisciplinary research and discussions on this fascinating new scientific challenge. It deals with both experimental and theoretical aspects. The scientific community and the interested layman are becoming ever more conscious of concepts such as self-organization, instabilities, deterministic chaos, nonlinearity, dynamical systems, stochastic processes, and complexity. All of these concepts are facets of a field that tackles complex systems, namely synergetics. Students, research workers, university teachers, and interested laymen can find the details and latest developments in the Springer Series in Synergetics, which publishes textbooks, monographs and, occasionally, proceedings. As witnessed by the previously published volumes, this series has always been at the forefront of modern research in the above mentioned fields. It includes textbooks on all aspects of this rapidly growing field, books which provide a sound basis for the study of complex systems.

Juval Portugali

Self-Organization and the City

With a Foreword by Hermann Haken

Includes chapters in collaboration
with I. Benenson, I. Omer and N. Alfasi

Two special chapters on "Synergetic Cities"
with Hermann Haken

With 120 Figures

Springer

Professor Juval Portugali, Ph.D

Department of Geography
and the Human Environment
Tel Aviv University
P.O. Box 39040, Ramat Aviv
Tel Aviv 69978, Israel
E-Mail: juval@post.tau.ac.il

Library of Congress Cataloging-in-Publication Data

Portugali, Juval.
Self-organization and the city / J. Portugali ; with a foreword by Hermann Haken.
p. cm. – – (Springer series in synergetics, ISSN 0172-7389)
Includes bibliographical references and index.
ISBN 3-540-65483-6 (hardcover : alk. paper)
1. Cities and towns. 2. City planning. 3. Self-organizing systems
I. Title. II. Series.
HT119.P68 1999
307.76 – – dc21

ISSN 0172-7389

ISBN 3-540-65483-6 Springer-Verlag Berlin Heidelberg New York

Typesetting: F. Herweg, Leutershausen
Cover design: *design & production*, Heidelberg
Printed on acid-free paper SPIN 10520418 55/3143AT 5 4 3 2 1 0

Foreword

Cities first came into existence more than five thousand years ago. How to deal with these partly fascinating, partly frightening creatures of mankind, both practically and intellectually, concerns all of us and, in particular, presents a real challenge to city planners. Each historical epoch has had its own particular attitudes associated with the "Zeitgeist". Accordingly, the planning and steering of cities were based on quite different criteria. But in spite of these differences, the concept of planning and steering was, and still is, the cornerstone of our dealing with cities. Nevertheless, the planner's dilemma is becoming more and more visible: cities and megacities seem to be unplannable.

In this book, Juval Portugali introduces a new idea: Cities are self-organizing systems. To substantiate his revolutionary concept, he uses several interlinked methods. On the one hand, and to my own delight, he employs in his arguments theoretical tools developed in the interdisciplinary field of synergetics. On the other hand, jointly with his co-workers, he has performed detailed model calculations on cellular nets. It has been a great pleasure and a wonderful experience for me to discuss these concepts with Juval Portugali over a number of years. I was repeatedly and deeply impressed by the way he established profound and often surprising links to other fields of science. Reading this book has fascinated me and I am sure that this fascination will also be felt by both professionals and laymen. Indeed, the text will provide the reader with new and deep insights. I am convinced that this book by my friend Juval Portugali will become a deserved success.

Stuttgart
April 1999 *Hermann Haken*

Preface

Cities are *par excellence* complex systems. This property of cities was always recognized and the study of cities and urbanism was from the start an interdisciplinary endeavor that involved disciplines such as geography, architecture, town-planning, engineering, economics, sociology, psychology, anthropology, archaeology, and more. Given the disciplinary diversity of the field, a major issue was to find a common language allowing discourse, co-operation and exchange of ideas between the many domains involved. For some time, during the 1950s and 1960s, it was thought that such a common ground could be found in a conjunction between the scientific method, as elaborated mainly in the economically oriented human geography, and the then newly emerging *General System Theory*. Very soon, however, this whole positivistic approach came under a strong Structuralist–Marxist and Humanistic–phenomenological criticism, and as a result the field of cities and urbanism has split into two, disconnected, parallel currents: one quantitative and positivist, often termed *regional science*, and the other qualitative–hermeneutic, which we'll call here *social theory of the city*.

Self-organization is the central property of, and a theory about, complex systems. Soon after its origin, during the 1960s, in physics and the exact sciences, it became a leading paradigm in the study of complex systems in general. Given the complexity of cities, it was not surprising that specialists in self-organization (mostly physicists) have found cities a fruitful field to apply their models, and that regional scientists have welcomed the new theory and made it the forefront of their research domain.

The central thesis of this book is that self-organization theories have much more to offer to the study of cities than being simply the modeling engine of regional science. The language and conceptual framework of self-organization resembles much of the language and conceptual framework of the non-quantitative social theory of the city. As such, it has the potential to provide a common language that will unite the two, currently split, urban currents.

Self-Organization and the City makes a first step in this direction and illustrates how the notion of self-organization can provide a common language and a conceptual and methodological framework for the study of cities and urbanism. The book is intended, therefore, for students of self-organization in general, and for students of cities and urbanism of all convictions: for regional scientists working with mathematical models of self-organizing cities and for urbanists in a variety of disciplines working within the frame of social theory: in social geography, architecture, urban and regional planning, economics, sociology, and so on.

I would like to close this preface with some acknowledgments. First and foremost I acknowledge Professor Hermann Haken, a great scientist, a marvelous person, a friend and a colleague: for his theory of synergetics and for our many conversations which have greatly inspired my work; for his encouragement and support in publishing this book, and for collaborating in the writing of the two special chapters on 'synergetic cities'. As an enthusiastic student of synergetics I very strongly believe in scientific co-operation and teamwork, that is to say, that the output of a good team can exceed the input of its members. This book is indeed the fruit of a synergistic team which includes Itzhak Benenson, who came all the way from Sverdlovsk to Tel-Aviv in order to construct and shape our FACS models and give them mathematical precision, Itzhak Omer, who was the first to suggest the usefulness of cellular automata as heuristic city models, and Nurit Alfasi, who joined the team just in time to elaborate on the implications of self-organization to urban and regional planning. Several chapters of the book were co-authored with them, and two special chapters (Chaps. 13 and 14) with Professor Hermann Haken. And while I bear full responsibility for any faults that might be found in the text, I would like to emphasize that without their collaboration and support the project of writing this book would not have been possible. I would also like to thank Y. Dorfman and O. Reuven-Safrir for doing several of the drawings.

Chapters 5, 6, and 7 of the book are based on research conducted, from 1992 to 1994, within the frame of an Israeli Academy of Science research grant (No. 891–171), on *The metropolitan space as a self-organizing system*, while Chaps. 8 and 14, are part of an on-going GIF (German–Israeli Foundation) research project (Grant No. I-458.224.07/95) on *Synergetics, inter-representation and cognitive mapping.* We are grateful to both research foundations. Since the early 1990s, as the study developed, several of its issues have been published in scientific journals. These papers provide some of the foundations for this book and are integrated in the following chapters. Chapters 1, 2 draw on my 'Notions concerning the nature of world urbanization', *Progress in Planning* **46**(3), 1996, 141–194; Chap. 3 on my 'Self-organizing cities', *Futures* **29**, (4/5), 1997, 353–380; Chap. 5 on my paper with Benenson and Omer on 'Socio-spatial residential dynamics: stability and instability within a self-Organizing city', *Geographical Analysis*, **26**(4), 1994, 321–340; Chaps. 6, 12 on my paper with Benenson on 'Artificial planning experience by means of a heuristic cell-space model: simulating international migration in the urban process', *Environment and Planning A*, **27**, 1994, 1647–1665; Chap. 7 on my article with Benenson and Omer, 'Spatial cognitive dissonance and sociospatial emergence in a self-organizing city', *Environment and Planning B*, **24**, 1997, 263–285; and Chap. 13 on Haken and my article 'A synergetic approach to the self-organization of cities and settlements', *Environment and Planning B: Planning and Design* **22**, 1995, 35–46.

Tel-Aviv
April 1999

Juval Portugali

Contents

Part V Self-Organization and Urban Revolutions

Introduction

The motivation for writing this book stems from two parallel, yet unrelated, scientific events that happened over three decades ago. The first was the rise of the self-organization paradigm in domains such as physics, mathematics and cybernetics (see Chap. 3 for details). The second, a split in the study of cities and planning that echoed, in fact, a split within human geography - a discipline that has played a major role in urban and regional planning and the study of urbanism. The split was between proponents of the positivist quantitative approach that dominated the field during the 50s and 60s and proponents of structuralist – Marxist and Humanistic (SMH) approaches that since the early 70s have dominated the field of human geography. From the early 70s onward, as a result of that split, the study of cities and urbanism is conducted in two separated domains: one quantitative, often termed *regional science* and influenced by the exact sciences and their scientific approach, and the other 'qualitative', often referred to as *social geography*, and dominated by social theory and philosophy and their hermeneutic approach.

Self-organization theories originated, as noted, in the exact sciences. Very soon, however, it was realized that the theories enfold general principles and methodologies that extend beyond the boundaries of their original research domains: Haken's theory of synergetics beyond the domain of laser and fluid, Prigogine's dissipative structures beyond physical–chemical systems, and so on. Self-organization, as is well established today, has captured the forefront of the system approach in science and as such became a paradigm relevant to phenomena in a wide spectrum of domains in the life sciences, social sciences and humanities. It was just natural, therefore, that when physicists went out of their discipline in search of new domains to apply their self-organization models, they have found the regional science approach to cities rather attractive. It was further natural that the quantitatively oriented regional scientists have welcomed the new theory and placed it at the research forefront of their field. Similar processes have happened, in fact, in other domains into which the self-organization paradigm has invaded and diffused.

The main drive behind the majority of self-organization studies in the social sciences and in regional science, is to examine the extent to which the concepts, methods and models of self-organization can be applied to processes and empirical data in the social and human domains. With few exceptions the question of how the theory of self-organization is, or can be, related to the rich interpretive literature that exists in social theory, social sciences and humanities, remained relatively untouched. One such exception is Haken's (1991, 1996) theory on the 'synergetics of cognition' who in the last two decades not only has applied synergetics to cognition, but has also integrated it with mainstream discourse in psychology and other cognitive sciences. This is not the case however, with studies about cities and self-organization. As a consequence of the above noted split between the qualitative and quantitative approaches to cities, and the conjunction between self-organization studies and regional science, most studies on the self-organization of cities are applicative in their scope. As such they form yet another line of research parallel to, and detached from, mainstream socio–cultural urban theory. The central thesis of this book is that the theory of self-organization has much more to offer to the study of cities, planning and urbanism than being regional science's sophisticated modeling approach.

My aim in *Self-Organization and the City* is thus to explore the possibility of a genuine discourse or even an integration between the notion of self-organization and socio–cultural theories of cities and urbanism. Not only to apply self-organization theories to the socio–economic domain, nor even to integrate them with the economic–quantitative regional science - this, as just noted, has been achieved. Rather, the aim here is to integrate the notion of self-organization with mainstream human geography and its social theory of the city. That is to say, to explore the similarities and differences between these two conceptual frameworks, as well as the possibility, that their integration in the context of cities and urbanism will enrich the three partners of this scientific endeavor: the study of cities and urbanism, the notion of self-organization, and social theory and philosophy.

Self-Organization and the City is, in fact, my second attempt to integrate social theory and self-organization theory. The first try was *Implicate Relations* – my study of *Space and Society in the Israeli–Palestinian conflict* (Portugali 1993). Implicate Relations – the theoretical framework of the above study – is a conjunction between David Bohm's (1980) theory of order and Haken's (1983, 1993) synergetic interpretation of self-organization. This conjunction, in particular its self-organization/synergetic component, provided the theoretical and conceptual nucleus and starting point for the research that we discuss here. This second attempt differs from the first in two respects. In focusing on the domain of cities and urbanism and, in including in the integration mathematical modeling and thus making the point that such models are not necessarily alien to the hermeneutic nature of social theory.

The split between the 'hard' regional science and the 'soft' social theory approaches to the city echoes, of course, the age-old split between Snow's (1964) 'two cultures': the culture of the 'hard' sciences such as physics and mathematics, and that of the 'soft' sciences mainly in the social sciences and humanities. The ambition to bridge the two streams of urban studies thus implies, in fact, bridging the urban manifestation of these two cultures. Is this a realistic aim? I think that the answer is 'yes' – to a large extent as a consequence of the rise of the self-organization paradigm. As elaborated elsewhere, a significant property of the notion of self-organization (and of Bohm's 1980 theory of order, which I do not discuss here) is "the discovery, in inanimate matter, of properties and behaviors hitherto found in living organisms and social relations" (Portugali 1985, pp. 232, 233). To my mind this discovery opens the way for a genuine discourse and dialogue between the two cultures that was not possible in the past. Not that from now on the mathematically oriented self-organization discourse of cities and the hermeneutic social theory of the city will become a new inseparable entity, but that both sides will recognize that beyond obvious differences there is also a large common ground. Large enough to allow the emergence of a dialogue that will benefit both sides. *Self-organization and the City* exposes, and elaborates on, this common ground in several ways.

First, by an explicit comparison between self-organization and social theory as two general conceptual frameworks. This kind of discussion forms the contents of the whole of Chaps. 4, 10, 15, and it appears occasionally when relevant in all the chapters of the book. This comparison focuses both on general similarities in the nature of the two theoretical bodies, as well as on specific similarities and often identities that exist between theories of self-organization and social theories. Thus, for example, I show that the synergetic concept of 'order parameter' is very similar to the notion of 'mode of production' in social theory, 'bifurcations' and 'phase transitions' of self-organization are not very different from 'revolutions' in social theory, 'circular causality' of synergetics is similar to 'reproduction' and 'structuration', and so on.

Second, by examining, from the perspective of self-organization, some classical issues, which because of their 'softness' have not as yet been subject to the self-organization perspective. These include discussions of the nature of cities and urbanism (Part I), of the question of the 'urban revolution' in ancient times as well as in the 20th century and on the verge of the 21st (Part V), and a discussion about the problematics of urban and regional planning in the age of post-modernity and 'hyper-modernity'. With the latter I refer to two claims made below in Chap. 3 and in Part III. One, that cities are self-organizing systems and as such to a large extent unpredictable and therefore unplannable. Two, that the perception of cities as chaotic and unpredictable entities is becoming more and more a basic sensation of life and urbanism on the verge of the 21st century. In these discussions self-organization is used mainly as a conceptual–theoretical framework and not, as usual, as a formal mathematical approach. Throughout these discussions we not only illustrate the strength of self-organization as a conceptual interpretive framework, but also illustrate and compare its relations to similar concepts and interpretations in social theory.

Third, by developing in Part II a new modeling approach which incorporates in its structure properties of both self-organization and social theory. We elaborate this new approach by gradually building a family of models (City, City-1, City-2, ..., City-6) and by applying them to what might be regarded as soft or humanistic issues of cities and urbanism: to the process of socio–cultural segregation, to the relations between international migration and the internal structure of cities, to planning and to urban socio–cultural morphogenesis. That is to say, to the emergence of new socio–cultural entities out of the very dynamics of the city. We term these models, or rather city-games, FACS-Cities (*Free Agents in a Cellular Space*) to indicate their elementary properties. They are built of two layers: an infrastructure cellular automata layer, describing the physical structure of the city, and a superstructure layer describing the spatial behavior of the free agents who make the population of the city. Our agents are *free* in the sense that they act intentionally on the basis of their personal history, memory and perception (i.e. cognitive map) of the city.

As just noted, the above FACS models incorporate properties of, and thus integrate, social theory and self-organization. The social theory properties enter these models in two ways. One, the rules of the games are based on 'ordinary', 'soft', non-formal knowledge and interpretations as accumulated during the years in the social sciences and the humanities. Two, while mathematical, our FACS-Cities are essentially *heuristic* models. Their aim is not to best-fit data to predictions, but to provide material and starting point for a free interpretation that is not different from ordinary hermeneutics as practiced in social theory. The difference is that in social theory the interpretation is made with reference to empirical data or observed events; in our FACS models the interpretation refers to the outcome of the simulations games – the virtual empirical results and events as they appear in our simulated, virtual, FACS-Cities.

The self-organization properties appear in these models in two ways too. One, in that the models are good simulators of self-organization. Given certain initial conditions and rules of the game, various aspects and phenomena of self-organization can be studied. Two, and as a consequence of the above, in that various self-organization theories can be employed as conceptual frameworks for the interpretation of the various simulation games. For example, in Part II, synergetics is employed not as a formal mathematical model, but as a conceptual framework for the interpretation of the simulations' results. As in Parts I, III and V of the book, here too, notions such as order parameter, slaving and so on, are not the explicit parameters of the models, but rather their interpretation tools.

Fourth, in most parts of the book, Haken's synergetic approach to self-organization is used as a conceptual framework to interpret events, empirical findings and results from the heuristic simulation games. However, it is only in Part IV, entitled *Synergetic cities*, that synergetics is explicitly employed as a formal mathematical theory and algorithm. Part IV indicates the direction, so we believe, for a possible integration between the "hard" formalism of synergetics as a leading theory of self-organization and the 'soft' social theory of the city.

Two chapters make the contents of the 'synergetic cities' of Part IV. One refers to the general dynamics of cities and settlements and the other to decision-making in the context of urban and regional planning. Both are based on the synergetic approach to cognition and brain functioning – a research domain which in the last two decades has formed the cutting edge of synergetics' scientific effort and in which considerable progress has been made. At the basis of this approach stands the algorithm of the so-called *synergetic computer* (Haken 1991) and its application in the context of cities and decision-making is rather novel. Synergetics has not as yet been applied to decision making and most, if not all, of its previous applications to cities and urbanism were made within the context of the more traditional 'master equation approach'. The advantage of this new pattern recognition approach is threefold. First, it has a very rich body of literature. Second, it is intimately related to psychology and hence potentially lends itself to integration with social theory. Third, it perceives the city in terms of an interplay between its physical structure and a cognized pattern (cognitive map), and as a consequence allows treating the city and decision-makers in it within the same theoretical framework.

The opening sentence of the book, in Chap. 1, is that 'Cities are with us for over 5000 years' (at least) – since the event, in 4th millennium B.C. Mesopotamia, that Gordon V. Childe has termed "the urban revolution". Starting from this perspective the main body of the book deals with the modern, postmodern and hypermodern/self-organizing cities of today. The last chapter of the book comes full circle and once again looks at the urban revolution of 5000 years ago. This time, however, from the perspective of self-organizing cities as developed throughout the book, and in conjunction with Lefebvre's (1970) claim that the urban revolution is only now occurring in front of our very eyes. This conjunction enables us, first, to look at the general phenomenon of urbanism as a self-organizing process. That is to say, as a sequence of urban cultures, or steady states, interrupted by chaotic urban revolutions which entail phase transitions and a move to new urban periods, or steady states, and so on. Second, it enables us to examine from this wide perspective the dramatic changes in society, cities and urbanism that seem to characterize the transition period from the 20th to the 21st centuries. That is to say, not in terms of *the* urban revolution, but from the perspective of a long self-organizing, space–time sequence of urban revolutions.

Part I **On Cities and Urbanism**

1. Cities as Concepts

1.1 Cities Have Been with Us for over 5000 Years

Cities have been with us for over 5000 years. This, of course, if we do not consider some early, premature, appearances such as 8th millennium B.C. Chatal Hüyük in Anatolia and Jericho in the Jordan valley. Since the urban revolution in 4th millennium B.C. Mesopotamia, cities form the largest artifacts ever produced by humans and their most dominant socio–spatial entity. Despite its archaic origin, urbanism survived all subsequent socio–cultural changes: It was the most dominant cultural, social, and political, spatial entity for the Sumerians, Acadians, Hittites, Egyptians, Greeks, Romans, Byzantines, Persians, Chinese, Muslims, Middle-Ages Europeans, and with the rise of modernity it played a major role in the industrial revolution. Cities such as Uruk, Ur, Jerusalem, Hatushash, Athens, Rome and so on, are considered as important landmarks in human evolution, and some of our most central terms and concepts are derived directly from cities – citizen, police, politics, bourgeois ...

And today, on the verge of the 3rd millennium A.D., urbanism is still at the center of our life. How is that possible? How is it that we employ the notions city, urban, and the like to refer to a tiny little place such as 2nd millennium B.C. Jerusalem, with its approximately 60 dunams and about 1500 inhabitants, as well as to New York, Mexico City, Rio de Janeiro, London or Paris with their huge territories and millions of inhabitants?

One way to answer this question is to say that cities are like triangles or squares – geometrical shapes that regardless of their variable quantitative sizes and properties, still share some structural–topological properties which constitute the necessary and sufficient conditions for their definition. Such a view would conform to what has been termed by students of cognition the Aristotelian view on concepts formation and categorization, and with what I'll call below 'the Aristotelian city'.

Another view which I would like to explore below is to say that cities should be seen in terms of networks stretching in time and space. This view is derived from Wittgenstein's theory of concepts formation and categorization and I'll call its urban implication 'the city of Wittgenstein'. In discussing linguistic concepts and categories Wittgenstein has likened our language to an ancient city. Christopher Alexander, per contra, in discussing architecture, building and cities, has likened the latter to a language. 'The city as a pattern language', thus forms a third view of the city.

A fourth view is to look at cities and urbanism in terms of a network, as in Wittgenstein, but as networks some of whose instances are more typical or more basic than others. Such instances of concepts and categories were termed by Rosch and co-workers as prototype instances and by Lakoff core instances. Accordingly, one can thus speak on the prototypicality, or the core, of concepts and categories. I'll term the urban implication of this view 'the prototypicality of core cities'. The notions of prototypicality and core remind one of Max Weber's 'ideal-type' which he has applied to his study of cities. The fifth view of the city will thus be termed 'the ideal-type city of Max Weber'.

The sixth and the last approach that I'll discuss here is the possibility to consider cities and urbanism in terms of what I've defined as Inter-Representation Networks (IRNs). That is to say, that cities are at the very same time an interactive network of internal (cognitive) representations of the external environment, and external (material) representations of internally represented concepts, categories and images. The implied city will be called below 'the IRN city'. As you can see I suggest the discourse of cognitive science on concepts formation and categorization with the implication that according to my view cities and urbanism are socio–spatial conceptual and material categories.

Cities as IRNs introduces to the discussion the property of self-organization. This follows from the very notion of IRN that suggests, in addition to what has been said above, that inter-representation networks are self-organizing systems. This is so with respect to concepts and categories and this is so with respect to the socio–spatial, conceptual and material categories called cities.

In the discussion below I first present the above six 'cities': The Aristotelian, Wittgenstein's, the city as a pattern language, the prototype core city, Weber's ideal-type and the IRN city. Then, in the chapter that follows I relate those cities to the discourse on urbanism in geography and planning.

1.2 The Aristotelian City

The classical view of concepts, which goes back to Aristotle's poetics, holds that concepts are monothetic, that is to say, that all members of a concept share at least one, or more, common properties. The latter provide the necessary and sufficient conditions to differentiate members from non-members of a concept (McShane 1991). From this specific viewpoint follows that all cities and urbanisms share some necessary and sufficient conditions which make them cities and urbanisms. This view is implicit in very many attempts to define urbanism and 'what is a city?', ranging from Gordon Childe's (1950) ten points definition in his seminal paper on 'The urban revolution' in ancient

Mesopotamia (Sect. 15.1.1 and Fig. 15.1 in Chap. 15 below) to the attempts in the Sixties to standardize and define metropolitan areas, such as Kingsley Davis' 12-page definition, the SMSA of USA in 1960, the FEA (Functional Economic Areas) and CUR (Consolidated Urban Regions) suggested by the Chicago geographers (see description in Haggett 1975, pp. 352–356), to David Harvey's (1973, 1985 and Sects. 2.7, 2.9 in Chap. 2 below) more recent discussion of 20th century cities. The Aristotelian view of cities is probably the most dominant one and its truism is usually taken for granted. Yet, it is also a most problematic and largely a false view, responsible, I think, for the failure of Childe's type of definitions whenever they were put to the Popperian test of falsification. That is to say, whenever they were extended beyond their original and limited case studies.

1.3 The City of Wittgenstein

One of the strongest (and nicest) challenges to the Aristotelian view is Wittgenstein's perception of concepts and categories in terms of what he described as *family resemblance*. His often-quoted example refers to the concept 'game':

> Consider for example the proceedings that we call 'games'. ... What is common to them all? – Don't say: 'There must be something common, or they would not be called 'games', – but look and see whether there is anything common to all. – For if you look at them you will not see something that is common to all, but similarities, relationships and a whole series of them at that. To repeat: don't think but look! – Look for example on board-games, with their multifarious relationships. Now pass to card-games; here you find many correspondences with the first group, but many common features drop out and others appear. When we pass next to ball games, much that is common is retained, but much is lost. – Are they all amusing? .... In ball games there is winning and losing; but when a child throws his ball at the wall and catches it again, this feature has disappeared.... And we can go through the many other groups of games in the same way; can see how similarities crop up and disappear.
>
> And the result of this examination is: we see a complicated network of similarities overlapping and criss-crossing: sometimes overall similarities, sometimes similarities of detail (Wittgenstein 1953, par. 66).

Now, if we replace the concept 'game' in Wittgenstein's paragraph by the concept 'city' it would read as something like this:

> Consider for example [the entities] that we call 'cities' ... what is common to them all? Don't say: There must be something common, or they would not be called 'cities' – but look and see whether there is anything common to all. – For if you look at them you will not see something that is common

> to all, but similarities, relationships and a whole series of them at that. To repeat: don't think but look! Look for example on North American cities, with their high rises, down town, suburbs and their specific architecture. Now pass to European cities; here you find many correspondences with the first group, but many common features drop out and others appear. When we pass next to the cities of the Occident, much that is common is retained, but much is lost. – Do they all have suburbs around their old city centers .... And we can go through the many other groups of cities in the same way; can see how similarities crop up and disappear.
>
> And the result of this examination is: we see a complicated network of similarities overlapping and criss-crossing: sometimes overall similarities, sometimes similarities of detail.

The city of Wittgenstein is thus a complex network of connections, extending in time and space, of entities that at a certain space–time point were perceived cities by virtue of some features that they shared with previous entities which at their time were perceived as cities by virtue of some properties they shared .... and so on. This is so with respect to the city as a concept, with respect to every single city and as the following paragraph testifies, this is so also with respect to language in general.

> ...ask yourself whether our language is complete; – whether it was so before the symbolism of chemistry and the notation of the infinitesimal calculus were incorporated in it; for these are, so to speak, suburbs of our language. (And how many houses or streets does it take before a town begins to be a town?) Our language can be seen as an ancient city: a maze of little streets and squares, of old and new houses, and of houses with additions from various periods; and this surrounded by a multitude of new boroughs with straight regular streets and uniform houses (Wittgenstein 1953, par. 18).

1.4 The City as a (Pattern) Language

Our language can be seen as a city, writes Wittgenstein, whereas Christopher Alexander (1979) says that buildings, neighborhoods, cities and metropolis, are products of a language of patterns. In their *A Pattern Language*, Alexander et al. (1977) have produced a lexicon of patterns, ranging from very large patterns of regions and metropolitan areas, through patterns of cities and neighborhoods, to patterns of alcoves, windows and door-handles. The book presents these patterns and shows how they are related to each other and thus form a whole language, structurally not very different from spoken or written languages.

Alexander's patterns are architectural entities of various sizes and scales. They can be compared to concepts, pictures, images, or schemata. The patterns of doors, windows, buildings, squares, neighborhoods and cities, are interrelated in a way similar to words, concepts, sentences, paragraphs, chapters and stories. The patterns are natural entities in the sense that they exist not only in physical structures in the environment, but also in "in peoples' heads and are responsible for the way the environment gets its structure" (Alexander 1979, pp. 49–50).

Alexander's patterns are

> 'something in the world' – a unitary pattern of activity and space, which repeats itself over and over again . . . each time in a slightly different manifestation. . . . these patterns are created by us . . . in our minds . . . [as] mental images of the patterns in the world: they are abstract representations of the very morphological rules which define the patterns in the world. However, unlike the patterns in the world . . . the same patterns in our minds are . . . generative. They tell us what to do . . . (Alexander 1979, pp. 181–182).

The pattern language is more complex than simple natural spoken languages. A natural spoken language has a set of elements (words), a set of rules which define the possible arrangements of words and in addition to these – "the complex network of semantic connections, which defines each word in terms of other words" (ibid, p. 184). A pattern language is still more complex in the sense that, like words, the patterns are elements and symbols, but unlike words, "each pattern is also a rule, which describes the possible arrangements of the elements – themselves again other patterns" (ibid, p. 185).

As in spoken languages, every person has his/her own personal pattern language, which forms a personal variant of the language of a larger social and cultural collectivity. And the artificially built environment is the product of a conversation between a large number of individual pattern languages, which are the means with which people act on the environment. 'And the enormous repetition of patterns, which makes up the world, come about because the languages which people use to make the world are widely shared" (ibid, pp. 209–210).

1.5 The Prototypicality of Core Cities

Wittgenstein's family resemblance perception of concepts and categories is today widely accepted with two important additions: The first is Rosch and co-workers' (1976) notions of prototypicality and basic-level which add that some instances of a concept are more typical than others. For instance, an orange is 'more' a fruit than a melon. The second is Lakoff's (1987) discussion in his *Women, Fire and Dangerous Things* in which he elaborated the idea that many concepts and categories have a radial structure with a core of the most basic exemplars related to other peripheral instances of the concept through similarities, metaphors and family resemblances.

Cities and urbanism, according to the above views, are thus a huge network of connected similarities, stretching in space and time from 4th millennium B.C. Mesopotamia, to all over the globe, until the here and now that we are discussing. In this huge evolving, diffusing, network one can identity core and periphery: cities, images, and urban phenomena which capture the central area of the net as the most typical or prototypical exemplars thus pushing aside, to the periphery, the other cities, images and urban phenomena. For example, according to Max Weber, despite the fact that the geo-historical record shows so many urban societies and instances of cities, it is only the city of the Occident which was an urban community in the fullest sense of the word. Like the orange which is 'more a fruit than a melon', that city is more a city that any other city in history. Weber calls this property 'ideal type'.

1.6 The Ideal-Type City of Max Weber

The *category concept*, or the *ideal type* as he later preferred to call it, was Max Weber's central methodological tool and "the key to that important element of his thinking which was derived from [Wilhelm Dilthey's] hermeneutics" (Frank's introduction to Weber 1909/1976). The idea is that in the study of culture and society the aim is not to subsume human action and phenomena under some general laws and so to explain them, as in positivism, but rather to grasp the individuality of each instance of a phenomenon. To do so you first posit the ideal-type of the category and then study the individuality of the various exemplars by looking at their links, similarities and differences with the ideal type and other instances. This method is applied by Weber (1958) in his study *The City.* He first posits the city of the Occident as an ideal type ("An urban 'community', in the full meaning of the word, appears as a general phenomenon only in the Occident", he writes (ibid, p. 80) and then proceeds by gradually weaving a whole network of similarities and differences between instances of cities in different periods and regions in the world.

Weberian cities are thus similar to the city of Wittgenstein; they too, form a complex network of connections, extending in time and space, of entities that at a certain space–time point were perceived cities by virtue of some features that they shared with previous entities which at their time were perceived as cities by virtue of some properties they shared .... As in Rosch's and Lakoff's cities, in them too, one can identify core and periphery: cities, images, and urban phenomena which capture the central area of the net as the prototypical or ideal-type exemplars thus push aside, to the periphery, the other cities, images and urban phenomena. The difference, of course, is that Weber's perception originated from his methodology of how one should study concepts and categories as they appear in the external social and cultural world, whereas Wittgenstein's, Rosch's and Lakoff's perception originated from their views as to how humans actually cognize such concepts and categories in their internal, individual worlds - in their minds.

1.7 IRN Cities

With the notion of IRN (*Inter-Representation Network*) I propose that the similarity between Weber's externally represented ideal-type city, and the internally represented city of Wittgenstein with Rosch's prototypicality and Lakoff's radial structure, is not accidental – the two simultaneously and sequentially represent each other. Weber's main concern were cities and urbanism as they appear in the (external) world. Rosch's, Lakoff's and other students of cognition main concern are concepts and categories as they 'appear' in the internal environment – in the mind. The notion of IRN proposes that the internal and external worlds are two aspects or representations of the very same network (Portugali, 1996). That is to say, that some of the elements which make the network associated with concept formation and categorization are internally represented in the mind–brain, while others are externally represented in the environment. And, that the network evolves and develops by an interactive sequential interplay between its internal and external elements, whose consequence is a sequential space–time, diffusion-like, process. More specifically, I suggest that with respect to concepts and categories such as cities, the process of their diffusion in space and time by means of this sequential external representation/internal representation interplay, is the process which has made and is still making, their family resemblance.

To see how the process of space–time diffusion makes family resemblances consider a new entity, which exists or is created at the external environment in a certain time ($t = t_0 + \delta t$) and place ($p = p_0 + \delta p$). This new entity will become a game (i.e. a new member of the category 'game') or a city (a new instance of the concept 'city'), if many of its properties resemble part of the properties we internally perceive as the properties of some existing games or cities (at $t = t_0$; $p = p_0$), more than they resemble the properties of any other existing concepts or categories. Once this new game or city joins the family of games or cities, the new properties it has added to the family now enable us to include new members hitherto not able to be qualified as games or cities. This interplay between the external and internal representations and properties of entities, is the motor behind their diffusion in space and time and the process, which makes their specific family resemblance network. The space–time diffusion of cities and urbanism from their core in 4th millennium B.C. Mesopotamia to our modern and postmodern cities and urbanisms thus entailed their specific network of similarities and differences. Such a process by its very nature cannot produce Aristotelian cities but only Wittgensteinian cities.

As the concept/physical entity 'city' diffuses in space and time in this sequential interplay between its internal and external representations, more and more instances of 'a city' come under its conceptual umbrella, more and more properties are assembled to the family resemblance network of cities, and with them the prototype or ideal type city at the core of the family resemblance network is also changing and moving in space and time. Just as an illustration: the Biblical text tells us that the meaning of what the first millennium B.C. Israelites have termed 'a city' (e.g. 10 dunams Tel Qashish at the Valley of Jezreel with a population of about 200 agriculturalists), would in today's Hebrew be probably termed 'a village' (a term which is almost absent in the Bible and by implication in 1st millennium B.C. Hebrew, despite the fact that this was an agricultural society – see Portugali 1984). Each time–space thus creates its own specific image of the city, each period, culture and society with its own ideal type, its own prototypical core city.

So what were the various ideal (proto)types of the city in the last 5000 years of its movement in space and time? This is a tempting question, but one which would take us far beyond the scope and context of the present study. What can be said here at this preliminary stage, however, is that if one looks at the overall form of this IRN process, by which the city as a mental and physical category diffuses and moves in space and time, one can observe a typical pattern. This pattern shows long periods of steady state, during which the city is dominated by a certain ideal (proto)type and core–periphery relations, interrupted by abrupt chaotic periods of urban change and revolution. This diffusion process is thus is not linear, smooth and gradual, but rather traumatic and punctuated. Such a process is typical, as we shall see below, of the dynamics of self-organizing systems, and indeed, in Chap. 15 below, I interpret repeated urban revolutions in ancient and modern times from the perspective of self-organization. Chapter 15 with its macro-scale view of cities and urbanism is the last chapter of this book and it closes the discussion. In the following we shall focus on a much smaller-scale view on cities. Our question here would be the following: 'what images of cities and urbanism have, in the last 50 years, captured a central prototypical position in the discourses of urban geographers and planners and pushed to the periphery many other instances'?

2. Prototype Urbanisms

What has been the nature of cities and urbanism over the last 50 years? What prototype images have captured the center stage of our perceptions and categorizations of cities and what instances and properties of cities and urbanism were pushed to the periphery? How has the IRN process – this on-going interplay between internal images and external appearances of the city – developed?

2.1 The Ecocity

One image or conceptualization which has captured the core-stage of cities at the middle of the 20th century is the Ecocity: a view of the city in terms of location theory with its economic principles, supported and supplemented by the ecological theory with its evolutionary principles. That is to say, for example, the concentric economic rings of von Thüen's (1826/1966) *Isolated State*, transformed into Alonso's (1964) *rbc* (*rent bid curves*), supplemented by Burgess (1925) ecological rings. Or, August Lösch's (1945) city-rich/city-poor landscape, supplemented and supported by Hoyt's (1939) ecological-sectorial morphology of the urban landscape. Let us have a look at a few examples of both these types of the ecocity, the ecological and the economic.

2.1.1 Eco-nomic Cities

2.1.1.1 From Thünen's Isolated State to the Isolated City

> Imagine a very large town at the centre of a fertile plain, which is crossed by no navigable river or canal. Throughout the plain the soil is capable of cultivation and of the same fertility. Far from the town, the plain turns into an uncultivated wilderness, which cuts off all communication between this State and the outside world.
>
> There are no other towns on the plain. The central town must therefore supply the rural areas with all manufactured products, and in return it will obtain all its provisions from the surrounding countryside.
>
> The mines that provide the State with salt and metals are near the central town which, as it is the only one, we shall in future call simply 'the Town'. (Thünen's *The Isolated State*, Chap. 1, Hypotheses.)

> The problem we want to solve is this: What pattern of cultivation will take shape in these conditions?; and how will the farming system of different districts be affected by their distance from the Town? We assume throughout that farming is conducted absolutely rationally.
>
> It is on the whole obvious that near the Town will be grown those products, which are heavy or bulky in relation to their value, and hence so expensive to transport that the remoter districts are unable to supply them. Here too we shall find the highly perishable products, which must be used very quickly. With increasing distance from the Town, the land will progressively be given up to products cheap to transport in relation to their value.
>
> For this reason alone, fairly sharply differentiated concentric rings or belts will form around the Town, each with its own particular staple product.
>
> From ring to ring the staple product, and with it the entire farming system, will change; and in the various rings we shall find completely different farming systems. (Thünen's *The Isolated State*, Chap. 2, The Problem.)

These two short chapters that open von Thünen's *The Isolated State in Relations to Agriculture and Political Economy* form a verbal description of The Isolated State – the model which provides the foundation to all his economic work. "This method of analysis", he writes,

> has illuminated and solved so many problems in my life, and appears to me capable of such widespread application, that I regard it as the most important matter contained in all my work (Hall, 1996, p. XXII).

At a later stage, so writes Thünen in an appendix to his book, this model was translated by a friend into the diagrams presented in Fig. 2.1. And despite Thünen's comment that these diagrams "are not essential to an understanding of the problem under discussion – and nowhere in the work have I referred to them", they have since become the symbol of his work – at least among students of geography, location and urbanism (especially Fig. 2.1a). Due to his *The Isolated State*, Thünen is regarded as the most important German economist of the 19th century; first, because it is the first formal economic model, and second, since Thünen has unconsciously invented here the economic principle of marginalism, some 50 years before Walras made it the basis for modern economic theory (see Portugali 1984, for further discussion and bibliography).

Due to his *The Isolated State*, Thünen is also regarded as the founding father of modern location theory; first, because his model enfolds all the ingredients of this theory – isotropic plane, spatial competition between land uses and the principle of marginal spatial utility – and second, and as a consequence of the above, since by changing a few key words in his verbal

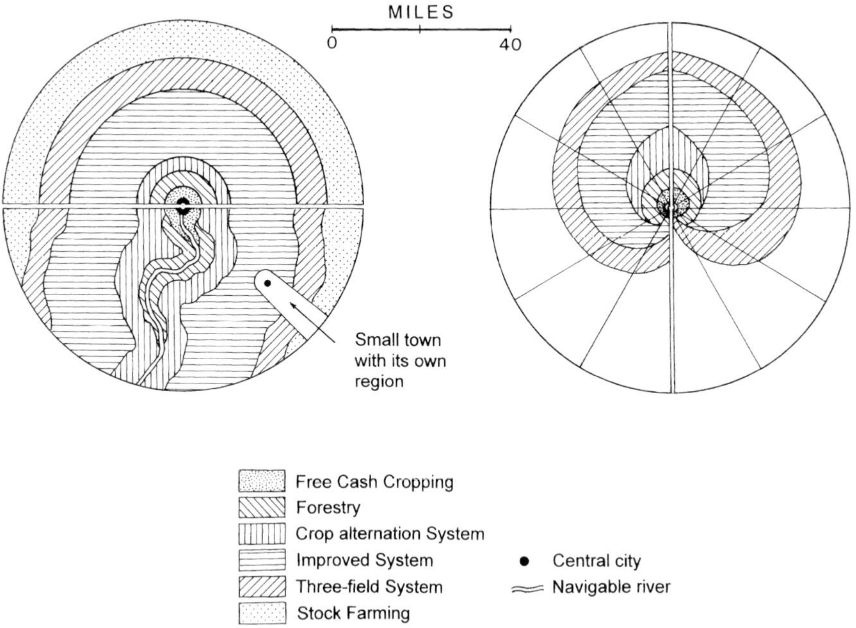

Fig. 2.1. "These diagrams", writes Thünen (Par. 384), "drawn by a friend of mine, are not essential to an understanding of the problem under discussion – and nowhere in the work have I referred to them. But since they afford a simple survey, ...I feel they might be welcome to the student ..." *Left, top*: "This shows the Isolated State in the shape it must take from the assumptions made in Section One ...". Left, bottom: "Here we see the Isolated State crossed by a navigable river. Here the ring of crop alternation become very much larger, stretching along the river ...The effect of constructing a highway is similar, ... " (Par. 385). *Right*: "The diagram illustrates the effect of the Town grain price on the extension of the cultivated plain" (Par. 386)

model one can get the standard urban land-use model as formulated some 100 years later by the economic land use theorists of the city:

> Imagine a very large CBD (Central Business District) at the center of an urban plain which is crossed by no navigable river, road or canal. Throughout the plain the urban land is capable of all land uses and with the same utility. Far from the CBD, the plain turns into wilderness, which cuts off all communication between this metropolis and the outside world.
>
> There are no other centers on the plain. The city center must therefore supply the other urban areas with all urban products and services, and in return it will obtain all its labor force from the surroundings.
>
> The mines and factories that provide the Metropolis with raw materials and industrial products are near the CBD which, as it is the only one, we shall in future call simply 'the Center'.

The problem we want to solve is this: What pattern of land use will take shape in these conditions?; and how will the urban system of different districts be affected by their distance from the Center? We assume throughout that decision making is conducted absolutely rationally.

It is on the whole obvious that near the Center will be allocated those land uses which are sensitive to the distance from the Center to the extent that they will not be supplied if located far from it. Here too we shall find services and products, which require an exposure to a very high threshold of potential customers in order to be supplied. With increasing distance from the Center, the land will progressively be given up to land uses cheap to transport in relation to their value.

For this reason alone, fairly sharply differentiated concentric rings or belts will form around the Center, each with its own particular land use.

From ring to ring the land use, and with it the entire metropolitan or urban system, will change; and in the various rings we shall find completely different land use systems.

Similarly to von Thünen's friend, we too can transform this verbal model into a visible diagrammatic model. All we have to do for this purpose it to take the standard image and model of a city as formulated in location theory. The latter is usually derived not verbally, as in Thünen's, but by means of an interplay between spatial demand curves and rent bid curves as in Fig. 2.2.

2.1.1.2 Cities as Central Places

Thünen's concentric rings image of the city was a source also of another image of the city, or rather of cities: cities as central places for their agricultural hinterlands, cities as mediators between their hinterlands and other cities, and cities as hierarchical systems of central places. *Central Place Theory* was developed by two people: Christaller (1933) in his work on *The Central Places in Southern Germany*, and August Lösch (1945) in his *The Economics of Location*. If Thünen has 'invented' the notion of marginal utility some five decades before its time, then Christaller and Lösch have suggested a genuine system theory several decades before Bertalanffy (1968) published his *General System Theory*.

Both Christaller and Lösch start with an 'isolated city' of a sort. Christaller starts with a kind of an isolated state which includes a geometrically central city and its dependent peripheral towns. From this starting point he then derives three basic hierarchies of central places (the $k = 3$ market, $k = 4$ transportation, and $k = 7$ administrative principles) from which the population of this imaginary state can consume goods and services and in which they can sell the products they produce. This is illustrated in Fig. 2.3. Lösch, on the other hand, starts with several independent isolated states, or cities, floating on an isotropic plane. He then assumes an increase of population

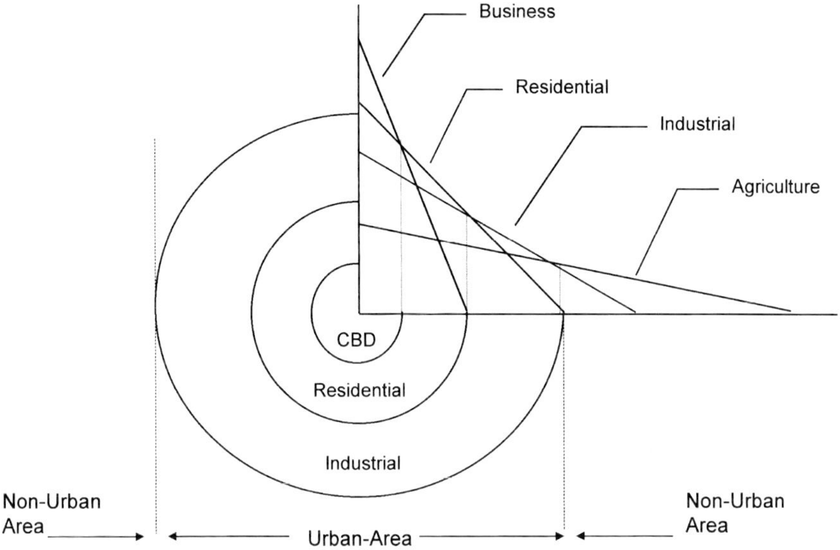

Fig. 2.2. Thünen's type of urban land-use system as derived 100 years later in location theory. Businesses are prepared to pay high rent at the center of the city, but are reluctant to 'live' far from it. Their spatial demand curves (or *rbc* – rent bid curves) are thus the highest and steepest. Industrialists, in this exposition, are exactly the opposite and residents are in between: they cannot afford to pay the high prices at the center, but are prepared to live far from it, and so on. Each land use thus occupies a ring were it can pay (bid for) the highest rent. Note that the principle of marginal utility which is implicit in Thünen's landscape, here appears explicitly as the central economic principle. The resulting spatial pattern is as in Thünen's, however

and economic activities which bring in more isolated cities, then by means of competition and general spatial equilibrium, the whole region becomes full and at a later stage as the process continues it reaches a spatial equilibrium in the form of a complex system of central places. Figure 2.4 describes the main stages in the process.

Lösch's theory is more ambitious and complex than Christaller's. While the latter's aim was confined to services, Lösch's aim was a general theory of location. As a consequence, his urban landscape is more complex, first with respect to the levels of the hierarchy and second, with respect to what he has termed as *city poor* versus *city rich* sectors, that is to say, sectors in his theoretical economic landscape, which because of the spatial distribution of cities in them, are better and worse served. The original model of Lösch is shown in Fig. 2.5, whereas Fig. 2.6 is a refinement of Lösch as suggested by Isard (1964).

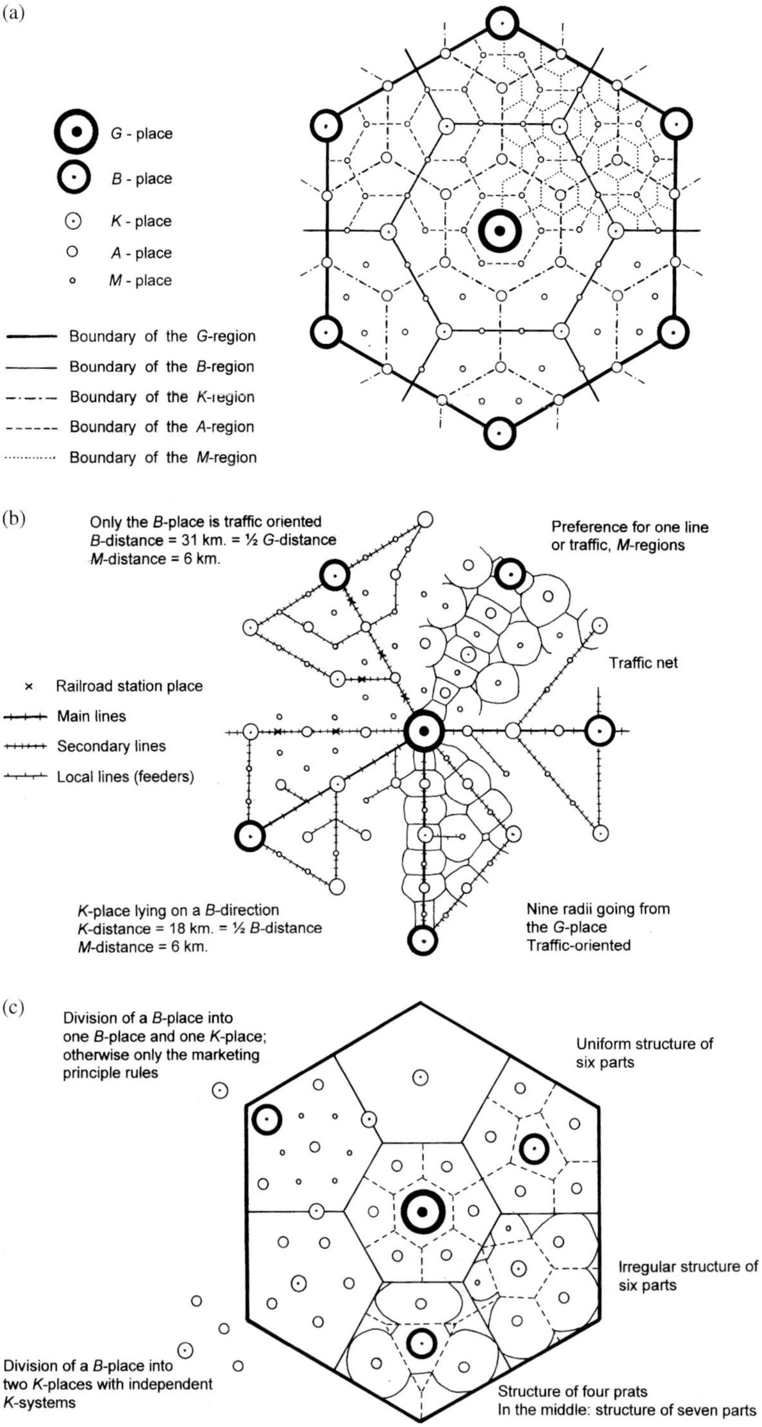
(a)
G - place
B - place
K - place
A - place
M - place
Boundary of the G-region
Boundary of the B-region
Boundary of the K-region
Boundary of the A-region
Boundary of the M-region
(b)
Only the B-place is traffic oriented
B-distance = 31 km. = ½ G-distance
M-distance = 6 km.
Preference for one line
or traffic, M-regions
Traffic net
× Railroad station place
Main lines
Secondary lines
Local lines (feeders)
K-place lying on a B-direction
K-distance = 18 km. = ½ B-distance
M-distance = 6 km.
Nine radii going from
the G-place
Traffic-oriented
(c)
Division of a B-place into
one B-place and one K-place;
otherwise only the marketing
principle rules
Uniform structure of
six parts
Irregular structure of
six parts
Division of a B-place into
two K-places with independent
K-systems
Structure of four prats
In the middle: structure of seven parts

Fig. 2.3. Christaller's systems of central places according to the three locational principles. (**a**) The marketing regions in a system of central places. (**b**) A system of central places developed according to the traffic principle. (**c**) A system of central places developed according to the separation principle. (Source: Christaller 1966, Figs. 2, 4, 6)

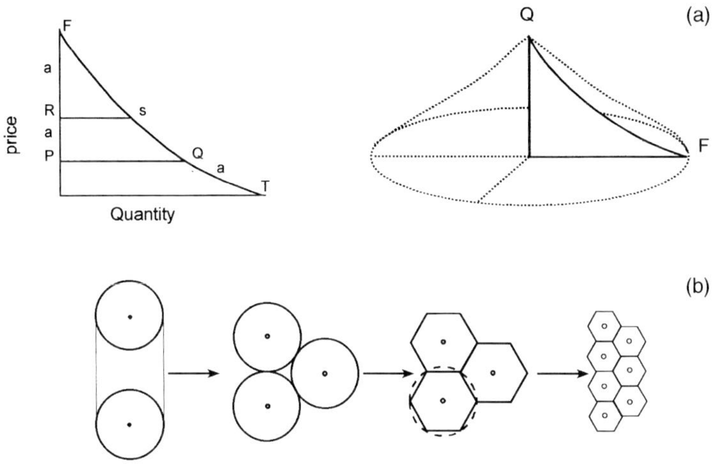

Fig. 2.4. The derivation of Lösch's system of central places. (**a**) The derivation of a spatial demand cone with its market area (*right*) out of an "ordinary" demand curve (*left*). (**b**) Development of market areas from the large circle to the final small hexagon. (Source: Lösch 1945, Figs. 20–23)

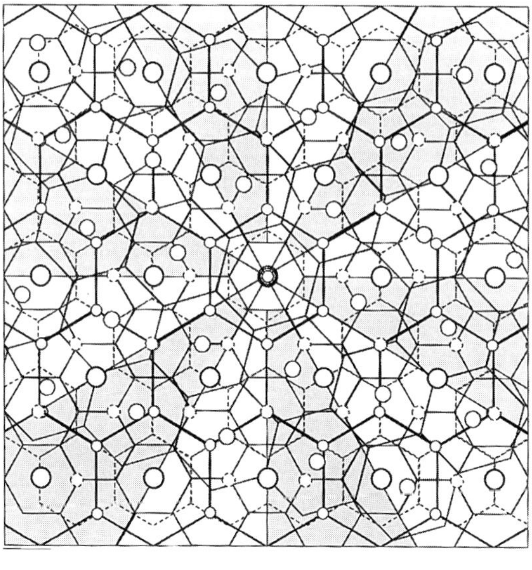

Fig. 2.5. Lösch's derived system of central places with their market areas, divided into city-poor, city-rich sectors. (Source: Lösch 1945, Fig. 28)

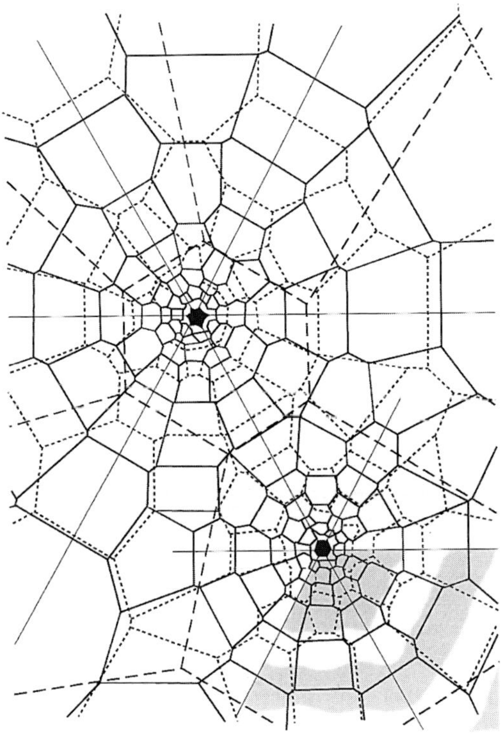

Fig. 2.6. A Lösch system of central places modified by Isard (1964) so as to be consistent with the resulting population distribution

2.1.2 Eco-logical Cities

Imagine an ecological system with a relatively high population density. The population is composed of several spatially segregated communities, each in its specific territorial niche. The individuals of which the population is composed are motivated by a simple aim – survival. For this purpose they have to form communities and interact among themselves, as individuals and collectivities, in various forms of symbiosis, competition, domination, invasion, succession, and the like. This on-going complex interaction, between individuals and communities, is the engine behind the dynamics of our ecological system. The latter might be a small water pool created after the rain, it might be a desert, a jungle, and also, so claim proponents of social and urban ecology, a city.

The ecological image of the city originated out of the Chicago school of *social ecology* and its product *urban ecology*. To be sure, social and urban ecology were not a direct, one to one, application of biological ecology to society and the city. In his seminal book *The City*, Park (1925) builds on top of the *biotic level* that describes the city in universal terms of symbiosis, competition, invasion etc., a *cultural level* that describes the city in unique human terms of social and moral norms, politics, religion and so on. In a paper on 'Urbanism as a way of life', his student and follower L. Wirth (1938) has further elaborated Park's view. He has, first, described society in terms of three orders: the ecological–biotic and on top, the cultural order and the political order. Then, from this general social order, he has delineated the urban from the rural on the basis of three ecological principles of *size*, *density* and *heterogeneity*.

As in biological ecology, where morphological analyses provide the basis to theorize about underlying mechanisms, here too, the formulation of general principles of urban ecology was associated with several detailed studies of *urban morphology*. The first and probably most influential ecological image/model was put forward by Burgess (1927) who, on the basis of his empirical studies in Chicago, has described the city as an entity that expands radially from its center and in the process forms a series of concentric zones (Fig. 2.7): a CBD zone, surrounded by a zone of transition characterised by high residential and business turnover, a working-class zone, a middle-class zone and an outer zone of suburban population-commuters.

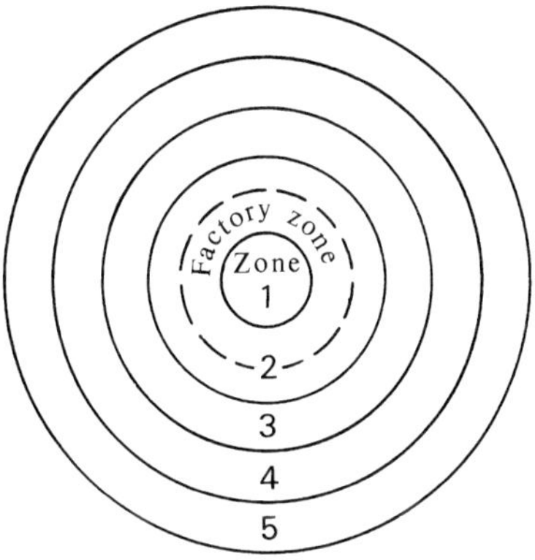

Fig. 2.7. Burgess's concentric zone model. (1) Central Business District, (2) Zone in transition, (3) Zone of working men's homes, (4) Residential zone, (5) commuters' zone

Burgess's model was followed by Hoyt (1939) who, on the basis of empirical studies on rent gradients in American cities, suggested a sectorial morphology of the city. In the latter, relatively homogeneous residential and non-residential areas grow outward from the city center probably along transportation routes, and in the process produce a sectorial pattern as in Fig. 2.8. Burgess and Hoyt models were integrated by Mann (Fig. 2.9) and at a later stage by Ullman and Harris (1945) in their *multiple nuclei model* in which urban growth starts not in one, but in several nuclei thus producing the morphology presented in Fig. 2.10.

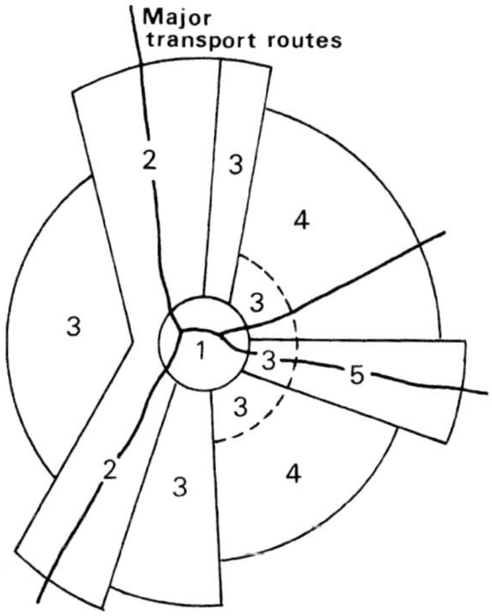

Fig. 2.8. Hoyt's sectorial model. (1) Central Business District, (2) Wholesale and light manufacturing, (3) Low-class residential, (4) Medium-class residential, (5) High-class residential

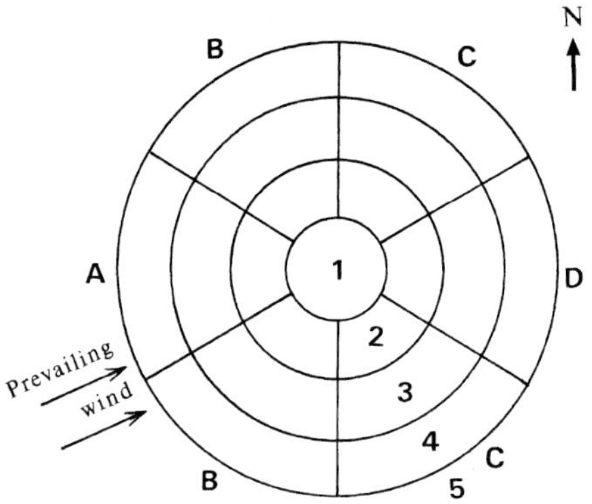

Fig. 2.9. Mann's model of urban structure. (1) Central Business District, (2) Transitional zone, (3) Zone of small terrace houses in sectors C and D; large bye-law housing in sector B, large old houses in sector A, (4) post-1918 residential area, with post-1945 development mainly on the periphery, (5) Commuting distance 'dormitory' towns, (A) Middlle-class sector, (B) Lower-middle class sector, (C) Working-class sector (and main council estates), (D) Industry and lowest working-class sector

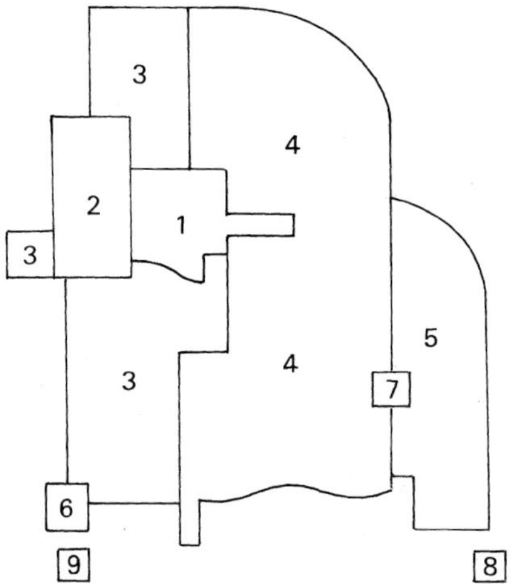

Fig. 2.10. Ullman and Harris's multiple nuclei model. (1) Central Business District, (2) Wholesale light manufacturing, (3) Low-class residential, (4) Medium-class residential, (5) High-class residential, (6) Heavy manufacturing, (7) Outlying business district, (8) Residential district, (9) Residential suburb

2.1.3 Their Common Origin

The morphological similarity between the ecological and economic cities is clear. The ecological city of Burgess is simply almost identical in its form to the von Thünen type concentric economic cities – a similarity which was noted by many. To this I would add that Lösch's city-rich/city-poor sectors are not very different from Hoyt's sectors, especially in light of our comment above that Christaller's and Lösch's central place theories were applied to the study of the internal structure of cities.

But the two eco cities are not only visually similar, but also with respect to their underlying mechanism. For both, reality is an arena where plants, animals, individuals and collectivities, compete and fight for survival, and in a similar way, for both the city is the arena for the urban process – the process by which people as individuals and collectivities compete over the urban land(use), either by means of an interplay of spatio-economic *rbc*'s, or by means of ecological invasion and succession processes identified by means of Chicago type factorial ecologies (Berry and Horton 1970).

These similarities between the two types of eco-: the economic and the ecological, are hardly surprising in light of the symbiotic relations that characterized the origin of the theory of economics and the theory of evolution when they first emerged in the first half of the 19th century. In his *Science, Ideology and World View*, Greene (1981) follows in some details the way several ideas about 'free', 'natural', competition and 'survival of the fittest', among individuals and collectivities, first appeared in liberal political economy and social philosophy in the writing of figures such as Malthus, Adam Smith and Herbert Spencer, and only at a later stage inspired Darwin. The impact of Darwin's biological *Origin of Species* was so strong, however, that it pushed to the shade its early historical origins in the human domain.

2.2 Sir Isaac Newton's Cities

Sir Isaac Newton's law of universal gravitation from 1687 states that two bodies in the universe attract each other in proportion to the product of their masses and inversely as the square of their distance. In direct analogy to Newton's law, geographers have suggested that two cities, regions, or districts in a city, interact with each other in proportion to the product of their masses and inversely according to some function of the distance which separate them, that is:

$$I_{ij} = kM_iM_jf(d_{ij}) \qquad (2.1)$$

Interaction (I_{ij}) between two locations i and j might refer to the flow of immigrants, goods, traffic, telephone calls, etc.; M_i, M_j (masses of cities i and j) might refer to population or size of shopping centers; whereas d_{ij}

might refer to geographical distance, economic distance (travel cost), social distance and so on and k is a normalizing constant.

The first models were formulated as a direct analogy to Newton's, with $f = 1/d_{ij}^2$, then, as a consequence of empirical studies it was realized that the power function is not always 2, and the model was thus generalized to $f = 1/d_{ij}^a$, where the power a is determined empirically. At a still later stage, to a larger extent as a consequence of Wilson (1970) work on *Entropy in Urban and Regional Modeling*, the model took the form

$$I_{ij} = A_i B_j M_i M_j \exp(-bd_{ij}) \tag{2.2}$$

where A_i and B_j are balancing factors, interpreted also as accessibility and potential terms. These two terms are interesting as they allow one to envision the city, on the one hand, as an accessibility surface, describing the accessibility of the inhabitants to the city's spatial distribution of goods and services. On the other, as a potential surface, describing the population potential of the city (i.e. spatial demand) to various commercial or service centers of the city. Figure 2.11 is a typical example.

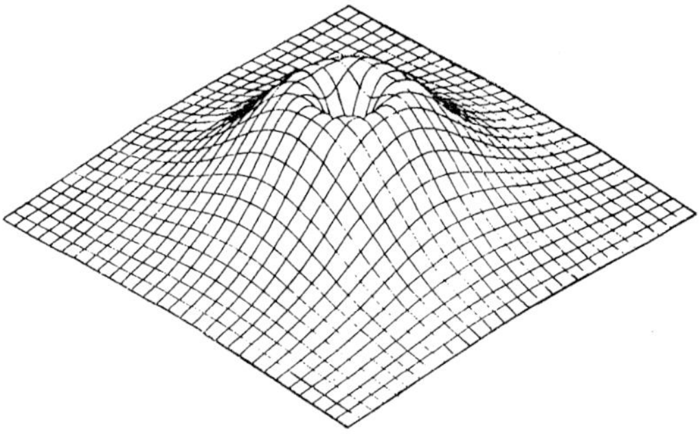

Fig. 2.11. The city of Manchester as an accessibility surface. Source: Angel and Hyman 1972

The family of gravity/interaction models is probably the most prominent form of 'physicalism' in the study of cities and their planning, but not the only one. As discussed by Ollson (1975), and exemplified in Figs. 2.2 and 2.5, $\exp(-bd_{ij})$, the negative exponential distance term, is at the roots of the economic cities of location theory just described, as well as of simulation models such as Hagerstrand's (1967) innovation diffusion and Morrill's (1965) ethnic segregation dynamics. At a more basic level physicalism is but one aspect of the Newtonian-mechanistic world view which forms the foundations for all positivist sciences (Portugali 1985).

2.3 Chicago

I know Chicago fairly well; not the real city of Chicago, but the ecocity of Chicago: its factorial ecology, spatial ethnic segregation, distance decay spatial population distribution, its land-use rings and so on. I know Chicago not because I've seen it – I've never been to Chicago – but because I've studied geography. From the various textbooks I know that there are nice and symmetric 'rings' around the center of Chicago – rings of land use, land value, of cultural groups and of population densities (though I never got rid of the thought that the symmetry of its rings had a lot to do with the location of the city on the lake – with the ease its rings on land could be imaginatively complemented in the water in a very symmetric way).

The whole of the eco-approach is strongly linked to Chicago. Not only to the Chicago School, but to the metropolitan field of Chicago which has become the ideal type of the city of the fifties and sixties: Its land use rings, spatially segregated ethnic groups, rent bid curves and distance decaying population densities appeared in most textbooks (e.g. Fig. 2.12, which is based on data from Berry and Horton 1970) and were forced on almost every

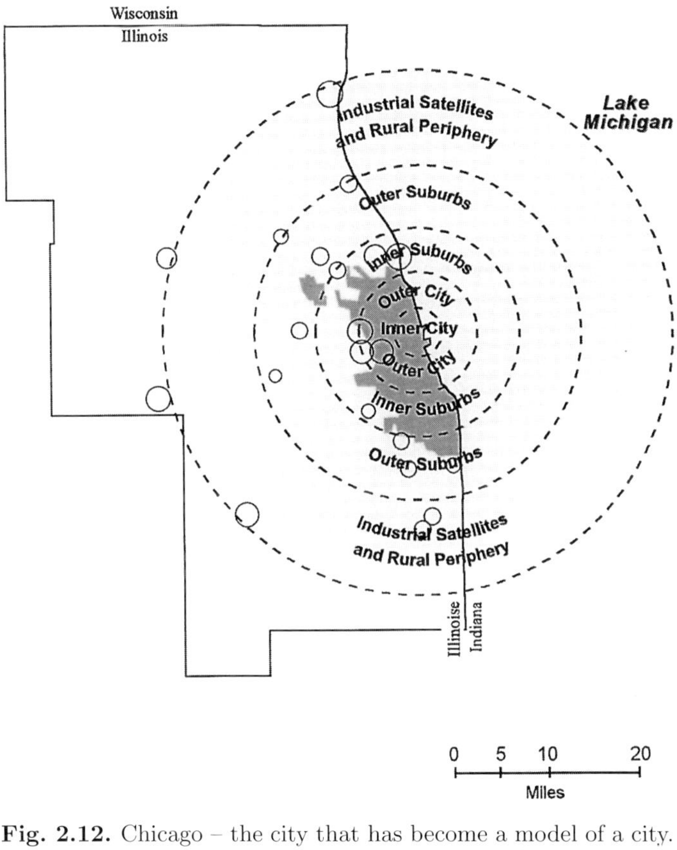

Fig. 2.12. Chicago – the city that has become a model of a city. In the figure, the model of a city is super imposed on the map of the city. Chicago, writes Haggett (1972, p. 342) in his *Geography: A modern synthesis*, "has had a special fascination for geographers. From the 1920s on, its spatial structure has been studied and dissected by a series of distinguished researchers. ..."

student of urbanism, planning and social geography. And when a certain instance of urbanism did not conform with the ideal type, as in the case of so many 'third world' cities, it was treated as exceptional, needing special explanatory maneuvers. The fact that the majority of the world population lives in exceptional cities whose structure and nature depart from Chicago never mattered much.

2.4 Monstrocity

> Will the city disappear or will the whole planet turn into a vast urban hive? – which would be another mode of disappearance? (Mumford 1961, p. 11).

Twentieth century city and urbanism were not the classical good old city of the noble Greeks and Romans, nor the city of the Occident which for Max Weber symbolized the ideal type of a city and for Mumford the high days of urbanism; it was no longer 'the city [as it used to be] in history', nor its ideal urban socio-cultural community. As painted by Mumford it now became a poor reminder of a great past, a decaying entity possibly doomed to deteriorate, disintegrate and maybe disappear. Mumford is the prominent figure in 'a group of extraordinarily effective and dedicated people' – planners who saw early 20th century cities and urbanism as the incarnation of all evil. In her *The Death and Life of Great American Cities* Jean Jacobs (1961, p. 31) presents a selective collection of superlatives by which these urban prophets have portrayed the American city of their time:

> The great city was megalopolis, Tyrannopolis, Nekropolis; a monstrocity, a tyranny, a living death. It must go. New York's midtown was 'solidified chaos' (Mumford). The shape and appearance of cities was nothing but 'a chaotic accident . . . the summation of the haphazard, antagonistic whims of many self-centered, ill-advised individuals (Stein). The centers of cities amounted to 'a foreground of noise, dirt, beggars, souvenirs, and shrill competitive advertising' (Bauer).

With the ideal type of the Occident city at the center of their minds, reinforced by Howard's (1965) *Garden Cities* and Geddes's (1968) notion of 'Organic Regions' they saw in early 20th century cities nothing but ill; nothing worth deeper study and understanding.

2.5 Taming Megalopolis

And yet this city was also a great challenge to a young generation of enthusiastic professionals for whom the more the city became a metropolis and then megalopolis and even monstrocity, greater was their professional enthusiasm to charge the monster and greater was the expected satisfaction to tame it. Read the preface to Eldredge's (1967) two volume, 1166 page book, *Taming Megalopolis*:

> In order to tame that sprawling beast, the metropolis fast becoming megalopolis, it is first necessary to understand his characteristics. ... Clearly this is a mountainous task and very close to the fringe of our present skills – and the number of skills to be enlisted in such an effort are legion. Traditional city planners – flanked by lawyers, housing experts, civil engineers, public health specialists, students of air pollution, conservationalists, social psychologists, real estate agents, sociologists, anthropologists, urban designers and architects, public administration scholars, computer programmers, political scientists, regionalists, economic geographers, economists, traffic engineers, recreation planners, surveyors of the higher culture, etc. – may be able to at least get on with the job. But the mere complexity of the matter demands that new higher levels of human thought and skills must be involved. These may take the form of an overarching synthesizing profession of planner of the total environment and of total systems analysis employing all professional specialists .... The logical plan of this book is simplicity itself: Part One, *What Is*; Part Two, *What Could Be*; and Part Three, *How to Manage an Urbanized World*, or how to get there.

With their backs to the ill cities of the past and to the prophetic melancholy of Mumford, they have pointed to a new brave and challenging urbanism. The city is nothing more than a big machine, they said; complex, problematic, full of 'noise' – yes, but at the same time controllable, socially-engineerable; just provide me an Archimedean point to lean on, to start with, and I'll move for you the whole city, metropolis, megalopolis.

2.6 Plan the Beast

And Archimedean point there was: From one side came the whole scientific arsenal of location theory and spatial analysis – their land use rings, rent bid curves, $k = 3, 4, 7 \ldots$ hexagonal central places, gravity/interaction models and all the rest. They provided explanation. From the other direction came *The Radiant City and The City of Tomorrow* as designed by Le Corbusier (1967) with its superblocks and wide urban highways. Le Corbusier and other architects provided the vision – the image of the city of tomorrow. Still from a third direction came the system people. Influences by Bertalanffy's (1968) *General System Theory* which goes beyond any specific disciplinary or professional domain, and equipped with notions such as 'open' and 'closed' and 'cascading' systems they have shown how to integrate and coordinate the location theorists with the architects and the many other experts which are listed in Eldredge's introduction (Chadwick 1971, McLoughlin 1969). At the front of this army of professional urbanists stood the planner – not anymore an utopist such as Howard or Le Corbusier of the first half of the 20th century, but a new modern(ist) planner – rational, comprehensive, with self-image of a professional, a technician; a planner which conducts his or her planning activities in line with a scientifically defined course of action (Fig. 2.13): First, identify the problems, then describe the system and formulate goals and objectives, then build models and use them to predict alternative futures, now evaluate the alternatives, choose the best one, implement it and send the outcome as a feedback to the starting point. All was prepared and ready to move and the taming of the shrew city was only a small step ahead. But the project failed. The beast refused to be planned and tamed.

2.6.1 The Disillusionment, or the First Planning Dilemma

By the end of the 1960s and early 1970s came the disillusionment. It became evident that 'rational comprehensive planning' (Meyerson 1956) is an irrational assumption, that planning is a political, incremental (Lindblom 1959) and essentially 'non-scientific' and non-technical process; it became apparent that planners cannot rationally tame the environment, not the global natural environment, nor the artificial (Simon 1979) environment – the towns, cities, metropoles and megalopoles – the biggest artifacts ever produced by humans; it became evident that science and technology can send expeditions to the moon, but cannot handle 'simple' sewage disposal problems in cities (Galdbraith 1975), nor traffic jams, not to speak of the really complex socio-spatial environmental problems such as pollution on a global scale, or poverty ghettos in Western metropoles, or poverty 'squatters in the megacities of the developing countries.

At the late 60s and early 70s urban, regional and environmental planning were in a dilemma which can be stated as follows: 'How can we account for the coexistence of spectacular scientific instruments to control the environment and for their failure in the efforts to tame the city, the metropolis, the megalopolis, the environment? How can we account for the fact that beautiful scientific instruments such as the gravity, interaction, or entropy maximization models (Wilson 1970) can hardly scratch the complexity of the urban scenario; and that so are the 'rent bid curves' of Alonso's (1964) urban land use theory, and the 'factorial ecology' of Chicago's 'urban ecology' (Berry and Horton 1970), and the 'location triangle' of Weber's (1929) industrial location theory, and the hexagonal geometrical landscapes of Christaller's (1966) and Lösch's (1954) 'central place theory'. All this scientifico–mathematical arsenal seemed "incapable of saying anything of depth and profundity about [the real problem of society and] ... when we do say something, it appears trite and rather ludicrous". So wrote David Harvey (1973, p. 129), the 'Danny la rough' of geography and planning of the early 70s, in his *Social Justice and the City* published in 1973.

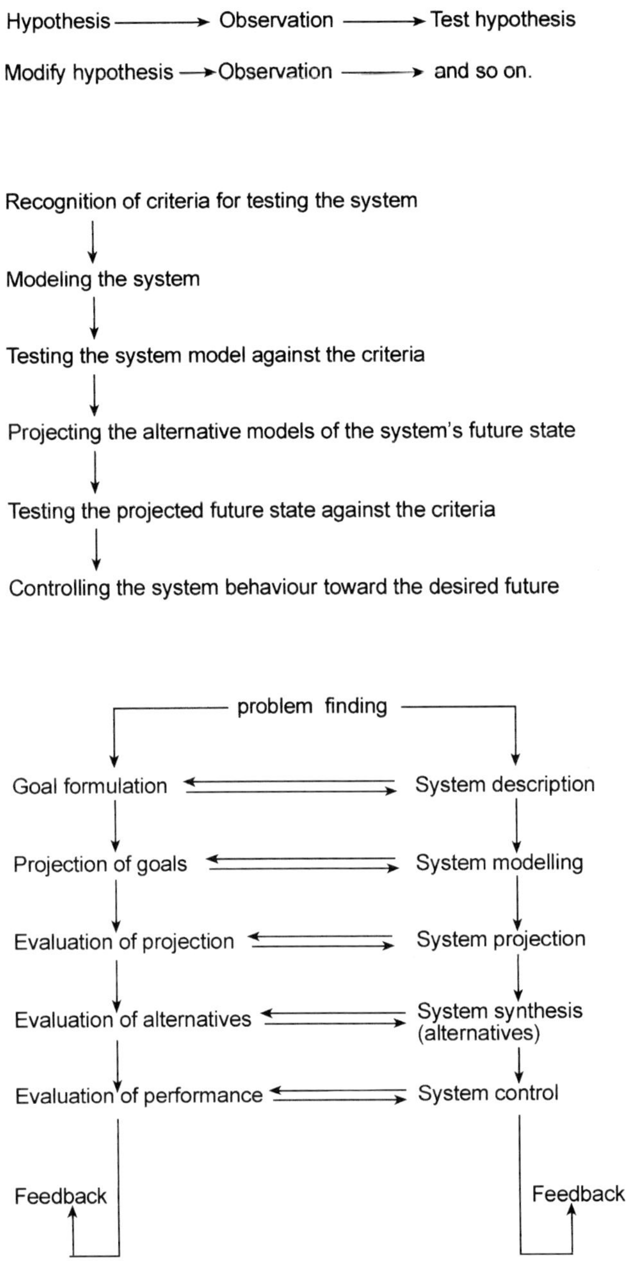

Fig. 2.13. Chadwick's (1971, pp. 66–68) derivation of *A System View of Planning*. The process of scientific enquiry of simple systems (*top*) is transformed into a process of scientific inquiry of complex systems (*middle*), and then into a system approach to rational planning (*bottom*)

2.7 The City of (In)Justice

David Harvey's (1973) *Social Justice and the City* pushed aside to the periphery the *rbc*'s, the land-use patterns, and all the rest. Suddenly the surfaces of our cities became transparent and through them we could clearly observe the deep structure of society, the ruling and the ruled classes and their relations of production as well as social relations of production, ... we started to see the hidden structure of the Capitalist engine which is the real power responsible for all that we see on the surface: the high rises, the suburbs, the movement of people, the rich and the poor, their wants, dreams, hates and loves ... all these, be they big and magnificent or small and ugly became secondary, peripheral, at best trivial surface representations, or a momentary configuration in a huge chess game played by the historical social forces and their relations of production.

Your efforts to tame this sprawling beast of Eldredge are Siziphean efforts, said Harvey to the legions of professionals listed in Eldredge's introduction; your land use rings, *rbc* and spatial interaction models are "incapable of saying anything of depth and profundity ..." (Harvey 1973, p. 129); not because you are bad guys or second-rate professionals, but because, like Don Quixote, you tilt at windmills, and by so doing not only do you not harm the real beast – the Capitalist mode of production – but you actually participate in its reproduction; not only that your sophisticated plans do not reduce injustice in the city, but they reproduce it; create it anew again and again. Capitalism with its city of injustice is advancing and flourishing behind the ideological false consciousness of your liberal science and planning.

Had there been a sub-title to Harvey's book it might have read 'the transformation of a social geographer from a liberal positivist into a structuralist-Marxist Social Theorist'. David Harvey's personal story is, in a sense, the story of a whole generation of geographers. As a participant in the 'quantitative revolution' of geography of the 1950s he gradually became one of the authorities in Positivist-quantitative geography. His book *Explanation in Geography*, published in 1969, is the best, and most comprehensive, geographical synthesis of Positivist theory, methodology and philosophy.

Harvey started to doubt his own 'Explanation' already when it was in press (Harvey 1970), and his 'Social Justice' is, an all-out Marxist-structuralist attack on his Liberal–Capitalistic 'Explanation'. 'Social Justice' has triggered a bombardment of anti-positivism, anti-quantification criticisms, mostly by the ex-leaders of the quantitative-positivist geography. Bunge the ex-theory and model builder, Gould the ex-behaviorist, King the ex-statistician, Smith the ex-neo-classical location theorist, to name but a few of the prominent figures; one by one they stand up, as in a methodist church, criticize their positivist-quantitative past and promise a new, socially relevant, geography (AAAG 1979).

If the early 1970s have been years of criticism, the late 70s were the years of re-formulations. The Journal *Antipode* became the major (though not the only) outlet for radical-Marxist geographers and planners in the US, and likewise the *International Journal for Urban and Regional Research* in Europe. Relph's humanistic–Heideggerian *Place and Placelessness* appeared in 1976, and a year later Yi-Fu Tuan's phenomenological monograph on experiential *Place and Space*. In 1977 Peet's Marxist *Radical Geography* was published. Ley and Samuels have edited *Humanistic Geography* and texts of forgotten geographers, such as the anarchists Elisee Reclus (Dunbar 1981) and prince Peter Kropotkin (Breitbart 1981), reappeared on the stage as a starting point for a renewed, socially relevant geography and planning. In 1980 Kirk came out with *Urban Planning in a Capitalist Society*, while in 1981 Dear and Scott with the edited *Urbanization and Urban Planning in Capitalist Society* – a collection of essays attempting "to define a general theory of urbanization and planning . . . a theory that generally insists upon the explicit derivation of contemporary urbanization processes out of the structure of the capitalist mode of production" (p. XIII). In 1982 once again came David Harvey, now with *The Limits of Capital* in which he attempts to go beyond Marx's *Capital* in order to fill some 'empty boxes' in the Marxist theory, especially those related to 'the urban process under capitalism'; a project which was completed three years later with the simultaneous appearance of his two new volumes *The Urbanization of Capital* (Harvey 1985a) and *Consciousness and the Urban Experience* (Harvey 1985b).

The above partial and by no means exhaustive list will suffice to give you a feeling for the spirit of geography and planning since the mid 1970s. You can see that the 'quotations' for the geographical and planning writings of Harvey, Peet, or Kirk, come from Marx, Engels, Lefebvre, Thompson, Althusser, Levi-Strauss or Habermas, while Tuan, Relph and Battimer, 'quote' Schutz, Husserl, Sartre, Buber, Gadamer and Heidegger. In short, these geographers and planners were inspired by 'social theory' or 'social philosophy' – that body of literature which takes the entity 'society' as its focal subject matter.

This is significant since 'geography as social theory' was not a side-stream of geographers interested in the philosophy of science and the role of the discipline in it, but it became the central current of the discipline affecting many planning schemes, almost every research, theoretical as well as empirical. This, I think, has no parallel in any other discipline not even in sociology whose source of origin was social theory itself. Twentieth century sociology, as noted by Frisby and Sayer (1986), has disentangled itself from social theory and from grandiose questions such as 'what is society' or 'how society is possible', or, what are the relations between the individual agent and the social structure. Contradictory as it may sound, geography and planning have done exactly this and became 'applied social theory'. Social theory has become a theoretical framework for studies on interregional migration, location of industries, regional or urban inequalities . . . geography and planning have taken social theory from the high spheres of philosophy 'down to earth'; literally.

2.8 The Humanistic City of Everyday Life

The humanistic city of everyday life is founded on two main columns: 'the city of daily routines' and 'the cognitive city in the head'. Both cities originated from the eco-physicalist cities of the 1950s and the 1960s and both mark their collapse.

2.8.1 The City of Daily Routines

"What about people in regional analysis" says the title of Hagerstrand's (1970) seminal paper which starts a new brand of geography – *time-geography*. In this paper Hagerstrand puts forward several innovative suggestions. First, that in addition to the study of cities by means of representative statistical samples, we might as well benefit from studying single individuals. Second, that this can be done by mapping the movement of individuals not only in space, as is usual in studies of the ecocity, but in space and time simultaneously – hence, time–geography. Third, when observing an individual's trajectories in space–time, one realizes that the individual is always in movement: when staying in a single location s/he is moving in time; when moving between locations, the individual is moving in space–time. Fourth, when observing the space–time movement of individuals in the city, one realizes that much of it is routinized (Fig. 2.14) – hence the notion of 'dance' in this context. Fifth, that by observing individuals' space–time movement, or 'dance',

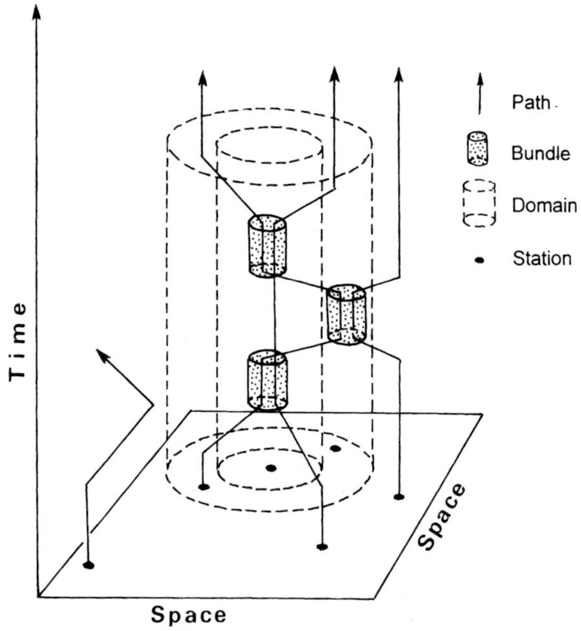

Fig. 2.14. Hagerstrand's web model of time-geography

in the city, and by focusing mainly on their routinized movement, one can learn about the nature of the urban environment within which people are operating. The city of daily routines opens a window into the significance of the 'weak force' of the city – the daily and the personal and the ordinary; not only as a methodology to identify and study the system of space–time constraints within which individals are operating, but also into the way individuals by their normal day to day routines create and construct the city as a humane place.

2.8.2 The Cognitive City in the Head

This is a city as human individuals, subjectively and inter-subjectively, perceive, remember, cognize, imagine, it to be: the visual shape of the city with its neighborhoods, streets and parks, as well as its non-visual properties of good/bad streets, prestigeous buildings, safe, dangerous, pleasant, friendly, alienated, neighborhoods, and the like. The city in the head was first designed to give urbanists a more realistic and refined basis for modeling human behavior in the city: In place of the 'homo-economicus' of the ecocity we now have the 'homo-imaginus' – an individual who takes decisions in the city according to what s/he thinks, imagines, perceives ..., the city is, or the distances and directions between locations are (Fig. 2.15). This subjective and inter-subjective cognitive map of the city replaces the measured objective map as a basis for decision making. Individuals decide to, and migrate, from one place to another in the city, or prefer to, and shop, in a certain area and not in another, or choose their commuting route, in line with a city as they see it in their mind's eye. But when they do so, when they migrate, shop, and commute, they participate in shaping and constructing the city and its patterns of activity, which they then perceive, imagine, cognize as their own subjective cognitive city, according to which they once again migrate, shop, commute and so on.

2.8.3 The Humanistic City

As you can see the cognitive city shapes individuals' daily activities and routines, which in their turn construct the city, and individuals' cognitive city, and so on in an ongoing process of reproduction and structuration. The heroes of this ongoing play of urban dynamics, are human individuals with their subjective near and far, big and small, pleasant and ugly – elements and properties which have no role to play in the theatre of the ecocity, but are the main actors of the humanistic city.

By looking at the city from within, from the perspective of people, of the individual, the humanistic city attempts to animate the skin and bones of the physico-ecocity; tries to capture the added existential, phenomenological, experiential, quality of the city; the *real* city as experienced by the people

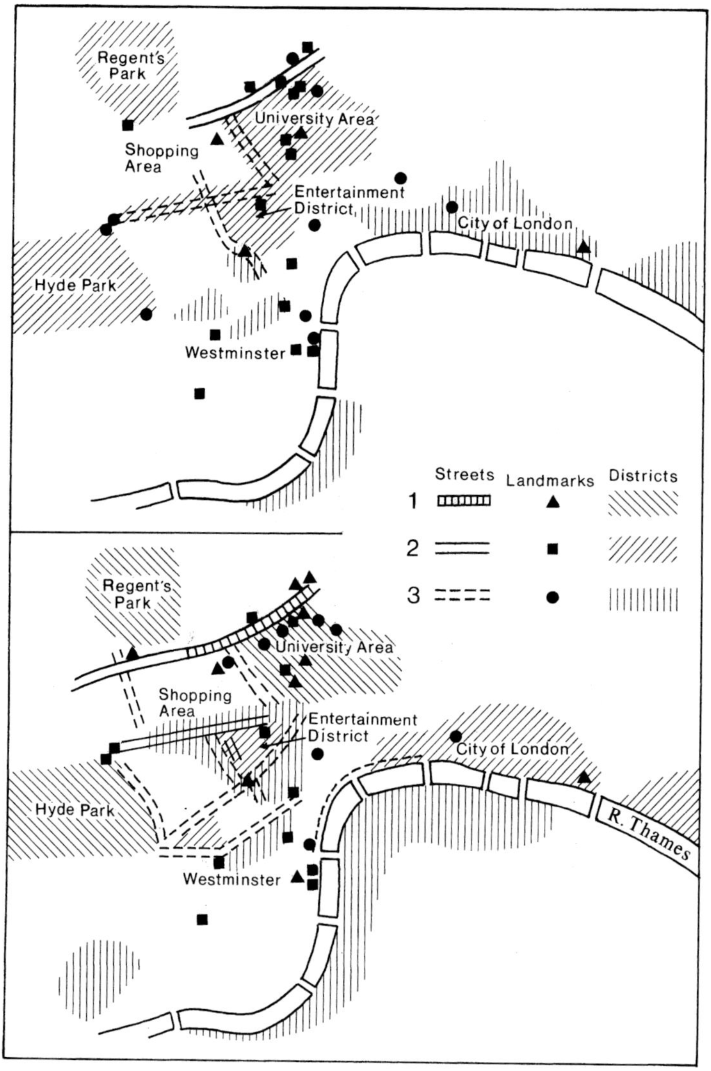

Fig. 2.15. An illustration of how a newcomer's familiarity with, and knowledge of, London changes with time. *Top*: The newcomer's cognitive map at the beginning, and *bottom*: after being some time in the city. Source: Ley (1983, 106)

who actually create and construct the city. It tries to capture not only the sense of *place* – those portions of the city which were created by people and thus directly transmit individuality, human scale and 'peopleness'; but also the sense of *placelessness* – those impersonal and alienated parts of the city created not by the personal wishes and activities of people, but by the impersonal interests of the multi-national, the big company, the institution, the system.

In this humane domain of qualitative and subjective properties, the positivist-quantitative-statistical methods of the old ecocity were of very limited use. Of limited use also were the formal check-lists of the rational-comprehensive. In their place humanistic geographers, urbanists and planners had to turn to "softer" methods of hermeneutics, free observation and conversation; and to the planners they have suggested learning, understanding and awareness: learn, understand and be aware of, the place and placelessness of cities, and this knowing, understanding and awareness will guide your planning decisions and actions (Johnston 1988).

2.9 The Marx(ist) City

Marx, as is well known, had very little to say about space and cities and yet the theory of cities and urbanism is intimately linked with his writting. This is so with respect to the discourse of ancient cities and urbanism which is strongly influenced by the Marxist interpretations of Childe's 'Urban Revolution' and of Whitfogel's 'hydraulic society'; and this is so with respect to discourse of urbanism in modern capitalist society in which central figures are people like Lefebvre with his (once again) 'Urban Revolution', Harvey with *The Urbanization of Capital* and Castells with his *The Urban Question.* (See further discussion in Chap. 15).

The Marxist (image of the) city is a city of big forces, of modes and relations of production, of class struggle between the oppressing and the oppressed, of the capitalists versus the working class, of the economic infrastructure and the political–ideological superstructure, all obeying the laws of history as revealed by Marx and elaborated by Marxism. Two such images, one of Castells' and the other of Harvey's, will suffice to give you the spirit of this new Marxist city.

Castells (1977) constructs his Marxist city by translating Althusser's non-spatial structuralist-Marxist conception of social structure to a spatial urban structure. Here the city is the spatial expression, or representation, of the structure of society as Marx and Marxism have revealed and elaborated. This translation is shown in Fig. 2.16.

In *The Urbanization of Capital* Harvey (1985a) portrayed another image of a Marxist city. It is constructed by showing how the very laws of capitalism as formulated by Marx(ism) entail, as a logical consequence, the specific capitalist urban landscape as we know it today. Capitalism as the dominant mode of production of our age is characterized by structurally inherent inner tensions and contradictions and is thus chronically unstable. As a consequence, the landscape of capitalism is full of tensions. One very basic tension

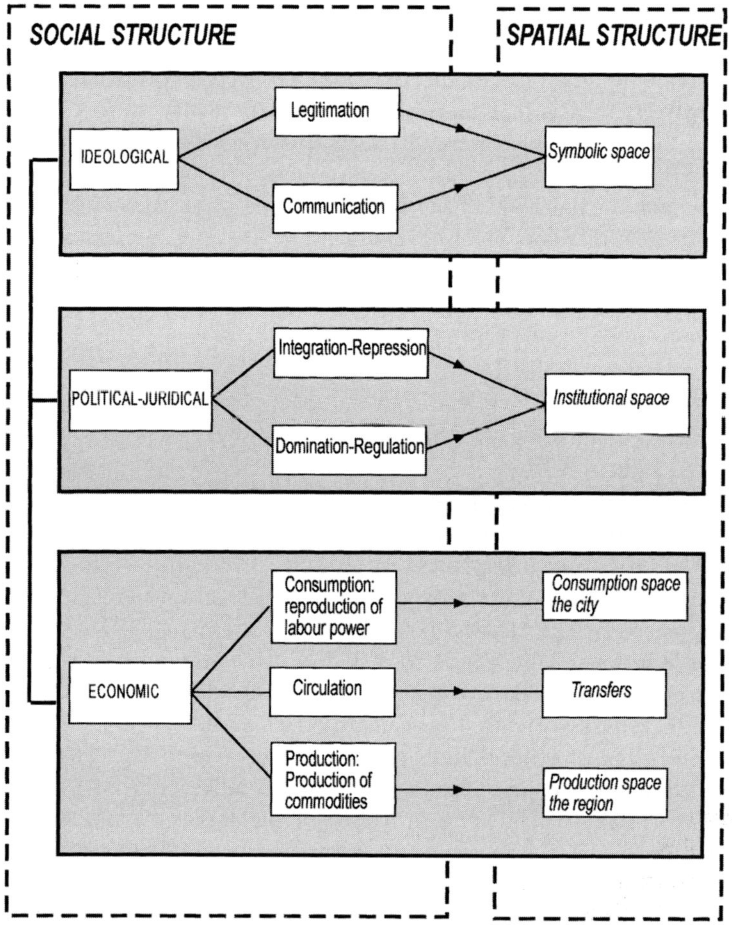

Fig. 2.16. The Marxist city as a spatial representation of social structure. After: Gregory 1994, 94

is between forces working toward spatial agglomeration and processes working toward dispersal over space (Fig. 2.17). The agglomeration processes pull clusters of economic activities into specific locations and regions, thus forming cities, and at the same time the inherently expansionary tendency of capitalism which demand that capital accumulation be also dependent upon time–space coordination between regions pushes toward spatial dispersion of economic activities. According to Harvey, this tension can be resolved only through 'the urbanization of capital' (Gregory 1994).

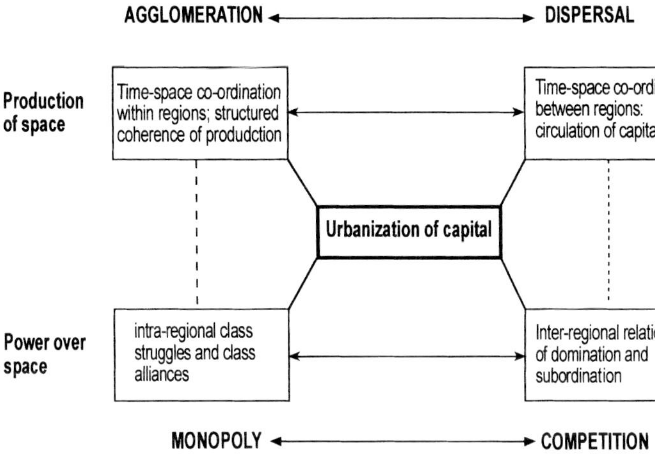

Fig. 2.17. The Marxist city as an outcome of basic tensions in the landscape of capitalism. After: Gregory, ibid, 92

2.10 De-Visualized Cities

Lösch, writes Gregory (1994, p. 58) in his beautiful chapter on 'geography and the world-as-exhibition', "wanted to disclose, to make visible, the systematic order of a rational economic landscape ...". This was characteristic, as Gregory shows, of the whole of the quantitative-positivist geography of the 1950s and 1960s and of what we have described above as the ecocity and as the city of Isaac Newton. The discussion of the two cities is very visualized in the sense that it is loaded with maps, diagrams of spatial order, geometries – all attempting to externalize to the surface, to expose and thus visualize, social, cultural, economic or political relations. Traditional geography and cartography provided society with the map – a medium with which one could literally see (in fact imagine seeing) a world, a city, otherwise invisible to the human eye; geography as a spatial science, with its ecocity and physicalist city, provided a medium with which one could practically see (imagine seeing) not only the physical landscape of cities, but also the dynamic and structure of socio-economic or cultural relations as they take place in the city. The visual became a language, and the very act of observing and seeing, an important component in the processes of explanation, analysis and interpretation.

This is hardly surprising if you realize that cities can be regarded as very large artifacts, and that it is inconceivable to study artifacts, small like needles, chairs or books, and large like cities and metropolitan fields, without reference to their detailed spatio-visual morphological properties. Yet SMH (structuralist–Marxist–humanistic) geography and planning did exactly this: "The way things appear to us in daily life", writes Harvey, and by that he means the way we visualize things, including cities, as relatively independent spatio-visual objects, "can conceal as much as it can reveal about their social meaning. This idea Marx captured in the concept of 'the fetishism of commodities'" (Harvey 1982, p. 17). And the word 'fetishism', as you know, carries with it an image of savages worshipping an inanimate object for its magical power – an image that a good (Marxist) scientist should stay away from. The visible city became a fetish and the study of its spatio-morphological properties, 'spatial fetishism'. Look at Harvey's the city of justice, or at the humanistic city we have portrayed above, and you'll see that the visual has gone, it is not living there anymore, not in the form of actual visual elements such as plans, maps and geometries, nor, as in Thünen's verbal description of his spatial model – *The Isolated State*, in the form of textual description of the visual-morphological elements of the city.

One of the first victims of SMH attack on positivism was the visual language of geography – the fetishism of the spatial – of the city as a spatial, relatively autonomous, artifact. Space, declared Gregory quoting Foucault, "is fundamental in any exercise of power" (ibid., p. 63). The visual became part of the oppressive. And what did we get instead? text! And is text more humane than the map than the visual? far from it! For one thing, from its very first appearance (together with urbanism and class society, see Chap. 15) text has been an instrument of control and oppression.

2.11 The Split

Harvey's 'Social Justice', 'The Limits' etc., supplemented and supported by Castells' (1977) 'Urban Question', from the one side, and the Humanist city, from the other, not only pushed all the above urban phenomena and entities to the periphery of urbanism, but in at least one instance out of the periphery, as in a Japanese Sumo fight: *Regional Science*, the successor of location theory plus quantitative geography, has to a large extent split from main-stream urban geographical discourse, and started its own, relatively independent urban trajectory: the core of the city and urbanism is here captured by the system – the city and the region as a system, mathematically tamed and controlled by graphs, equations and the hard data that support them. Social Justice plus the Urban Question, and the regional science urban system have developed almost in parallel. And if there has been some interaction between the two it was mainly of regional science taking notice of social justice (as in the recent *Modeling The City* by Bertuglia, Clarke and Wilson 1994).

2.12 The Second Planning Dilemma

This new social geography has enriched human geography, and the study and planning of cities immensely; there can be no doubt about that. It has also enriched social theory itself, as the writings of the social theorist Antony Giddens (1984) would testify. And yet it also has its drawbacks which only recently started to show up; after the first waves of enthusiasm started to calm down, and after social theory itself entered a period of deconstructivism and postmodernism.

What you see today is a general atmosphere of discomfort or disillusionment from structuralist-Marxist-humanist (SMH) geography, urbanism and planning; very similar to the disillusionment of the late 60s. This new social geography and planning which has transplanted 'understanding' in place of 'explanation', and phenomenology in place of spatial analysis, social relevance and connected theory/praxis in place of abstract mathematical games/modeling... somehow have not delivered the goods it promised to deliver. We have been promised a socially relevant geography, linking theory with the daily praxis of commuting, migration, spatial interaction, or environmental perception. And yet it seems that the concepts of class, infrastructure, superstructure, labor, modes of production, relations of production, use-value and so on, are no less abstract and remote from daily human life then the previous positivist *Central Business District*, *Rent Bid Curve*, hexagonal market area and the like. As in the late 60s this became specifically evident when the new socio-spatial theory was applied to the real environment. Here SMH planners found themselves in a frustrated situation: Positivism has at least given you tools for social engineering; SMH geography, urbanism and planning teaches that the tools are false (consciousness), that the game is sold to the politicians, the big money, and the multinationals. As a planner this insight gives you, at first, a great feeling that you know and understand, that you get into the deep structures, expose the real/hidden forces, that intellectually you control the situation? but then what? what are you to do with it when as a planner you have to make a decision about urban renewal, or road networks; what would you say? Start talking about base and superstructure? The labor-process? How this beautiful theoretical insight becomes praxis? Gradually it became evident that SMH geography and planning are remote from reality and social relevance even more than positivism. Thus, since the mid 80s, we hear once again the very same question: 'how can we account...'; but this time not only for the coexistence of great scientific achievements, on the one hand, and the failure to apply them to society, on the other, but also 'how can we account for the failure of the alternatives'.

This general feeling of discomfort is the ground on which 'postmodern geography' – the latest trend to originate from the production line of 'geography as social theory', is currently growing (Harvey 1989, Soja 1989, Cooke 1990). Postmodern geography not only admits the impotence of positivist and SMH geography to control and plan the environment, but makes it its ideological platform. Not only that science cannot control society and its shrew environment, but that it should not attempt to do so. Let society, and its artificial products the city, the metropolis, the megalopolis, be what they have come to be: uncontrollable, unpredictable and unplannable. Let us urbanists and planners make them more so by deconstructing all that has been constructed; let us deconstruct disciplinary boundaries inside science and between it and art.

2.13 The Postmodern City

Today's city – the city of end 20th century – is the postmodern city: untamed, shrew, capricious, ever-changing; actually it is not a city but a text; a text written by millions of unknown writers, unaware that they are writers, read by millions of readers, each reading his or her own personal and subjective story in this ever-changing chaotic text, thus changing and recreating and further complicating it. Today's urbanism is a big theatre at the center of whose stage we see a kaleidoscope of shapes and forms, high-tech science-fiction structures mixed with Phareonic lotus capitals from second millennium B.C. Egypt (e.g. Sterling's Staats Galerie in Stuttgart), and in and around it a similarly pluralistic kaleidoscope of cultures and sub-cultures of Italians, Chinese, Japanese, Jews, Indians, Gays, Lesbians ... Somewhere at the side of this end-2nd-millennium urban stage, way in the periphery, one can still observe the base and the superstructure, with their modes and relations of production, the capitalists versus the proletariat, talking, shouting in vain trying to be heard. In this postmodern city nothing is stable, nothing is true nor matters for more than a second, not the Marxist urban categories, nor any other grand theory or truth: all must go, must move, clear the way to the new next whatever it is.

For the pessimist positivist and SMH urbanist or planner this sounds chaotic; for optimists – highly desirable and creative: an ever changing, ever moving reality. Yet this is not what takes place in the currently emerging postmodern reality; at least not up till now. Take the emerging urban landscape and architecture, probably the most visible representation of postmodernism. Indeed it started with free and creative quotations from the futurist high-tech Archigram architecture and 2nd millennium B.C. Egyptian columns. However, very quickly it turned into a uniform style with the double column, the gable and the circular window as its trademarks. Walk around in Tel Aviv or many other cities and you'll see that you can't avoid instantly identifying the uniform, postmodern, style. And yet, the words 'uniform' and 'style' are the very opposite of postmodernism.

So the postmodern trend toward plurality and coexistence, in art, science, urbanism and society, is at least problematic – and this is indeed the deadlock of the 1990s: You can't tame, plan, engineer, the environment, since you are trapped in its chaos, and you cannot participate in its chaotic play since you are trapped in its structure, fashion and style. This deadlock shows up in a debate in the geographical journal *Society and Space* (1987) on modern and post-modern geography and planning. Postmodern geographers accuse David

Harvey that his Marxist-structuralist geography is totalitarian, and Harvey responses that there is nothing more totalitarian than such an accusation. Both are right, of course, since both are playing a zero-sum game: there is no room for chaos in the highly ordered Marxist world and there is no room for stable structures in the highly chaotic postmodern city.

2.14 Los Angeles

I've been to LA once, for a short period of time, and remember very little of it. These days I am getting to know LA once again; not by being there, however, but by reading what geographers like Soja (1989) and Gregory (1994) have to say about urbanism in general and urbanism at the end of the 20th century in particular. I learn that LA is *the* city of end 20th century, the 'quintessential postmodern place', a city which needs no deconstruction because it is deconstructed with no center, no Chicago in its structure. Is LA the most typical city of end 20th century? Far from it. It is unique; unlike any other city in the world. But, quote Soja (1989) in his *Postmodern Geographies* and Gregory (1994) in his *Geographical Imaginations*, "it all comes together in Los Angeles". People like Soja, Gregory and others work hard these days to 'Chicago' LA, to make it the ideal type, the prototypical city at the core of the huge network of cities which exist in the here and now of late 20th century, to make it a measure against which the rest of the cities of the world will have to be evaluated. Will they succeed? Will LA become a 'paradigmatic city', 'a prototopos', 'a megocosm', as Soja wants it to be? "an ordered world in which the micro and the macro, the ideographic and the nomothetic, the concrete and abstract can be seen simultaneously in an articulated and interactive combination"? (Soja, ibid, p. 191). Do we see here the beginning of the postmodern LA School which will replace the passe modern Chicago School? In order to get an answer to these questions we'll have to wait for a while. And while waiting we can look and see how an ideal type of city is being created or at least attempted.

2.15 The Hypermodern Self-Organizing City

Strangely enough, a picture of urbanism very similar to the postmodern city is emerging out of the regional science vision – that branch of geography which since Harvey's 'social justice revolution' in 1973, was pushed to the periphery of the geographical ring or altogether out of it. Today's cutting edge of regional science is the theory of self-organization as formulated by people such as Hermann Haken (1983, 1987) the inventor of the theory of Synergetics and Nobel prize laureate Illia Prigogine who developed the notion of 'dissipative structures' (Nicolis and Prigogine 1977; Prigogine 1980; Prigogine and Stengers 1985). Both persons are mathematicians and physicists, both originated in the sciences, yet both have found in matter properties hitherto assigned to humans and humanity: history, evolution, unpredictability, irreversibility, nonlinearity ... (Portugali 1985). Such properties, they have found, are typical to open, complex and thus self-organizing systems. 'Open', in the sense that they exchange matter, energy, information and people, with their environment; 'complex', in the sense that their parts are so numerous, and changing, that there is no way to describe them in terms of cause and effect (as did the urbanists of the 1950s and 1960s), nor in terms of probabilities (as did the urbanists since the end 1960s and the regional scientists of the 1970s and 1980s). Such cities are thus chaotic and unpredictable and they self-organize themselves independently of our scientific predictions and planning rules. All that is left for us to do, as scientist and planners, is to sit and watch, or at best become participants in this huge self-organizing process. Here too, life in general and urbanism in particular, are like one big theatre in which we the audience–scientists–planners sit and look and often respond to the dynamics of this changing complexity, and by so doing we – like the audience in the theatre – also play our role in this on-going 'game of life'.

The city of self-organization is related to the many cities described above in three important ways. First, as just noted, it has many similarities to the postmodern city. Second, in the domain of cities and urbanism it evolved mainly out of the discourse of regional science – the successor of location theory and quantitative geography with their Ecocity and Newtonian city. Third, it has many resemblances with SMH geography and urbanism and through them to social theory. In particular to the structuralist-Marxist trend: The Marxist city begins with the perception that the urban system under capitalism is 'chronically unstable' – in the domain of self-organization one is usually dealing with systems 'in far from equilibrium conditions'. In Harvey's city this chronic instability is a property of capitalism. Postmodernists claim that this is a property of the postmodern city of our age. In the self-organizing city this is a property of all complex, open and thus self-organizing systems. Marxists see social change in terms of a sequence of revolutions; hence Childe's urban revolution of antiquity and Lefebvre's urban revolution of modernity (see Chap. 15). Self-organization theory sees systemic change in term of a sequence of bifurcations by means of which the system jumps from one steady state to another. Furtheremore, as we shall see in the next chapter, according to Haken's synergetics approach to self-organization each such steady state is governed by one or a few 'order parameter(s)' – a notion which is not very different from the 'mode of production' which governs society's development during its non-revolutionary steady-state periods; and in both social theory and synergetics individuals by their daily routine activities obey the laws of the dominant mode of production and by so doing they also support its continuation. In social theory this process is called social reproduction and/or structuration, in synergetics – circular causation.

The above properties of the notion of self-organization – its association with quantitative approaches, on the one hand, and with qualitative-hermeneutic approaches, on the other – make self-organization a paradigm which has the potential to provide a common integrative ground to the various cities described above, and thus to reconcile their seemingly irreconcilable nature. Showing this is a central task of the remainder of this book.

3. Self-Organizing Cities

Self-organization, that is to say, the phenomena by which a system self-organizes its internal structure independent of external causes, is a fundamental property, as we have seen in the preceding chapter, of open and complex systems. Such systems exhibit also phenomena of nonlinearity, instability, fractal structures and chaos – phenomena which are intimately related to the general sensation of life and urbanism at the end of the 20th century. On the other hand, self-organization is a formal theory. It is, in fact, a general umbrella for several theoretical approaches which, while agreeing on general principles, differ in their treatment of such systems, in the emphasis they give to the various processes and properties, and in the subject matters they refer to. In this chapter I discuss some of those theories and methodologies of self-organization which were applied to the domain of cities and urbanism. The discussion proceeds under the title of seven categories of cities which are related to general theories or specific methodologies: dissipative cities, synergetic cities, chaotic cities, fractal cities, cellular automata cities, sand-pile cities, and FACS & IRN cities. The discussion of each category of cities starts with a short introduction to the general principles of the approach and then elaborates its self-organizing city. However, before starting with the self-organizing cities let us say a few introductory words about self-organization as a general theory.

3.1 A Concise Introduction to Self-Organization

While the concept of self-organization appeared already in the early years of cybernetics, implicitly in the writings of McCulloch and Pitts (1943), and explicitly in studies such as Ashby's (1947) psychological discussion of the nervous system, and Yovits and Cameron's (1959) and von Foerster and Zopf's (1962) studies in the domain of system theory, its modern form is related to several theories developed since the mid sixties. In particular to Haken's (1983a, 1983b, 1987, 1990, 1996) theory of *synergetics*, to Prigogine's *dissipative structures* (Nicolis and Prigogine 1977, Prigogine 1980, Prigogine and Stengers 1984), to Eigen's (1971) *catalytic networks*, as well as to Lovelock (1979), Maturana and Varela (1980), and Margulis (1995). Of the latter, Haken's synergetics and Prigogine's dissipative structures were probably the most influential in the domain of cities and urbanism and our concise introduction will lean mainly on their interpretations. Further notions, such as the mathematics of *complexity*, Mandelbrot's *fractal geometry*, the mathematics of *chaos*, models of *cellular automata* (CA) and *self-organized criticality*, will be discussed in subsequent sections.

The authors who first coined the notion of self-organization were fascinated mainly by the property of non-causality of the systems they have confronted. That is to say, by the finding that in certain situations external forces acting on the system do not determine/cause its behavior, but instead trigger an internal and independent process by which the system spontaneously self-organizes itself. The authors of the second wave of studies, led, as noted, by physicists such as Haken and Prigogine, were attracted by an even more complex process; the latter can be nicely illustrated by the Bénard experiment which has since become a classical way to convey the notion of self-organization.

Bénard, a French physicist working at the beginning of the century, found the following about a liquid in a vessel heated from below: At the beginning of the process, when the temperature difference between the heated bottom and the cool top is low, the heat is being transferred by conduction and no macro-motion can be observed in the liquid. However, as the temperature difference increases and a certain threshold is reached, the movement in the liquid becomes unstable, chaotic and then a strikingly ordered pattern appears: The molecules of the liquid which at the beginning were moving at random, suddenly exhibit a coherent macro-movement in rolls which are millions times larger than the molecules. And if the liquid is contained in a round vessel, the motion of the rolls form a hexagonal pattern on the surface of the liquid (Fig. 3.1). This pattern is in fact an outcome of the movement of the hot liquid which rises through the center of the honeycomb cells, and of the cooler liquid which falls along their walls. All this happens as if by an external force. Yet no such force exists – the spatial order appears spontaneously, by means of self-organization.

Fig. 3.1. Top view of a liquid in a circular vessel (From Haken 1996). When the liquid is heated from below and the temperature gradient exceeds a critical value, hexagonal cells are formed. In the middle of each cell the liquid rises, sinking back down at the edges of the hexagon. Compare to Figs. 2.3–2.7, above

The Bénard experiment, which has been repeated and elaborated by others, exhibits the main features of self-organization. First, that a system that is open and is thus part of its environment can attain a spatio–temporal structure and maintain it in far from equilibrium conditions; not in spite of, but as a consequence of, a sufficient flow of energy and matter. This contradicts the traditionally held view in physics, that systems must be looked at as essentially isolated from their environment. According to the second law of thermodynamics, such systems tend to move toward molecular disorder, that is to say, toward an increase of entropy.

Second, that this flow of energy and matter through its boundaries allows the system not only to spontaneously self-organize itself, attain a certain structure and maintain it in far from equilibrium conditions, but also to 'create' or 'invent' novel structures and new and novel modes of behavior. Self-organized systems are thus said to be 'creative'.

Third, self-organized systems are complex in two respects. One, in the sense that their parts (e.g. the molecules in the liquid) are so numerous that there is no technical way to establish causal relations among them. Two, in that their parts and components are interconnected in a nonlinear fashion by a complex network of feedback loops. Mathematically this form of behavior can be described by a set of nonlinear equations.

All the above implies a new view of science with a potential to bridge the traditional division between Snow's (1964) *Two Cultures*. That is to say, that in some cases inanimate material systems – the subject matter of the 'hard/exact' sciences – exhibit evolution, history, creativity, irreversibility and so on, that is, phenomena hitherto assigned to the subject matter of the 'soft' life sciences and even to the 'softer' social and human sciences. This property of the new theory came out from the start, as in the case of Nicolis and Prigogine (1977, p. 4) who in order to convey the notion of self-organization to their fellow physicists and chemists, wrote that "An appropriate illustration would be a town that can only survive as long as it is a center for inflow of food, fuel ... and sends out products and wastes".

From the start it was felt by the proponents of self-organization that what one faces here is not just a new, more elaborated, physical–chemical theory, but a new paradigm relevant to all domains of science studying systems which are open and complex. And indeed, soon after its first appearance in the domains of physics, chemistry and mathematics, the new theory was discussed, applied and elaborated in many other scientific domains, including life sciences, brain studies, cognition, medicine, economics, sociology, philosophy and also in the domain of cities and urbanism.

3.2 Dissipative Cities

Dissipative cities are the product of Prigogine's theory of *dissipative structures* and its application, by Allen and co-workers, to the study of cities and systems of cities (Allen 1981, Allen and Sanglier 1981, Allen et al. 1985). As the name indicates, Prigogine's theory of self-organization puts specific emphasis on the process of dissipation. A good starting point for his argument might be the conjunction between Boltzmann's order principle and the Bénard experiment as described above. The latter, as noted, exhibits a coherent motion which 'means that many molecules travel with nearly the same speed'. According to Boltzmann's principle (which relates entropy to probability), there is almost no chance for this coherent self-organized motion to occur. 'Yet it occurs'. The reason is that in

> far-from-equilibrium condition, the concept of probability that underlies Boltzmann's order principle is no longer valid .... Classical thermodynamics,

explain Prigogine and Stengers,

> leads to the concept of 'equilibrium structures' such as crystals. Bénard cells are structures too, but of a quite different nature. That is why we have introduced the notion of "dissipative structures", to emphasis the close association, at first paradoxial, in such situations between structure and order on the one side, and dissipation or waste, on the other. We have seen ... that heat transfer was considered a source of waste in classical thermodynamics. In the Bénard cell it becomes a source of order. ... The interaction of the system with the outside world, its embedding in nonequilibrium conditions, may become in this way the starting point for the formation of new dynamic states of matter – dissipative structures. ... Bénard cells, like all dissipative structures, are essentially a reflection of the global situation of nonequilibrium producing them (Prigogine and Stengers 1984, pp. 142–144).

For anyone familiar with location theory the close resemblance between the hexagonal structure of the Bénard cells and the Christaller's (1966/1933) and Lösch's (1954) hexagonal landscapes of central places (above, Chap. 2, Figs. 2.3, 2.5b) not just invites, but almost demands, a comparison. And indeed, this challenging similarity was taken up by Allen and co-workers who in a series of studies have reformulated the static central place theory of Christaller and Lösch in terms of Prigogine's dissipative structures.

Allen and co-workers have developed a sequence of several models which elaborated their theoretical treatment of hierarchical landscapes of central places, first with respect to systems of cities in a given region and later at the intra-urban scale in connection with a single city. At a later stage they have also applied their models to real case studies of Brussels and the Belgian provinces (Sanglier and Allen 1989. See also Pumain, Saint-Julien and Sanders 1987).

A typical model of Allen's starts with an infrastructure of localities in a region, each with its residents and jobs. The actors are individuals who migrate in order to get employment, and employers who offer or take away jobs depending on the market's situation. The migration (or interaction) between localities and the introduction and extraction of economic activities (i.e. employment opportunities), create for each locality a kind of local 'carrying capacity' and for the system as a whole nonlinearities and feedback loops which link population growth and manufacturing activities. For example, in Allen and Sanglier's (1981) study, the positive feedback is due to common infrastructure, economies of scale, etc. and the negative feedback is due to crowding, pollution and so on (ibid). An example for a simulated scenario produced by the model is Fig. 3.2. As can be seen in Fig. 3.2a, this specific scenario starts with a hypothetical region characterized by a rectangular lattice of homogeneous localities. Then, the mere play of chance factors, such as the place and time where different enterprises and migrations start, produces symmetry breakings which entail an uneven distribution of population, employment and so on (Figs. 3.2b–d). The result is an evolutionary process by which new urban centers emerge, grow, and form the whole of the regional system of central places; as the system evolves, some old localities grow, others decline or even disappear, thus constructing the specific history of this region.

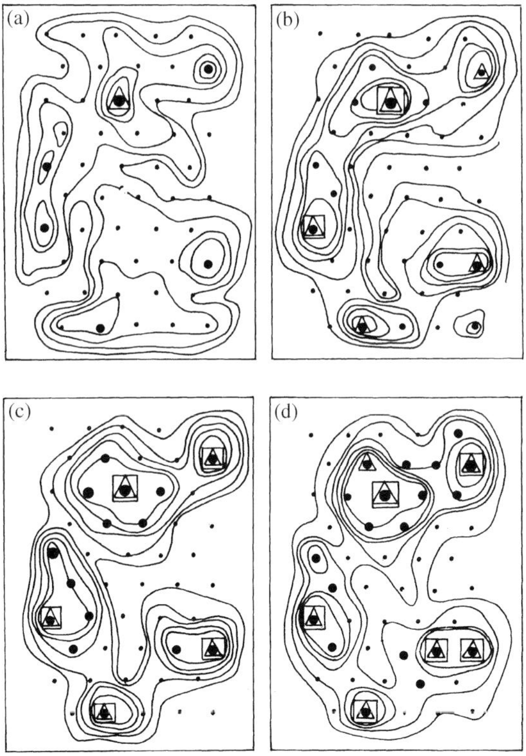

Fig. 3.2. Allen and Sanglier's simulated evolution of a dissipative system of cities. (**a**) at time (t) $t = 4$; (**b**) at $t = 12$; (**c**) at $t = 20$; (**d**) at $t = 34$

The application, by Allen and co-workers, of the theoretical principles of dissipative structures to the question of the emergence of a hierarchical landscape of central places, immediately exposes the similarity and difference between the 'old' static approaches of Christaller and Lösch, and the new treatment by means of self-organization. In both the old and new approach, economic activities and interactions give rise to central places which are usually urban centers. However, while in the old formulations the landscape reflects an equilibrium state which is the optimized sum of the properties of the various economic forces, the new landscape is 'more than the sum of its parts' – it reflects a far-from-equilibrium situation in which the spatial hierarchical order among the central places is obtained, maintained and then transformed, by means of an interplay between interaction and fluctuations, on the one hand, and dissipation (as in the Bénard cells), on the other.

3.3 Synergetic Cities

Synergetics – the working together of many parts, individuals, subsystems, groups, ... – is the name assigned by Hermann Haken to his theory of self-organization (Haken 1979, 1983a, 1983b, 1985a, 1985b, 1987, 1990, 1993, 1996). As the name indicates, the emphasis here is on the interrelations, interaction, and synergy among the many parts of the system and its overall structure and behavior.

Though synergetics originated in physics, it is by no means a physical theory that tries to reduce complex phenomena to the laws of matter. From the start Haken emphasized that the physical systems he was studying are similar in their behavior to phenomena of collective behavior in a variety of disciplines, and indeed, many of the notions which now form the theory were revealed and developed by a close study of case studies in a variety of domains, including sociology, psychology, cognition, AI (artificial intelligence), and also cities and urbanism. In all these studies Haken's (1996, p. 39) central methodological guide was to 'look for qualitative changes at macroscopic scales'. Some of these case studies, in particular those of the laser, pattern formation in liquid, pattern recognition and the finger movement experiment, have become paradigm cases and a convenient way to convey the principles of synergetics. For the purpose of the present comparative discussion it will be useful to present three such cases – the laser paradigm, the pattern formation paradigm, and the pattern recognition paradigm.

3.3.1 The Laser Paradigm

The process that produces the laser can be regarded as the generic paradigm of synergetics. The gas laser consists of a glass tube filled with a gas of atoms or molecules. The gas tube has two mirrors at its ends, which serve to reflect the light running in the axial direction rather often so that this kind of light wave can interact sufficiently strongly with the individual atoms. Because at least one of the mirrors is only semi-transparent, the laser light can eventually emerge through this mirror.

In a conventional lamp, when an electric current is sent through the gas, the individual atoms may get excited and emit individual independent light waves. In the laser, the individual electrons in the atoms correlate their movements and generate a beautifully ordered coherent light wave (Fig. 3.3). This is a typical act of self-organization. There is nobody who tells the laser system how to behave in such a coherent fashion; it finds its well-ordered behavior by itself.

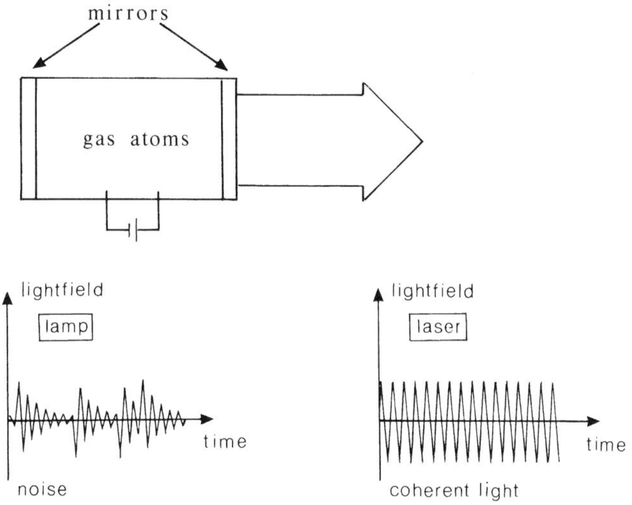

Fig. 3.3. The laser paradigm. *Top*: Typical set up of a gas laser. A glass tube is filled with gas atoms and two mirrors are mounted at its end faces. The gas atoms are excited by an electric discharge. Through one of the semi-reflecting mirrors, the laser light is emitted. *Bottom, left*: In a conventional lamp, the light field generated by the gas atoms consists of individual wave trains that represent noise. *Right*: In the laser, a highly ordered, i.e. coherent, light wave is generated

The interpretation of synergetics starts with Einstein's observation that when an excited atom emits a light wave, this light wave may cause other excited atoms to deliver their energy to it so that it is enhanced in its intensity. In the laser, initially quite a number of atoms emit their light waves independently of each other and with somewhat different wave lengths. Each of these might get support from the other excited atoms. In this way a kind of a Darwinian competition among the light waves for the energy resources of the excited atoms begins. This competition is won by an individual light wave which grows fastest. The winning light wave describes and prescribes the order in the laser and it is thus called the *order parameter*. It dominates the movement of the individual electrons as if by enslaving them and forcing them to move in its own rhythm. In the language of synergetics this is called the *slaving principle*.

The transition from the state of a lamp with its microscopically chaotic light field and the state of the laser with its well-ordered light field is quite sharp and occurs at a critical strength of the power input by the current into the laser. Thus the change of a single, rather unspecific, parameter, the power input in the case of the laser, may cause a systemic phase transition. This parameter is termed the *control parameter*.

When the laser wave is perturbed, it adjusts rather slowly compared to the adjustment time of the individual electrons. This is quite a general feature: the order parameters are slowly varying quantities compared to the enslaved subsystems. Consequently, when excited atoms are shot into the tube, they are enslaved by the order parameter (i.e. the light field in that tube) first by delivering their own energy to the order parameter, and then by acquiring its rhythm. This interplay between the rhythms of the system and its subsystems is analogous to many cases in human life: Language can be regarded as a slow moving order parameter. When a baby is born, it is subjected to the language of his or her parents and the other people. The baby learns the language and in technical terms is thus enslaved by the language. But by so doing, he or she eventually 'emits' their personal energy into the language and in this respect supports the language. As we shall see in some detail below, the same happens in city dynamics: the individual who immigrates to a new city has to learn the city and adapt to its rhythm. The individual is thus enslaved by the city's order parameter. But by adapting to the global movement of the city the individual's energy enters into, and supports, the order parameter of the city. This phenomenon is called *circular causality*.

Quite clearly, the concept of order parameters and their relationships to the individual parts of the system, a relationship governed by circular causality, applies to a great variety of phenomena in society. On the one hand, the individuals are the parts of a human society and determine its macroscopic manifestations, such as language, religion, form of government, culture, educational system, city structure and so on. On the other, the behavior of the individuals is determined by these macroscopic manifestations or institutions, which play the role of order parameters. Order parameters may compete with each other, or may coexist, or may cooperate. Such phenomena are well known in laser physics with respect to the electric field strength acting as order parameters. In social life these phenomena also occur, e.g. languages may compete and one language may win the competition as it happened in the United States, where two main streams were competing with each other, namely English and German. In other nations languages may coexist as in Switzerland and in Israel. Finally, in a way languages may cooperate, e.g.

when terms originally generated in one language are adopted by another language. This happens quite often with technical terms taken from English, for example.

3.3.2 The Paradigm of Pattern Formation

The case study here is the Bénard instabilities. As we've seen, as the temperature difference between the heated layer at the bottom and the cool layer on top exceeds a critical value, quite suddenly a macroscopic motion of the liquid becomes visible (Fig. 3.1 above). The temperature difference thus controls the macroscopic behavior of the system and in the language of synergetics it is called the *control parameter*.

As the control parameter grows, the liquid start its motion, rolls are created, their rolling speed increases, and the initial resting state becomes unstable. *Instability* thus shows up. Slightly above the instability point, the system may undergo quite different collective motions of roll configurations (Fig. 3.4). At the beginning the amplitudes of these roll configurations are small and independent of each other. When they grow further, they start to influence each other – in some cases they compete until one configuration suppresses the others, in others, they co-exist and even stabilize each other. "The amplitudes of the growing configurations are called *order parameters*. They describe ... the macroscopic structure of the system" (Haken 1996, p. 39).

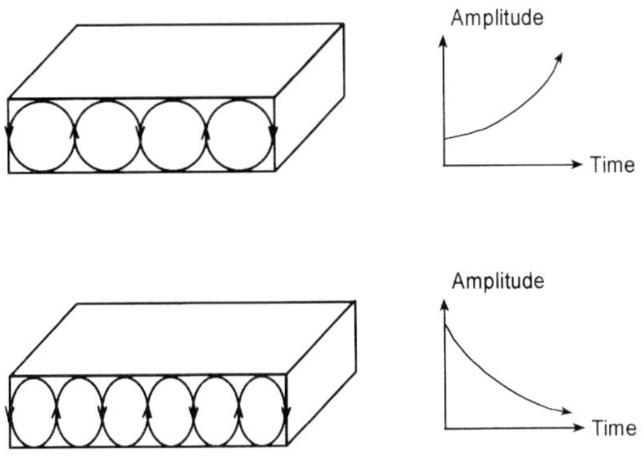

Fig. 3.4. *Left*: two different roll configurations in a fluid. *Right*: The behavior of the amplitudes of these configurations in the course of time. While in one case the amplitude increases, in the other it decays

The order parameters not only determine the macroscopic structure of the system, but also govern the space–time behavior of its parts. By winning the competition the order parameters enslave the many parts of the system to their specific space–time motion. This is a basic theorem of synergetics and it is called the *slaving principle*.

In some cases, for example when the fluid is enclosed in a circular vessel, all directions for roll systems are then possible, each being governed by a specific order parameter. Which pattern will eventually be realized, depends on initial conditions. It is as if the system internally stores many patterns. This repertoire of patterns is not stored in a static fashion, but is dynamically generated anew each time. This property is termed *multistability*.

3.3.3 The Paradigm of Pattern Recognition

A typical experiment of pattern recognition can start as follows: a test person (or a computer) who has many patterns, of faces, city maps etc., stored in memory, is presented with a portion of one of the patterns. The task is to recognize the pattern – to decide what face, city map, etc. it is. According to synergetics, what happens is a process analogous to pattern formation as described above: at the start, the cognitive system of the person (or computer) is in a state of multistability as it enfolds many patterns which coexist. When a few features or part of the pattern is offered, several pattern configurations and their order parameters are formed by means of associative memory. The order parameters enter into competition and when a certain order parameter wins the competition and enslaves the cognitive system, the task of recognition is implemented. This analogy which is illustrated in Fig. 3.5, was first demonstrated by Haken in 1979 and has become the basis for an intensive study of synergetics of cognition and of brain development and activity (Haken 1979, 1990, 1996). Figure 3.6 is a typical implementation with respect to face recognition, by means of the so called synergetic computer. Further discussion on this issue is given in Chaps. 13, 14, while an implementation with respect to city maps, is presented in Chap. 14, Fig. 14.5.

As suggested by Portugali and Haken, the above conceptualization offers also an appropriate framework for the study of cognitive maps of cities, regions and large-scale environments (Portugali 1990, Portugali and Haken 1992). The basic idea is that cities, regions etc. can be regarded as large-scale patterns which can never be seen in their entirety. As a consequence, the cognitive system constructs a whole cognitive map on the basis of only a partial

Fig. 3.5. Haken's (1979, 1991) analogy between pattern formation and pattern recognition. *Left*: a configuration of some parts of a system gives rise to an order parameter which enslaves the rest of the parts and brings the whole system to an ordered state. *Right*: a few features of a pattern shown to a person (or a computer) generate an order parameter which enslaves, and thus complements, the rest of the features, so that the whole pattern is recognized

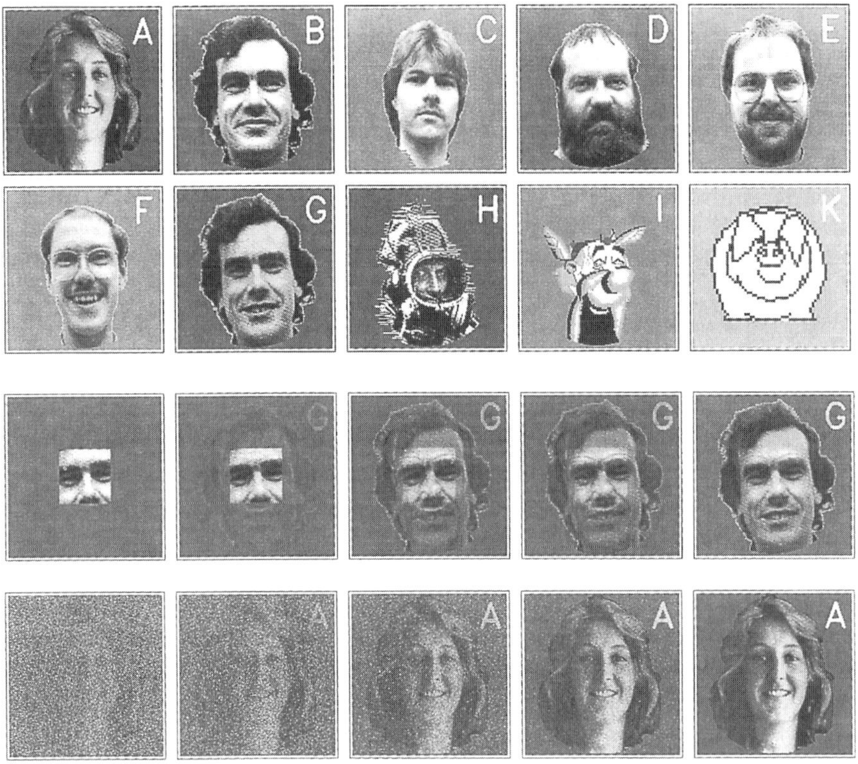

Fig. 3.6. Pattern recognition of faces by means of the synergetic computer. *Top*: examples of prototype patterns stored in the computer, with family names encoded by letters. *Middle*: when part of a face is used as an initial condition for the pattern recognition equations, their solution yields the complete face with the family name. *Bottom*: when a noisy face is used as an initial condition for the pattern recognition equations, again their solution yields the complete face with the family name

set of features available to it. Thus it can be said that a partial set of environmental features offered to a person triggers a competition between several configurations of features and their emerging order parameters, until one (or a few) win and enslaves the system so that a cognitive map is established.

Synergetics have developed from the start two approaches to the study of phase transition and qualitative change in self-organizing systems. One is by means of probability distributions and direct numerical solutions. This line of research has used the conceptual framework of order parameters and the slaving principle in a rather implicit manner; its main instrument was the so called *master equation*. The second approach developed by focusing on the state variables of the system and by an explicit consideration of the order parameters, the slaving principles and the other tenets of synergetics as presented above. These two lines of research are characteristic also of the approaches of synergetic cities. One approach, led by Weidlich and co-workers, followed Weidlich's sociological and economic applications of synergetics, and the other by Haken and Portugali, was inspired by Haken's elaboration of synergetics in the domains of cognition, pattern recognition and brain activities.

3.3.4 Slow Cities and Fast Regions

One way to look at Haken's synergetics and its slaving principle is in terms of an interplay between slow and fast processes:

> If in a system of nonlinear equations of motion for many variables these variables can be separated into slow ones and fast ones, a few of the slow variables ... are predestined to become 'order parameters' dominating the dynamics of the whole system on the macro-scale (Weidlich, 1999).

This perspective stands at the basis of Weidlich and co-workers' studies on sociodynamics and more recently on cities and urbanism (Weidlich 1987, 1994, 1999).

According to this perspective, fast and slow processes are easily identifiable in processes of settlement and urbanism. The fast ones typify the local urban microlevel of building sites, streets, subways, etc., whereas the slow processes typify the macrolevel of whole regions which are often described as systems of cities. The relations between the slow and the fast processes are described by the slaving principle: on the one hand, the regional system

> serves as the environment and the boundary condition under which each local urban microstructure evolves. On the other hand, the ... regional macrostructure is ... the global resultant of many local structures (Weidlich 1999).

This circular causality between the local and the global, allows one to study global regional systems by assuming that local processes adapt to the slow regional ones, and to study local urban processes by treating the regional context as given, and of course to study the complex interplay between the local and the global. In all three cases Weidlich has recently prescribed a four stage approach: stage 1 concerns the configuration space of the variables; stage 2, measures the utility of each configuration; stage 3, defines transition rates between configurations which are in fact utility differences; stage 4, derives stochastic or quasi-deterministic evolution equations for the system under consideration. The central evolution equation is the master equation which defines the probability that the configuration under examination is realized at a certain time.

The above theoretical procedures have recently been used to study the role of population pressure in 'fast and slow processes in the evolution of urban and regional settlement structures', and in urban evolution. Fig. 3.7

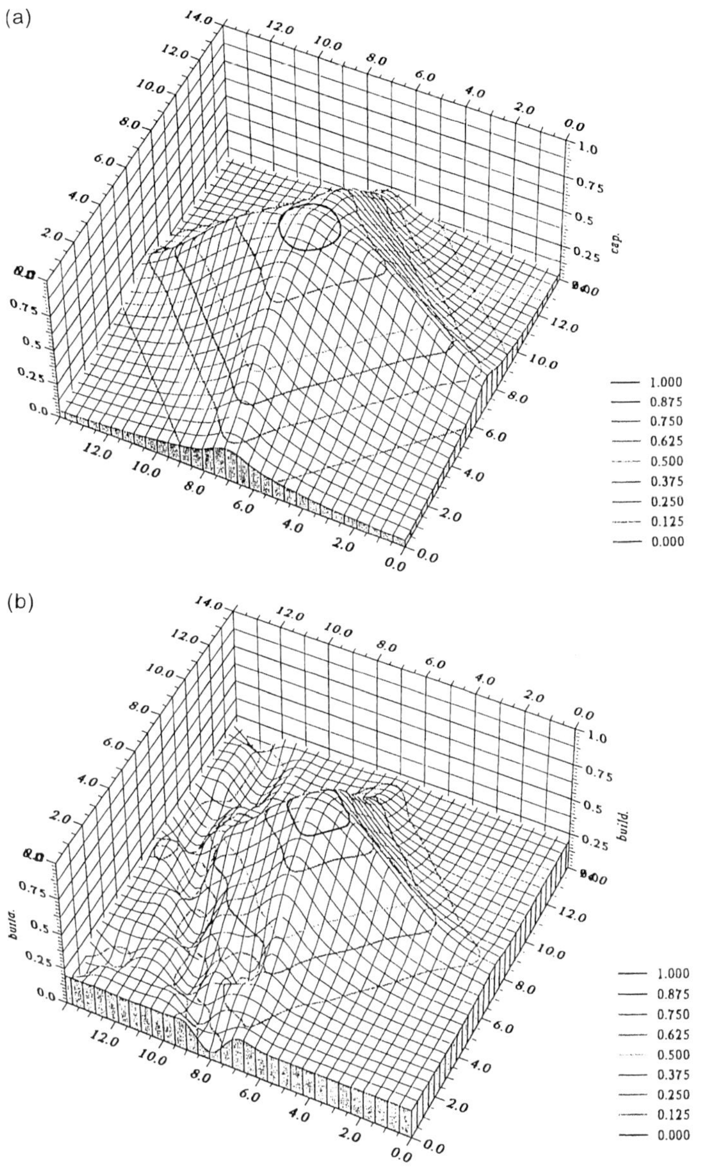

Fig. 3.7. Building and development under population pressure (Weidlich, 1999). *Top*: on a uniform urban plain. *Bottom*: on an urban plain with disturbances

brings some results from these studies, in which the city capacity for building and development is related to population pressure. It shows the evolving city capacity when the urban plain is uniform (*Top*), and when it is disturbed in one of its sites (*Bottom*).

3.3.5 Pattern Formation and Pattern Recognition in the City

Thus far, most applications of synergetics to the domain of cities and regions were in line with Weidlich's approach as described above. That is, they have employed mainly the very early and basic notions of synergetics – the order parameters and slaving principle – and usually in a rather implicit way. On the other hand, much of the advance made in the last two decades in synergetics was done in connection with issues of cognition and brain functioning. In fact, the above analogy between pattern formation and pattern recognition provided the foundation for these advances.

In a recent study, Haken and Portugali (1995) suggested that this analogy is specifically attractive to the study of cities. The latter can be perceived as self-organizing systems which are both physical and cognitive: individuals' cognitive maps determine their location and actions in the city, and thus the physical structure of the city, and the latter simultaneously affects individuals' cognitive maps of the city. In their preliminary mathematical model Haken and Portugali construct the city as a hilly landscape which is evolving, changing and moving as a consequence of the movement and actions of individuals (firms etc.). The latter give rise to the order parameters which compete and enslave the individual parts of the system and thus determine the structure of the city. The significant and new feature of this exposition is that the order parameters enslave and thus determine, two patterns: one is the material pattern of the city, and the other is the cognitive pattern of the city, that is to say, its cognitive map(s). This is exemplified diagrammatically in Fig. 3.8.

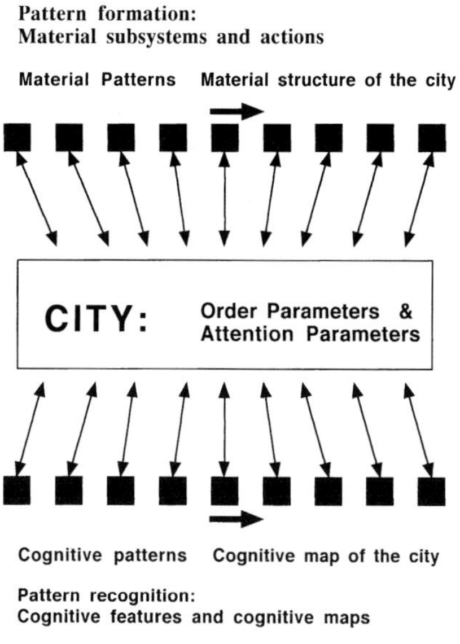

Fig. 3.8. The city as a self-organizing systems which is at the same time both physical and cognitive. Its emerging order- and attention-parameters enslave the city's cognitive and material patterns

One of the more interesting outcomes of the model are the *attention parameters* which emerge by means of self-organization. The latter can be seen as the order parameters of specific sub-systems composing the pattern. In a state of multi-stability, or in the case of an ambivalent pattern (e.g. vase/face) they determine which aspect of the pattern is seen (i.e. attracts attention) first. This is of utmost importance in city dynamics. The city is full of patterns, yet individuals are attentive to only a few of them. The latter form the cognitive maps of the city and it is according to them that individuals and firms take decisions and act in the city. In the model we investigate cases where one attention parameter dominates the dynamics and cases where no cross attention is paid; that is, when two or more urban communities are cognitively not aware of each other. This situation entails the emergence and persistence of an urban cultural or socio–economic mosaic where a few coexisting attention parameters govern the dynamics. In Part IV on 'Synergetic Cities' (Chaps. 13, 14), we present the above approach in full and elaborate it in connection with city dynamics and decision making in the context of urban and regional planning.

3.4 Chaotic Cities

Self-organization can be regarded as a theory about chaos and order – of the way chaotic systems self-organize themselves and attain order. The theory of dissipative structures suggests that this is due to dissipation; synergetics suggests that this is due to cooperation between the many parts of the system. From this perspective might follow that self-organization is a theory about *Order Out of Chaos* as the title of Prigogine and Stengers' (1985) book declares, and, that the various self-organized cities are ordered cities which the theory of self-organization explains just how their order state was created. This conclusion is only partially correct, however, for it is but a special case of a more general situation in which the ordered self-organized cities might still be chaotic. To see how, we have to look at two forms of chaos: local or *microscopic chaos*, versus global, macroscopic or *deterministic chaos*.

Local chaos stems from the irregular motion or behavior of the very many individual parts of a complex system. Examples might be the motion of the molecules of a gas, the movement of cars on an uncrowded freeway, or of people in an uncrowded piazza, and so on. In all such cases the individuals are moving in irregular and uncoordinated trajectories. This situation is a typical starting point for a process of self-organization.

Deterministic chaos arises when as a consequence of self-organization, for example, the many individual parts are suddenly attracted by a few attractors, or enslaved by a few order parameters, and as a consequence exhibit a coordinated motion. On the face of it this new state is the exact opposite of chaos; yet, it is not. Quite often in these cases, the system is dominated by order parameter(s) which are macroscopically chaotic: for some time one order parameter dominates the system, then suddenly another; the point is that these jumps back and forth occur irregularly in a chaotic manner. Or, put differently, the trajectories of the enslaved parts or subsystems jump irregularly and abruptly between the two attracting forces. Such a behavior was given the contradictory name *deterministic chaos* to emphasise the fact that the conditions for such a chaotic behavior can be determined externally and mechanically.

The connection between chaos theory and socio–spatial dynamics has been scrutinized recently by Dendrinos and Sonis (1990) in a very general and comprehensive way. Their work paves the way for subsequent studies which will explicitly apply the various notions of chaos to the specificity of cities and urbanism. While such studies have yet to come, it might be useful to indicate a few possible directions.

As self-organizing systems, cities are chaotic in several respects. First, like open and complex systems in general, the evolution of self-organizing cities exhibits a very distinct and routinized path: a long period of 'steady state', followed by a short period of strong fluctuations or chaos, from which the system re-emerges to a new level of steady state and structural stability, and so on (Chap. 4, Fig. 4.2). A common way to describe this process of self-organization is by means of a bifurcation diagram as in Figs. 4.3, 4.4 in Chap. 4. As can be seen, the system, i.e. the city, moves from one structurally stable state to another, via bifurcations, when every evolutionary move is a transition from a microscopic chaotic state to an ordered, macro, steady state. This property is of special importance as it helps to illustrate the difference between the 'old' stable urban structure and form discussed in location theory (above Chap. 2, Figs. 2.3–2.5), and the new notion of far-from-equilibrium structural stability of a self-organized city.

Second, as just noted, when in a steady-state, the city might still exhibit phenomena of deterministic chaos in the form of abrupt, chaotic jumps between the order parameters that dominate the city. For example, between residential versus commercial activities, or between a few socio–cultural order parameters that enslave the city's cultural landscape and so on. An interesting variation in this connection concerns our finding that often, when the city as a whole evolves stably, local areas within it exhibit unstable or chaotic behavior. It is as if a few unstable chaotic areas are found captive within the otherwise stable city. In studying this situation by means of a cellular automata model we suggest that these local islands of instability are needed in order to maintain the overall global stability of the city. The city thus maintains global structure by sociospatially imprisoning local chaotic elements that threaten its global stability. We have termed it *the captivity principle* and suggested that it might play a supplementary role to Haken's slaving principle (Portugali et al. 1994, Chap. 5, Sect. 5.8 below and Haken and Portugali, in preparation).

Third, the play between chaos and order might show up not only in the long-term evolution of the city, but also in its daily routines. The movement of cars on the roads, of pedestrians on pavements and the like, are characterized by shifts between unstable and stable motions and might thus be candidates for this kind of interpretation. For example, it might be interesting to look at the movement of cars on the freeway, from the randomly distributed speeds during nights to the almost uniformly distributed speeds during rush hours, in terms of self-organization and a shift from chaos to order; and then, during rush hours, to examine the abrupt changes of speed of almost solid bodies of thousands of cars, in terms of deterministic chaos.

It is important to note that given the differences between the notion of chaos as used in common language and chaos as a formal mathematical theory, implementation of the ideas just mentioned must be made with special care.

3.5 Fractal Cities

Fractal Cities is the title of Batty and Longley's (1994) book in which they show how Mandelbrot's (1983) theory of fractal geometry can be applied to the study of cities, their structure and evolution. My aim here is not to summarize Batty and Longley's book, but rather to examine in a rather preliminary way the interface between fractals, chaos and self-organization, and the implications thereof to the study of cities.

Mandelbrot's theory is based on two innovative notions – self similarity and the fractal dimension, and, on the idea that a rather simple iterative process might produce highly complex geometrical shapes. Using these principles, Batty and Longley show how complex geometries of urban form, growth and evolution, on intra-urban and inter-urban regional scales can be generated by means of Mandelbrot's type of fractals. (For an illustration see Fig. 3.9, below, in the next section on cellular automata cities.)

Fractal geometry is related to self-organization via the notion of chaos. As shown by several authors (Mandelbrot 1983), an important property of the Boltzmann's strange attractor noted above, is that the trajectories of its many parts are self-similar and can be described by a fractal whose dimension is between 2 and 3. More generally, some of the attractors or order parameters which govern a self-organized system in its steady-state might be fractals. That is to say, they have a fractal dimension, and they generate complex, self-similar shapes by means of simple iterative rules.

The above property is significant to our intuitive understanding of the meaning of steady-state and order parameters in self-organizing cities: To say that a city is in a steady state and that it is governed by one or more order parameters does not mean equilibrium and stability, as is the case of Christaller's and Lösch's central place theories, for example, but rather a rich and complex evolution and change according to a given ordering principle. The fractal cities assembled and generated in Batty and Longley's book provide an implicit example for this. Explicit examples come from more recent studies of Batty and co-workers in which they generate fractal cities by means of cellular automata models and interpret them in the context of self-organized criticality. These cities are discussed in the following two sections.

3.6 Cellular Automata Cities

A standard two-dimensional cellular utomata (CA) model is a lattice of cells, where each individual cell can be in one of several possible states (empty, occupied, etc.) and have one out of several possible properties (developed, underdeveloped, poor, rich, and so on). The dynamics of the model is generated by an iterative process in which in every iteration the state of each cell is determined anew by some transformation rule(s). The rules are local and they refer to the relations between the cell and its immediate neighbors. The name of the game is to see how, what, and in what circumstances, local interrelations and interactions between cells entail global structures, behaviors and properties of the system as a whole. As it turned out, simple CA games are capable of generating very complex global structures and behaviors – a property which made CA models a very attractive research tool.

The origin of CA models goes back to Alan Turing and his ideas concerning self-reproducing machines, to John von Neumann's elegant demonstration that such machines, or *automata*, are in principle possible, and then to John Conway's *Game of Life* which was an explicit CA game (Gardner 1971).

CA models have been used quite extensively to simulate and study processes of self-organization. For example, Langton, in biochemistry, Manneville et al. (1989) in physics with respect to fluid dynamics and turbulence theory, Demongeot et al. (1985) in the domain of neurobiology and computer sciences, and Langton (1986) in the study of Artificial Life. From a wider perspective it is important to mention Eigen and Winkler's (1983) book *Laws of the Game*. (For further examples see Toffoli and Margolus 1987, Wolfram 1984).

The attractivity of CA models to the study of self-organization stems from the fact that the more conventional tendency to use differential equations in models of self-organization, often makes computation

> very tedious ... In this situation (especially if we are interested mainly in qualitative results) we can abstain from a numerical integration of exact differential equations and turn to the analysis of much simpler systems represented by networks of cellular automata (Mikhailov 1990, p. 40).

The main difference between studies of self-organization by means of differential equations and by means of CA, is that in the first, the basic concepts of self-organization, such as attractors or order parameters, are explicit mathematical elements in the models, whereas in the second case they are implied or derived from the simulation as its interpretation concepts: you set up a CA game, observe its evolving scenario, and then, post factum you derive a phase-space that describes the evolution of an attractor or an order parameter.

The attractivity of CA models to the study of cities is almost self-evident. Real cities are built of discrete spatial units such as houses, lots, city-blocks and the like. CA models are also built of discrete spatial units – the cells. In real cities the properties of local spatial units (e.g. land value) are determined, to a large extent, in relation to their immediate neighbors; so are the properties of the cells in CA models. These resemblances make CA models, intuitively and mathematically, natural tools to simulate urban processes. This urban and geographical potential of CA models was noted by Tobler (1979) who suggested calling their usage 'Cellular Geography', and indeed, in the last few years we see a growing number of studies which employ this type of modeling to the study of cities. (For full bibliography see the special issue of *Environment and Planning B: Planning and Design* 27, 1977, on cellular automata.)

From what has been said above, it is clear that CA is also a very useful tool to study self-organizing cities. From this perspective one can divide the various models of CA cities into implicit and explicit self-organized CA cities. The first group refers to studies the aim of which is to use the simulation capabilities of the CA city in order to best-fit a certain simulated pattern to an existing one. The studies of Batty on the Buffalo–Niagara frontier provide a good example (Fig. 3.9; see also the study of White and Engelen, 1993 on Cincinnati). The general motivation here is to explain an existing or historical pattern, or alternatively to predict a future one for the purpose of planning. The fact that the model has properties of self-organization, just adds more realism and sophistication to the simulation.

The second group concerns explicit self-organized CA cities. Here the central motivation is to use the model as means to investigate the self-organization properties inherent in cities and urbanism. For example, how micro decisions and behaviors of individuals and firms, taken at the local scale, are related to the global behavior and structure of the city as a whole. Such models are essentially heuristic and they regard the simulated CA city as essentially a learning device. Examples here are Chaps. 4–10 below in which we examine how the location of individuals from different cultural groups, varying in their attitude toward strangers, entail different spatial segregations of the global cultural–urban landscape, and how the latter affects their movement in a kind of circular causation.

Because of their iterative structure, CA models can be used as convenient tools to generate fractal structures. In this connection Batty (1997) has recently demonstrated some such applications in which CA models generate

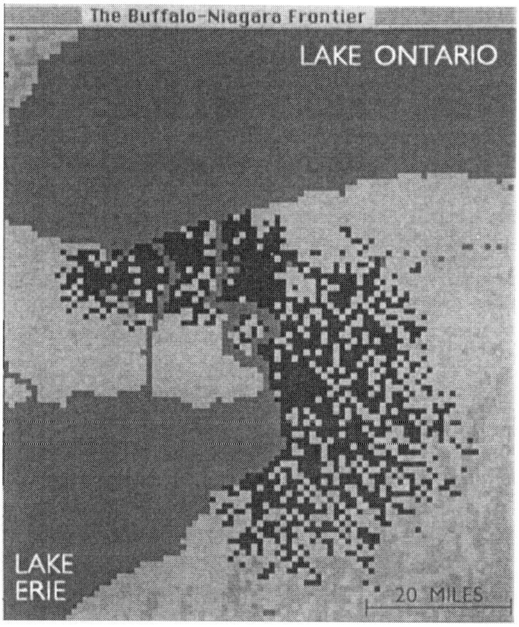

Fig. 3.9. Simulation using cellular automata on the Buffalo–Niagara frontier. Source: Batty (1997)

'fractal cities', both abstract and realistic cities (Fig. 3.9, see also Fig. 3.10). This is significant as it provides an interface between the rich worlds of fractal cities, and CA cities, making the two different facets of the various self-organizing cities.

3.7 Sandpile Cities

> Imagine building up [a sand] pile by slowly adding particles. As the pile grows, there will be bigger and bigger avalanches. Eventually a statistical stationary state is reached in which avalanches of all sizes occur, that is the correlation length is infinite. Thus, in analogy with equilibrium thermodynamical systems, the state is 'critical'. It is self-organized because no fine-tuning of external fields was needed to take the system to the critical state: the criticality is unavoidable.
>
> The sandpile has two incongruous features: the system is unstable in many different locations; nevertheless the critical state is absolutely robust. On the other hand, the specific features, such as the local configurations of sand, change all the time because of the avalanches. On the other, the statistical properties, such as the size distribution of the avalanches, remain essentially the same (Bak, Chen and Creutz 1989).

The sandpile is the canonical example of *self-organized criticality* (Bak, Chen and Creutz 1989, Bak and Chen 1991, Batty 1995, 1996, Batty and Xie 1999). Compared to 'grand' theories of self-organization such as synergetics and dissipative structures, the theory of self-organized criticality is more specific and thus limited, but at the same time more detailed. It is a kind of a zooming-in to the internal dynamics of self-organized systems in their steady state periods – when the system is controlled by what in synergetics is called order parameters: given a control parameter – the flow of sandgrains – beyond a certain critical threshold, the system is controlled by a certain order parameter, which in the above canonical example can be described by the slope of the sandpile. The interesting part of the theory starts here: it shows how complex and rich the internal dynamics of a steady state situation can be.

For human geographers and students of urbanism the above statistical observation that 'the size distribution of avalanches remain essentially the same', immediately reminds us of the old 'rank–size rule', according to which the size-distribution of cities remains essentially the same under circumstances of ongoing population growth, for example. The rank–size rule is one among several phenomena in human geography that exhibit similar statistical regularities. The implication is that self-organized criticality might have the potential to provide a theoretical foundation for geographic regularities, such as the rank–size rule, which were often criticized for not having any theoretical basis. (The first criticism that I know about goes back to the mid 1930s when Christaller (1933), in his book has criticized Auerbuch's 'rank-size rule' (from 1913) as being nothing more than playing with numbers.)

The above similarity indicates a potential, yet the essence of self-organized criticality is not the final statistical size-distribution, but the process behind it. And here students of urbanism are in a problem: unlike in many physical domains, in the human domain of cities we simply do not have the detailed data that is needed to describe the dynamics. This challenge was taken-up by Batty and Xie (1999) who suggested a way out of this dilemma by following applications of self-organized criticality to the 'Game of Life' – the simple CA game noted above (Bak, Chen and Creutz 1989, Alstrom and Leao 1994). These applications showed not only that the CA model generated self-organized criticality, but also that its chain reaction is a fractal process which in the Game of Life is of approximately dimension 1.7.

The general strategy suggested by Batty and co-workers with respect to the use of the theory of self-organized criticality is that the statistical data which is needed to apply the theory to the study of cities can come from empirical studies of reality, whereas the data and information about the fine-grained dynamics, can be obtained from the conjunction of CA and fractal cities. Batty's project of applying the theory to the study of cities is just at its beginning. From preliminary attempts to look at the evolution of Buffalo from 1750 to 1990, and by studying more abstract simulations, he suggests that 'there is evidence that real cities in their evolution over time display this characteristic'; yet he also indicates that there are many technical limitations which have to be overcome before a proper application of the theory can be implemented (Batty and Xie, 1999). Figure 3.10 is an example of fractal cities and systems of cities, generated by means of a CA model, and interpreted by reference to the theory of self-organized criticality.

3.8 FACS and IRN Cities

Consider a city game constructed of two layers: an infrastructure layer which is a usual CA city with its 2-dimensional cellular space, and on top, a superstructure layer of individual free agents. They are free agents in that they can move from one cell to the other all over the 'city', they have past, they have plans for the future, and they act intentionally; they can see beyond local situations, their seeing is subjective and is captured by the notion of cognitive maps. The latter determine their actions and behavior in the city. In short, each free agent is a self-organizing system, each is a virtual human.

FACS (free agents on a cellular space) is the name of this city model, and IRN (Inter-Representation Network) describes its dynamics: a circular causation between two self-organizing sub-systems, at two different scales, that together form a single network of internal and external representations. At the local scale we have the individual free agents; they are the 'parts' of the city, but unlike ordinary parts, each of them is itself a self-organizing system. Their activities determine the city which can thus be seen as the external representation of their actions and behavior. At the global scale we have the whole city as a self-organizing system which in its turn determines the internal representations (that is to say, cognitive maps) of individuals and through these their action and behavior in the city, and so on in a circular causality.

In Part II that now follows, we use this model to study how, by means of self-organization, the city dynamics entails the emergence of a new urban cultural group – a phenomenon which is typical of current postmodern and hypermodern cities. We start the game with two groups of individuals, Greens and Blues, belonging to two cultural groups. They enter the city as immigrants, intending to find a proper location in it, when the Greens' intention is to live among Greens, and the Blues', among Blues. If such a free agent finds a satisfying location, it will live there and will become an inhabitant, if not, it will try to move to another location in the city, thus participating in creating the city's intra-urban migration. Some agents who cannot find a location according to their intentions, leave the city and thus create its out-migration, and some get stuck in a location that they do not like. Because of various systemic situations they cannot migrate and are thus enforced to behave counter their intentions. The outcome is that they enter a situation of *spatial cognitive dissonance*. With time, in order to resolve this cognitive

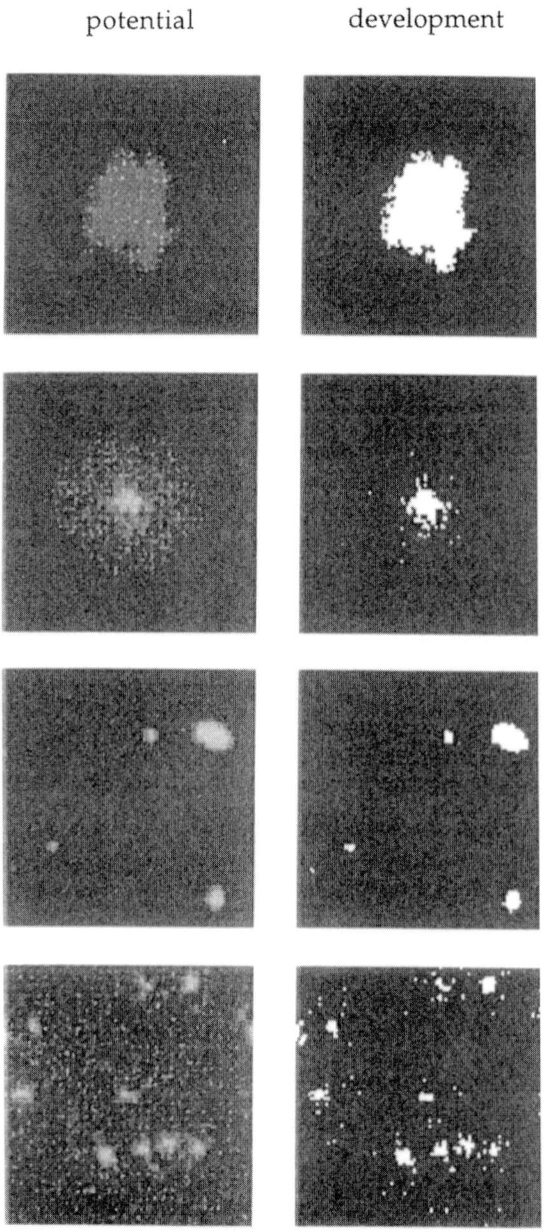

Fig. 3.10. Cities and city systems generated through feedback. Source: Batty (1996)

conflict, some might change their intention, and in certain situations this change of intention gives birth to a new cultural identity in the city.

What we try to achieve by this new kind of modeling is to be able to examine, simultaneously, self-organization at the local level of the individual, and self-organization at the global level of the city: to see how the city dynamics might create a self-organization process at the individual level, and how the latter might entail self-organization at the city level. This approach departs from the usual procedure in self-organization studies which tends to study macro-scale city self-organization by ignoring micro-scale self-organization at the level of the individual, and to study micro-scale cases, as in cognitive studies concerning the behavior of individuals in the city, by ignoring the global self-organization process of the city, or by assuming it as fixed.

Part II **City Games**

4. Free Agents in a Cellular Space

4.1 Free Agents

From what has been said above follows, first, that we suggest viewing the city as an open, complex and thus self-organizing system, exhibiting phenomena of non-linearity, bifurcation and phase-transition. Second, that cellular automata and cell-space models provide a specifically appropriate infrastructure upon which to construct models of cities as self-organizing systems, and, as heuristic tools to learn their properties. With these suggestions we in fact join the cities surveyed above in Chap. 3 ; yet our attempt is to go a few steps further. We further suggest that, by and large, most of the 'cities' presented in Chap. 3 were constructed by applying to the human domain of cities, urbanism and society, the various notions of self-organization as developed mainly in physics, chemistry and mathematics. What we propose to do instead, in the chapters that form the second part of the book, is not to apply, but to *adapt*, the notion of self-organization to the human domain of cities and urbanism. As already noted at the end of Chap. 3, we propose to do so by adding to the models human intentionality, hermeneutics and memory – notions which are absent from most of the the cities of Chap. 3, but form the foundations of the cities of social theory in Chap. 2.

Chapter 2 described in some detail the discourse on cities and urbanism as evolved in social theory. It followed the development by which the prevailing Marxist and humanistic perceptions of cities have pushed some parts of the eco–physical perception of cities out of the territory of their discourse. In particular, how regional science, the successor of the eco–physical cities, had to depart from mainstream human geography to start its own urban trajectory, parallel to the trajectory of the SMH cities. The most recent products of these two parallel trajectories are the postmodern city – the latest product to originate from the production line of geography as social theory, and the self-organizing city – the latest product of what might be seen as the post- or rather 'hyper-modern' version of geography as social physics (see Chap. 15, especially Sect. 15.1.6). The discussion in Chap. 2 was concluded by listing some properties, which the cities of self-organization share with postmodern and Marxist cities. In light of the discussion of the self-organizing cities in Chap. 3, you are now in a better position to judge and see the nature of these similarities and resemblances.

4.1.1 Between Atoms and Free Agents

You can see that the self-organizing city is indeed similar to the postmodern and Marxist cities in several important respects; but you also see that there are differences. You can see that they are similar in being inherently unstable, far from equilibrium, and so on, but that they differ with respect to the source of these resemblances. In the self-organizing cities of Chap. 3 it is the systemic openness and complexity of these cities; in the cities of social theory it is systemic openness and complexity too, but more importantly and in addition to these – the nature of the elementary parts of these human systems.

In dissipative structures, fractal and CA systems, and in most (though not all) applications of synergetics, the system is complex, but its elementary parts – atoms, molecules and the like – are simple, homogenous and objective entities. The complexity of such systems thus arises at the systemic level. When applied to organic, say, biological systems, the same property is retained – the animals in a herd, the ants in a society of ants, the birds in the flight, are treated as if they are atoms, with no personality, subjectivity or individuality. And the same with the human individuals, families or households, which form the elementary parts of the self-organizing cities surveyed in Chap. 3. With few exceptions and reservations, they too are treated like atoms, or as 'one dimensional people', to rephrase the title of Marcuse's book.

In the Marxist and humanistic cities of social theory the elementary units – individuals, families, households and so on, are themselves open, complex and thus self-organizing entities. They are free agents who have personality, and subjective individuality. These free agents have intentions, they see the world and their cities in their own peculiar subjective way, and they have their own personal memory. And consequently the fact that in certain urban situations they act and behave collectively, in concert, cannot be taken for granted; it has to be understood, interpreted, explained.

Collective behavior is also characteristic of the self-organizing cities of Chap. 3. But there the phenomenon to be explained is 'how it is possible that entities that have no individuality and consciousness behave in concert as if they 'know'? With free agents the puzzle is exactly the opposite: 'how is it possible that free agents such as humans, who do not know each other, who are so different from each other in their intentions, plans and background, behave collectively as if they have decided in advance to do so'? In most of the 'cities' of Chap. 3, the answer is that they are economic persons, that is, they react in a specific, pre-determined way, to certain stimuli. The problem is that apart for a few limited situations, the assumption of the *homo-economicus* does not hold water. This was shown in a very convincing way by the criticism of the proponents of the Marxist and humanistic cities on the positivist ecocities. And as we show below in Chap. 7, the assumption of the economic person stands in opposition to recent psychological theories of human behavior, as well as to the very notion of self-organization.

4.1.2 Intentionality, Hermeneutics and Memory

The subjectivity and personality of human individuals and families have three important aspects, which are absent, in most of the self-organizing cities of Chap. 3: intentionality, hermeneutics and memory. Individuals in the city act and behave intentionally in line with their wishes, wants and so on. In order to act and behave intentionally, they need information about the city. This information they subjectively extract from what they see and experience in the city. They extract this information by means of logic, imagination, past experiences, knowledge and other 'tools' commonly assumed to form the content of the individual's memory. This process by which the individual extracts information by means of memory, and by so doing in fact creates and constructs his/her own and others' city, is termed *hermeneutics*. And to complete the picture and the feedback loop one should add that memory is also the place were intentions are created, represented and stored. In the discussion here and in the 'cities' that we develop below, in Chaps. 5–10, we try to take into account these properties and we thus term our models *Free Agents in a Cellular Space*, or in short FACS.

4.1.3 Internal and External Information

Much of the city dynamics is created by individuals' intra- and inter- migration movements. That is to say, by inhabitants or firms who change their residential or firm location in the city, and by various sorts of immigrants who enter the city and try to find a home or a working place in it. As a free agent, the spatial decisions such an individual takes in this process are based on two forms of information: internal and external. Internal information is derived from the individual's memory as related to the spatial consequences of his/her cultural identity and socio–economic status. For example, in the models/games we develop in the following chapters we postulate individuals with Green, Blue or Red cultural identities and with varying levels of socio–economic status. A person whose cultural identity is Blue might prefer to spatially locate among Blue neighbors, whereas his/her socio–economic status might constrain his/her choice to certain rich/poor areas in the city. External information refers to actual spatio–cultural and spatio–economic configurations in the city at a certain moment. For example, a Green individual might find him/herself at a certain time living among Blue neighbors in an area which is beyond his economic status; or, a Blue immigrant might find that at a certain time all vacant houses in the city are surrounded by Green neighbors. The location decision of the individual and his/her migration behavior in the city thus reflect this dialectical tension between internal and external information. This tension appears in all the chapters of Part II: implicitly in the models of Chaps. 5, 6 and explicitly in the games of Chaps. 7–9. A more detailed discussion on decision making in the context of cities is given in Chap. 14.

4.2 In-Formation

It is important to emphasise that in the above and in what follows we use the notion of information in a rather informal, intuitive and literal manner. Literally, the term information can be read either as to *in-form* (Bohm and Hiley 1993), or as *in*(ternal) *form* or *form*(ation). That is to say, a certain form or pattern that is stored or enfolded in a given element. The latter might be the brain, and it might also be a book, a building or a city. In the first case, one would speak about internal, or internally represented, information, while in the second case, about external, or externally represented, information.

As is well known the term 'information' refers also to a formal theory of information, originaly formulated by Shannon, in order to study the capacity of a communication channel to transfer information (Shannon and Weaver 1959). For this purpose he constructed a concept of information which is not related to any meaning – it is information with "meaning exorcised" – as Haken (1988a) phrases it in his *Information and Self-Organization.* In that book Haken shows that Shannon information refers to closed systems and that when perceived from the perspective of the synergetic approach to open, self-organizing systems, the formal concept of information might include semantics, that is to say, meaning. This happens "if the response of the receiver is taken into account" (Haken, ibid, p. 16), in other words, if intentionality, memory and hermeneutics, as well as internal and external information, that we have discussed above in Sects. 4.1.2, 4.1.3, are taken into account. This is interesting because it implies that when looked at from the perspective of self-organization and synergetics, the formal meaning of information comes close to its literal, intuitive usage.

4.2.1 Between the Local and the Global

Information about the city, internal and external, comes in a number of spatial scales. The extremes are the local scale, concerning a house and its immediate neighbors, and the global scale, concerning the cultural–socio–economic spatial configuration of the city as a whole. In between, individuals also have information about the distribution of certain phenomena in the city, for example, the spatial distribution of vacant houses. The spatial structure of the city as a whole is constantly evolving and changing, to a large extent as a consequence of the inter- and intra-urban migratory movements of individuals, whose migration decisions are based on their internal and external information from local- to global-scales. This local–global urban process of self-organization is only one theme in the wider context of the relations between the local and the global.

The above externally represented information about the city can only partly be subject to direct sensation or experience by individuals. Because of size, usually, people cannot see (hear, smell, ...etc.) the various spatial scales of the city in their entirety. Yet, despite of this limitation they usually do have an internally represented image, or a cognitive map, of the city at its various scales. They construct such cognitive maps by complementing and interpreting the externally represented information to which they are exposed when moving and acting in the city, by information internally represented in their mind. Information about the city at its various scales is thus constructed by this on-going interplay between internal and external information. This interplay provides the basis for the notion of IRN as introduced by Portugali (1996b) and cast into the formalism of synergetics by Haken and Portugali (1996), as well as to the FACS and IRN cities of Sect. 3.8 in the previous chapter.

The whole dynamics of self-organizing systems concerns the relations between the local and the global: between the local and global information available to the individual, between the local properties of the individual elementary part and those of the global properties of 'the system' (i.e. the city), between the local behavior of the individual and the global behavior of the city, between the properties of the local neighborhood and the global city, or region of cities, and so on. The basic perception in the various theories of self-organization is that of *circular causality* or *simultaneous determination*: The behavior, action, properties and interaction of the local and/or the individual level, determine the properties, structure and behavior of the global, and the latter simultaneously determine the behavior, action, properties and interaction of the local, and so on in an ongoing process.

This tension between the local and the global is also central to social theory and philosophy as well as to the social theory of the city. The long and complex discourse on this issue, in social theory, has been summarized, in an oversimplified diagrammatic manner, by Gregory (1981, p. 11) in terms of four basic positions (Fig. 4.1): reification, voluntarism, dialectical reproduction and structuration. The view we develop in this study in the context of our self-organizing cities, which is strongly connected with Haken's synergetics, is close to Giddens' structuration as presented in Fig. 4.1. This can be seen by comparing dialectical reproduction and structuration, in Gregory's diagram with Haken's diagrammatic representation of his paradigms of pattern formation and recognition (Figs. 3.5–3.6 in Chap. 3). In Haken's conceptualization the order parameter(s) determine the behavior of the subsystems; in our cities, as in Giddens' structuration, not only behavior, but also the personal properties of the individual are determined through his/her interaction with other individuals and the city as a whole, which is simultaneously determined by the activities etc. of the individuals. We shall come back to this issue in Chap. 7 below.

4.3 Self-Organization with Free Agents

Self-organizing systems, be they cities or the individuals that form their subsytems, are open, complex and are characterized by circular causality, repro-

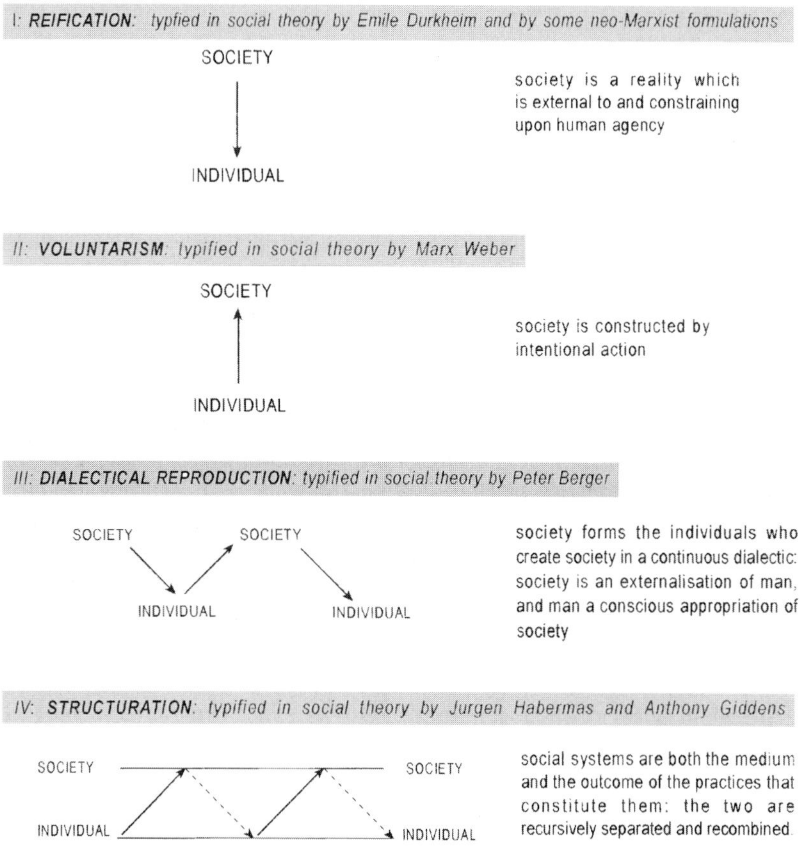

Fig. 4.1. Gregory's (1981, p. 11) diagrammatic representation of four basic positions concerning individual–society relations: reification, voluntarism, dialectical reproduction and structuration

duction and structuration. These properties constrain, in fact, the possibility to treat such systems mechanically and causally. As we have seen above (Chap. 3), Haken has suggested that in such circumstances it is operationally more useful to examine the overall morphological or global behavior of the system. When this is done it has been found that the evolution of open, complex systems exhibits a very distinct and routinized path: a long period of 'steady state', followed by a short period of strong fluctuations or chaos, from which the system re-emerges to a new level of steady state and structural stability, and so on. This rhythm is represented graphically in Fig. 4.2. A substantial portion of the research effort in the study of self-organization is thus directed to the way the system moves from its unstable chaotic stage to a steady state.

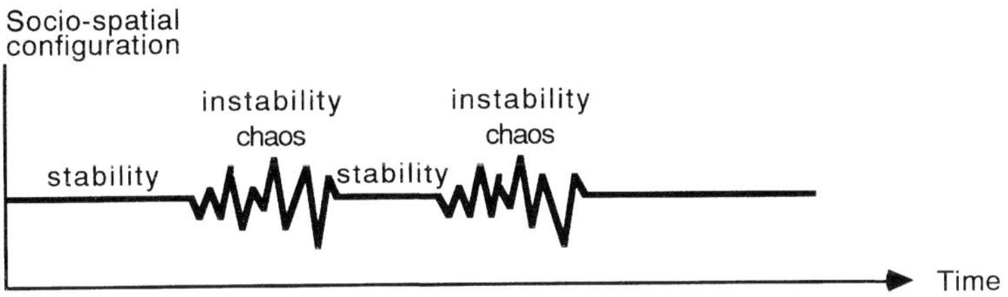

Fig. 4.2. Stability and instability in the evolution of open, complex, self-organizing systems. Source for Figs. 4.2–4.5: Portugali (1993)

The transition from instability or chaos to steady-state forms also the port of departure to Haken's (1987) synergetics approach to self-organization described above. As we have just seen, according to synergetics, during the period of systemic chaos or instability several 'order states' compete, until one 'wins', enslaves the system and brings it into a new steady state. The amplitude of this winning state is termed the order parameter. During a steady state the system is thus governed by a single or a few order parameter(s). The elements or parts of the system thus give rise to an order prameter which then enslave the parts, which then by their enslavement support the order parameter (s) and so on in a process of circular causality.

When the elementary units of the system are one dimensional entities, such as atoms, molecules, bees, or economic persons, the main challenge is to identify the order parameter(s) that dominate the dynamics of the city. But when the elementary units are free agents, when they are themselves self-organizing entities, then one faces a triple challenge: to identify the order parameter(s) that dominate the city, those that dominate the individual free agents, and the relations between them. This is in fact the same issue we've referred to above as the relations between the global and the local: the way the global order parameter(s) of the city are related to the local order parameter of the individual.

4.3.1 Stratigraphic Change

It is common to describe the process of self-organization by means of a bifurcation diagram. This is exemplified in Fig. 4.3. The diagram of Fig. 4.3 exhibits what has been termed (Portugali 1993) *stratigraphic change*: the system moves from one stable state to another, via bifurcations, and with every evolutionary move the previous and alternative steady states die and disappear. Thus the system moves, via bifurcations, from one stratum to another. Stratigraphic change is the usual form of change discussed in the literature on self-organization and the main form of change considered in the self-organizing cities of Chap. 3: Every new bifurcation entails a new city – a new urban stratum. Such a process implies a unique evolutionary path to every city and the urban process might thus be said to be historical. Past

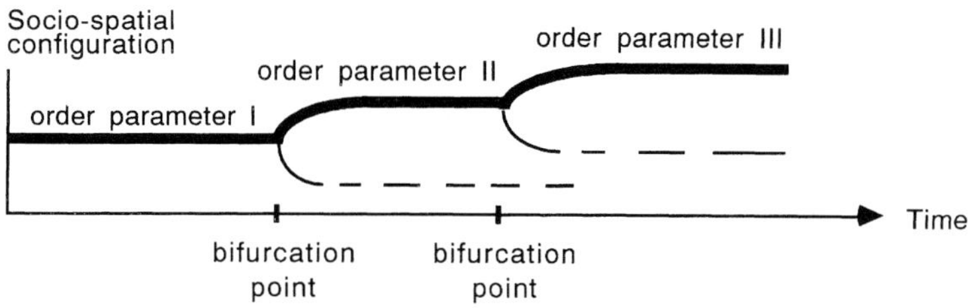

Fig. 4.3. Stratigraphic change

cities are important in the sense that they played a significant role in creating the new city, but they are not part of its present and the city thus has no memory and needs no memory to operate.

The theory of self-organization originated, as noted, in the natural sciences and only at a later stage was it applied to the social domain; and as we've seen in the discussion throughout Chap. 3 and above, usually this was done from a natural sciences orientation. Since our aim is not to apply, but to adapt the notion of self-organization to the human domain of free agents and cities as perceived by social theory, we have also to allow for new concepts of change. As a first approximation this can be done by introducing two additional forms of systemic change: furcative and hermeneutic.

4.3.2 Furcative Change

In a furcative form of change (Fig. 4.4), with every stage of furcation the system becomes more complex in the sense that the old or alternative states (or order parameters) do not die but continue to live, enslaved by the dominant order parameters. An immediate example for a furcative change is the individual human being, the development of which is assumed to proceed in pre-programmed stages or phase transitions. For example, the oral, anal and oedipal 'steady states' of Freud, or the cognitive–developmental stages of Piaget, might be interpreted in terms of a sequence of order parameters.

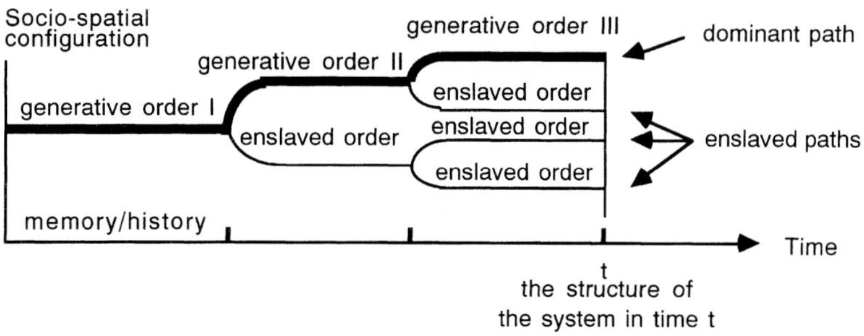

Fig. 4.4. Furcative change

Each phase transition thus entails a new order parameter which adds a new quality to the person, but does not totally eliminate the qualities of previous order parameters. An example in the cultural–social domain might be the sequence of socio–spatial revolutions identified in the evolution of society: for example, the Hominid, agricultural (Neolithic), and urban revolutions. From the perspective of self-organization it can be said that each of these revolutions can be interpreted as a bifurcation from which society emerged into a new (and more 'progressive') phase. The evidence we have on those revolutions suggests that the process was not that of a stratigraphic, but a furcative change: agriculture indeed enslaved hunting and gathering, but did not eliminate it – it made hunting and gathering a marginal non-dominant form of production within agriculture. In a similar way urbanization was in fact a reconfiguration within which agriculture production was globalized (Portugali 1993, 1994a and Chap. 15 below) and hunting–gathering was further pushed to the margin of social productivity.

A very recent example is the events associated with the collapse of the USSR. Following the 1917 furcation, Communism enslaved religious, nationalistic and ethnic aspirations. The latter did not die (as many thought), but were enfolded into the implicate order – into the collective memories of the various socio–spatial groups. From this point of view in the recent furcative change – in the collapse of the USSR – the various national and ethnic groups with their specific beliefs and aspirations once again unfolded to become entities in the explicate domain. Furcative change thus implied memory.

With respect to the human individual as a free agent this implied memory raises no problems – the individual is born with a ready made memory, so to say. But at the level of society, or a city, this is a question: what is the meaning of a collective social, urban, or city, memory?, and where about in the city it is located or distributed? We shall come back to these issues in the next chapters, but here in a preliminary way one might look for an answer in the consequences of furcative changes in cities. The landscape of most modern cities is, in fact, a mosaic of ancient, old and new buildings, streets, roads, quarters, streets' names, styles, each of which is a memory trace which participates in creating the present landscape of the city. In Chap. 1 we quoted Wittgenstein who used the metaphor of a city to convey the notion of a language. With few changes we can rephrase Wittgenstein's passage and use the metaphor of a language (and by implication of a memory) to convey the notion of a city:

> ... ask yourself whether our city is complete; – whether it was so before the industry, the roads for automobiles, and the high rises were incorporated in it; for these are, so to speak, new concepts and words of our city. (And how many words and sentences does it take before a language begins to be a language?) Our city can be seen as an ancient language: a maze of voices words and expressions, of old and new concepts, and of concepts with additions from various periods; and this surrounded by a multitude of

new dialects with specific regular words and uniform schemata (rephrasing of Wittgenstein 1953, par. 18)

(Bohm's notion of holomovement and the continuous process of enfoldment/unfoldment is thus integrated in a natural way with the notion of furcative change and as such with the theory of self-organization, see discussion below.)

4.3.3 Hermeneutic Change

With a hermeneutic form of change (Fig. 4.5), the system furcates as above, but not by eliminating or enslaving competing order parameters, but by expanding the boundaries of the dominating and thus increasing their internal complexity. Hermeneutics is 'the art of interpretation' and it was and is the usual approach in religious writing. The name of the game is to interpret the old (anachronistic) law in such a way that it will adjust to the new reality and at the same time will preserve its old generative principles. Hermenuetics was introduced to modern scientific discussion by Dilthey, Heidegger, Habermas and Gadamer, with a 'soft' claim that it is a property of every textual analysis (Levi 1986), and with a 'hard' claim (that we've already adopted in our discussion of the properties of the free agent) that it is typical of human activity in general (Giddens, 1976). At the individual level, hermeneutics is the process by which, using an existing set of order parameters, the individual extracts, or creates, in-formation from the ex-formation that s/he sees and experiences in the city. The resultant hermeneutic change thus adds new meaning and content to an existing personal perception of a given city structure without changing it. A case in point that we discuss below concerns a person belonging to a certain cultural group, say Blues, who is changing his attitude toward people from another group, say Greens, as a consequence of living a long time among them.

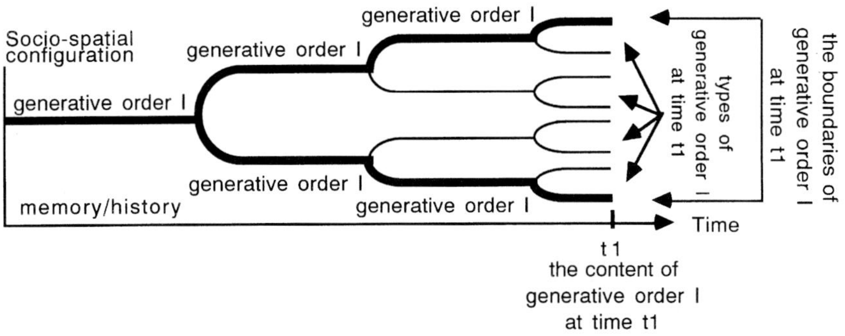

Fig. 4.5. Hermeneutic change

4.4 Free Agents in a Cellular Space (FACS)

In order to allow for intentionality, hermeneutics and memory to be basic properties of the individuals that operate in the models, and for the interplay between internal and external information, as well as between the local versus global, we transform below the CA self-organizing cities into FACS cities, and treat the various simulations as 'games'. Let us start with the first task.

Like the CA models described in Chap. 3, the FACS models we present below are means to study self-organization in the city. And similarly to them, we, too, start from the standard CA models. Our models, however, differ from regular CA city models in that they make explicit the role of the human agents in the self-organizing urban process. The full details of these FACS models are elaborated in Chaps. 5–10 below. Here we present their general structure and properties.

4.4.1 Cellular Automata

Our starting point, as noted, is the standard two-dimensional cellular automata (CA) model. As originally formulated (Toffoli and Margolus 1987), Cellular Automata $\mathcal{A}$ is a lattice of cells, each of which can be in one of k possible states $S = \{S_1, S_2, \ldots, S_k\}$. The states of cells are changing in time, depending on the state of the cell and on the states of the neighboring cells:

$$S^H(t+1) = F_t(S^H(t),\ S^{U(H)}(t)), \tag{4.1}$$

where $S^H(t)$ is the current state of cell H, $U(H)$ is a neighborhood of cell H, $S^{U(H)}$ denotes the set of states of the neighboring cells (neighborhood's configuration), and F is the transformation rule according to which the new state of H is selected out of S.

As already noted, this formulation is formally and intuitively attractive to urban and regional modeling: like real cities which are built of discrete spatial units, so also are cellular automata models; like houses and lots in real cities, the properties of the spatial units in cellular automata models (the cells) are determined in relation to their immediate neighbors.

4.4.2 City

CA city models as developed in line with (4.1), are indeed very attractive, yet at the same time they are also limiting. This is so since they were originally designed with reference to physical and biological systems whose main actors are spatially fixed (tissue cells, atoms of crystal lattice, meadow plants, trees in the forest, etc.). As discussed above, such open and complex systems interact with their environment by exchanging matter and energy through their boundaries. The city as an open complex system exchanges with its environment not only matter and energy, but also human population, that is to say, individuals that act and interact intentionally, perceive and interpret the urban scene in their own specific way, and so on. The movement of these free agents, between and within cities, creates interregional (or inter-cities) migration, which forms the interactive links between a city and its environment, and intraregional migration, which plays an important role in the internal dynamics of cities. Immigrants are indeed free human agents, but their activities take place on an infrastructure which is a cell space of houses, parcels of land, network of streets and so on. In formulating our urban models we've attempted to take into consideration both the cellular and the human components of the city structure. Our first attempt is a CA model that we term City:

$$S^H(t+1) = F_t(S^H(t),\ S^{U(H)}(t),\ S^{\mathcal{A}}(t),\ E(t)). \tag{4.2}$$

Formally, City is a generalization of a standard CA model in two respects. First, we suppose that the transformation rule depends on the parameters of the CA as a whole, and denote this dependence as $S^{\mathcal{A}}(t)$. Below we interpret it as a dependence on the fractions of cells in each of the possible states $\{S_1,\ S_2,\ \ldots,\ S_k\}$. Second, we suppose that the transformation rule can depend on an external parameter $E(t)$.

Using City, we start, in Chap. 5, to examine phenomena of residential socio–spatial segregation in a city and the existence and role of local regions of instability within an otherwise stable urban system. In the various simulations played on City, the actors are defined in terms of their intention of locating close to members of their own socioeconomic, ethnic or cultural group. This formulation enabled us to isolate the impact of the preference for culturally or socially homogeneous neighborhoods from other complex location factors operating in the city, and trace the implications of such preferences on residential patterns. One important implication was the finding that the emergence of local pockets of socio–spatial instability is governed (among other things) by a self-organization principle we term the *captivity principle.* The latter is essential to the systemic stability and reproduction of the city as a whole.

4.4.3 City-1

The above formulation attempts to take account of the human agency properties and at the same time to remain as close as possible to the original CA formalism. This is achieved by the fact that human properties appear in the model either implicitly, or as external input to it – as the rules of the game.

In our next model 'City-1' the human properties of the actors appear explicitly as integral elements of the model. This is achieved by constructing City-1 as a two-layer model: a population model of human agents describing the migration and interaction activities of individuals, superimposed on a CA infrastructure describing various domains of the urban landscape (Fig. 4.6).

The general structure of the model (we still denote it as $\mathcal{A}$) can be written as follows:

$$
\begin{aligned}
S^H(t+1) &= F_t\big(S^H(t), A^H(t), S^{U(H)}(t), A^{U(H)}(t), S^{\mathcal{A}}(t)\big), \qquad (4.3)\\
A^H(t+1) &= G_t\big(S^H(t), A^H(t), P^A(t), S^{U(H)}(t), A^{U(H)}(t), S^{\mathcal{A}}(t), E(t)\big),\\
P^A(t+1) &= L_t\big(S^H(t), A^H(t), P^A(t), S^{U(H)}(t), A^{U(H)}(t), S^{\mathcal{A}}(t), E(t)\big),
\end{aligned}
$$

where $P^A(t)$ is the current location of an agent A and, in addition to (4.2), $A^H(t)$ is a state of a human agent A occupying cell H, $A^{U(H)}(t)$ is a set of states of the human agents in the $U(H)$, and G and L are additional components of the transformation rules: G defines the state of an agent, occupying cell H at the next time step and L defines the location of A at $t+1$. In City-1 the state of a cell is represented by means of its value, as a continuous characteristic varying on [0, 1]. It is normalized at each time step by dividing the cell value by the maximum over the City. The state of an agent is represented by means of two characteristics. First is a Boolean cultural identity, allowing us to study the interaction of two cultural groups in the city space. Second is an economic status, varying on [0, 1], that can change during the time an agent is in the city. The vector $E(t)$ describes immigration inflow, and is thus included in the second and third of equations (4.3).

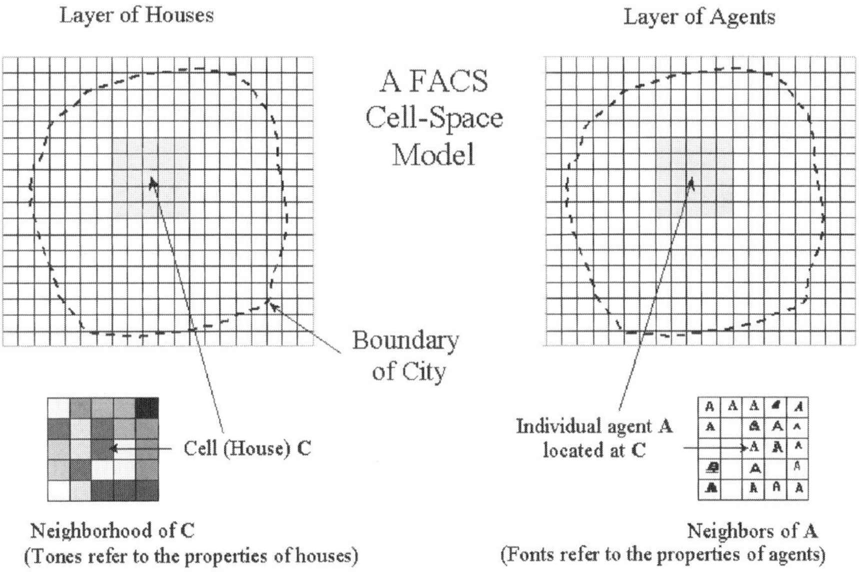

Fig. 4.6. A typical FACS model is constructed of two-layers: a population layer of human agents describing the migratory and interaction activities of individuals (*right*), superimposed on a CA infrastructure describing the urban landscape (*left*). This structure is typical of the City-1, City-2, ... City-6 models developed below

By means of City-1 we consider, several interrelated questions: (a) in Chap. 6, the impact of international migration on intra-urban migration, the urban space economy, and socio–spatial segregation – the empirical case being the recent waves of Jewish immigrants from ex-USSR to Israel (b) in Chap. 12, the implication to planning theory and praxis of the central property of cities as self-organizing systems: that they are not predictable and controllable and in this respect not plannable (in the conventional way of planning). City-1 is an elaboration of City in several respects: (a) it takes into consideration not only cultural affiliation and behavioral intentions of individuals, but also their actual and potential economic situation, as well as the urban land value with respect to both rented and owner occupied houses (b) It is constructed also as a planning device – a heuristic computer game whose aim is to enable planners to gain 'artificial planning experience' on the basis of which they will be better qualified to take intuitive planning decisions.

Another issue we study by means of City-1 concerns the interplay between two order parameters which are central to the urban dynamic: the cultural order parameter and the economic order parameter. In particular we examine the non-linear consequences of self-organization in the city in relation to socio–cultural spatial segregation. It was found that when individuals are motivated by economic preferences only, the result is a socio–spatial integrated city: when cultural preferences dominate, the city is first segregative but with time spatial segregation disappears; and when individuals' decisions are motivated by both cultural and economic preferences the city evolves from the start as a segregated city and persists being so.

In terms of synergetics the above findings may be interpreted in the following way: Assume, first, a control parameter which refers to individuals' motivation and changes from an economic motivation to a mixed economic and cultural motivation, and then to a cultural motivation. Now, in the usual way it can be said that as the control parameter grows and crosses a certain threshold, the economic order parameter emerges and enslaves the urban system. Then, as the control parameter grows further and crosses a second threshold, two order parameters emerge (the economic and the cultural) in relations of interdependency. As the control parameter grows even further and crosses a third threshold, the cultural order parameter takes over with the consequences noted above. (I am indebted to Professor Haken for suggesting this interpretation to me.)

4.4.4 City-2

The third model, City-2, is structurally similar to City-1 as formulated above. It differs from it in that it is specifically designed as a heuristic tool to examine, first, the circumstances by which spatial cognitive dissonance in the city may arise and lead to changes in individuals' intentions, cultural affiliation and identity. Second, the implications thereof extend to residential spatial segregation in the city. City-2 is thus generally similar to (4.3) with the addition that the cultural identity of an agent occupying cell H is a continuous variable on [0, 1] that can change during the time the agent is in City-2. State $A^H(t)$ of an agent in cell H is therefore a two-component vector $A^H(t) = (A^H_{\mathrm{e}}(t),\ A^H_{\mathrm{c}}(t))$, where $A^H_{\mathrm{e}}(t)$ represents economic status and $A^H_c(t)$ cultural identity of an agent. The system is governed now by means of the following system of equations:

$$\begin{aligned} S^H(t+1) &= F_t\big(S^H(t), A^H(t), S^{U(H)}(t), A^{U(H)}(t), S^{\mathcal{A}}(t)\big), \qquad (4.4)\\ A^H_{\mathrm{e}}(t+1) &= G_{\mathrm{e}t}\big(S^H(t), A^H(t), P^A(t), S^{U(H)}(t), A^{U(H)}(t), S^{\mathcal{A}}(t), E(t)\big),\\ A^H_{\mathrm{c}}(t+1) &= G_{\mathrm{c}t}\big(S^H(t), A^H(t), P^A(t), S^{U(H)}(t), A^{U(H)}(t), S^{\mathcal{A}}(t), E(t)\big),\\ P^A(t+1) &= L_t\big(S^H(t), A^H(t), P^A(t), S^{U(H)}(t), A^{U(H)}(t), S^{\mathcal{A}}(t), E(t)\big). \end{aligned}$$

Like City-1, City-2 is a two-layer model: a population layer composed of individuals (persons, families or households) with their cultural and economic properties, and a housing stock layer of lattice of cells (i.e. houses) with their properties. The individuals who enter the city (the immigrants) and those living in the city (the inhabitants) interact with each other and the system of cells/houses, and this interaction results in (a) intra- and inter-city migration dynamics, (b) changes of the properties of individuals and (c) changes in the properties of the cells.

In both City and City-1 the socio–cultural properties of individuals (Greens vs. Blues and veterans versus 'Olim' – Hebrew for 'Jewish new immigrants') and their socio–spatial intentions/wants (segregatives or neutrals) were determined externally, outside the models. The aim in City-2 is to allow for these properties to be determined internally, inside the model. Such a model will enable us to examine the very dynamics of (a) the tendencies, values and wants of individuals, (b) the way a new socio–spatial group with its own identity is born, and (c) the implications thereof to residential spatial segregation in the city.

It is important to note that the questions of human intentionality and socio–spatial emergence are central to social theory. In this context our usage of City-2 in Chap. 7, in fact, suggests and adds our own theory or view on the matter. It differs from the usual discourse in social theory (and in social geography) in the means employed for that purpose: in addition to, and in place of, the 'usual' hermeneutic discourse which is typical of social theory, we'll make an extensive use of City-2 which is the specific heuristic–hermeneutic model we construct below.

4.4.5 City-3

City-3 is, in fact, a refinement of one aspect of City-2. In the latter individuals' decisions are made in relation to local information only. Individuals decide to move to and from a certain place by looking at the properties of their nearest neighbors. In the model this included the neighbors in the 5×5 cells around the cell under consideration. In City-3 we include also the role of global information. That is to say, individual agents make their location decision by taking into consideration both the properties of the nearest neighbors to the cell/house they consider, and in addition to this, the cultural properties of the city as a whole. For example, the existence of a large, spatially distinguished (i.e. segregated), cultural entity at the global scale of the city as a whole, balances the effect of the structure of the local (5×5 cells) neighborhood. Thus, a spatially distinguished and segregated Green community at the global city scale might affect a Green individual to retain his/her cultural identity in face of a spatial cognitive dissonance at the local scale of the neighborhood (the 5×5 cells around the house). This is achieved mainly by adding to the measure of sensitivity to local cognitive dissonance a sensitivity measure to the global, or overall, cultural and economic structure of the city. The general form of City-3 is the same as City-2 (see (4.4) above), but the interpretations of each of the equations, regarding dependence on the $S^{\mathcal{A}}(t)$, i.e. the state of the system as a whole, is different. Compared to City-2, in City-3 the state of an agent A, its migration decisions, and the state of the cell it occupies, depend on the *degree of segregation* of the socio–spatial cultural groups that might emerge out of the dynamics. This enables us to explicitly consider the role of global external information in the evolution of cognitive dissonance and the subsequent emergence (or not) of a new cultural group.

4.4.6 City-4 and City-5

As one can see, the main methodological tool of Part II is the family of City models or games that we have just surveyed. We close this survey, by introducing two more models/games: City-4 and City-5. We introduce them despite the fact that they are at a rather preliminary and still experimental stage and, as such, still limited in their application. Rather than to summarize what has been done, these two models are included to indicate where we intend to go to. The two models depart from the previous ones in two respects: in their definition of the individual agents, and with respect to the definition of the infrastructure of the modeled city. City-4 defines the individual agent by means of a cultural code (c-code) or memetic code (m-code), whereas City-5 moves from CA-city to GIS-city, that is to say, from an infrastructure which is based on an abstract cell space, to one which is based on the 'real' structure of the city as encoded by means of GIS (geographical information systems). Let us introduce the two of them.

City-4. While 'free', the virtual agents that we play with in the models surveyed above are culturally uni-dimensional. Each agent's identity is defined as Green, Blue or in between (Yellow, Red), along a single, uni-dimensional cultural scale. Our cultural reality is much more complex in several respects, one of them being the fact that a person's identity is multi-dimensional and as such cannot be stretched along a single dimension. In City-4 we therefore make a first attempt to construct a new city game with culturally more complex, multi-dimensional, agents. Formally, each agent's cultural identity A_c is defined by a cultural code, which is a multi-dimensional Boolean vector of length K namely, $A_c = (A_{c1}, A_{c2}, \ldots A_{cK})$, where $A_{ci} \in \{0,1\}$ for all i.

The general form of City-4 expands that of City-2 and City-3. Formally, in City-4, instead of one equation that describes the dynamics of an agent's cultural identity A_c, we have a set of K equations describing the dynamics of each component of A_c. As a result, the general form of City-4 is the following:

$$\begin{aligned}
S^H(t+1) &= F_t\big(S^H(t), A^H(t), S^{U(H)}(t), A^{U(H)}(t), S^{\mathcal{A}}(t)\big), \qquad (4.5)\\
A^H_{\mathrm{e}}(t+1) &= G_{\mathrm{e}t}\big(S^H(t), A^H(t), P^A(t), S^{U(H)}(t), A^{U(H)}(t), S^{\mathcal{A}}(t), E(t)\big),\\
A^H_{\mathrm{c}i}(t+1) &= G_{\mathrm{c}it}\big(S^H(t), A^H(t), P^A(t), S^{U(H)}(t), A^{U(H)}(t), S^{\mathcal{A}}(t), E(t)\big),\\
&\qquad i = 1, 2, \ldots, K.\\
P^A(t+1) &= L_t\big(S^H(t), A^H(t), P^A(t), S^{U(H)}(t), A^{U(H)}(t), S^{\mathcal{A}}(t), E(t)\big).
\end{aligned}$$

In Chap. 8 the interpretation of (4.5) is such that all components of an agent's cultural identity are equally important and do not correlate.

As elaborated in some detail in Chap. 8, the preliminary results of City-4 are quite interesting. First, the formulation of City-4 is more general, in the sense that our previous City models appear to be a special case of it. Second, while the number of groups that can potentially be constructed might be large (2^K), the actual number of groups that emerge in the city is much smaller than the potential. Furthermore, their number is changing in time and does not stabilize. Third, and as a consequence of the above, the cultural groups that emerge and co-exist in the city are not pre-determined as before, but emerge out of the dynamics.

City-5. This is our first attempt to construct our city game on an infrastructure which is not a cellular space as before, but which is a GIS space. That is to say, a real city map as captured by a Geographical Information System. There are several ways to implement this task. The one employed here is to represent a city's infrastructure by a GIS-based vector map consisting of a number of layers of streets, houses, open spaces, etc. and to put on top the layer of free agents as defined in the five City models discussed above. The resulting model can still be represented by (4.5), but the definition of $U(H)$ is essentially different. In the previous City models the neighborhood of a house is always a square of 5×5 cells with the house at the center. In the GIS-based version, $U(H)$ is determined by the configuration and types of streets around the house, visibility of the neighboring houses, and so on. Preliminary experiments with City-5 demonstrate the role of the heterogeneity and variety of the neighborhoods on the city dynamics.

4.4.7 Modeling Groups Internal Complexity

Parallel to the City-4 model, we present here another exposition of cultural identity as a multi-dimensional characteristic of an individual. We call this model City-6. It differs from City-4 in its interpretation of the relations between the different dimensions of the individual's cultural identity. While in City-4 we do not assume a-priori dependencies between any two components of cultural identity, in City-6 each individual has a *hierarchical identity.* That means that an agent might be a member in more than one identity group. For example, in addition to being, say, religious (highest level of hierarchy) an individual also might have a community identity (second hierarchical level), family identity (third level), and so on. Cultural identity of the individual in City-6 does not change in time and in this respect City-6 continues and expands City and City-1 models, with their permanent Blue, Green, Neutral, etc., identities. The purpose of City-6 is to specifically study the effects of groups' internal complexity on the dynamics of residential distribution in the city space. Its general form, therefore, represents migration changes:

$$P^A(t+1) = F_t\big(A^{II}(t),\ P^A(t),\ A^{U(H)}(t),\ E(t)\big), \tag{4.6}$$

where, as above, $P^A(t)$ is the current location of an agent A, $A^{U(H)}(t)$ denotes states of agents within neighborhood of A's current location and $E(t)$ represents external parameters (immigration).

In City-6 the identity of an agent A is denoted by means of a vector $(a_i,\ b_{ij},\ c_{ijk},\ \ldots)$, where the first component, a_i, designates a group at the highest hierarchical level, the second component, b_{ij}, designates a group at the second level of hierarchy, and so on. If $a_i \in \{a_1,\ a_2,\ \ldots,\ a_I\}$ (that is, there are I different identities at the highest level of hierarchy), then for an agent with identity a_1 at the highest level its identity b_{1j} at the second level belongs to the set $\{b_{11},\ b_{12},\ \ldots,\ b_{1J(1)}\}$, for an agent with $a = a_2$ its identity b_{2j} at the second level belongs to the set $\{b_{21},\ b_{22},\ \ldots,\ b_{2J(2)}\}$, etc., where $J(1),\ J(2),\ \ldots,\ J(I)$ is the number of sub-groups within each highest-level group, and the values b_{ij} are all different. Similar relationships hold for second and third levels, etc.

An a-priori hierarchy of identities implies a nested process of agent's housing search and a hierarchy of resulting residential distributions. Let us denote by the symbol D_i cells, which are occupied by individuals having identity a_i at the highest level of hierarchy, by D_{ij} let us denote cells, which are occupied by individuals having identity b_{ij} at the second level of hierarchy, etc. The following condition is always true: $D_i = D_{1i} \cup D_{2i} \cup \ldots \cup D_{iJ(i)}$, and it also true for second and third levels, third and fourth ones, etc.

The dynamics of residential distributions in the City-6 model is studied in Chap. 10, where we intensively use Atkin's (1974; 1977, 1978) Q-Analysis to analyze the emerging hierarchical residential structures.

4.5 Urban Games with FACS

'The game of life' is one among several heuristic games or models 'played' by Eigen and Winkler (1983) in their *Laws of the Game* as part of their attempt to study processes connected with the emergence of life and other emergencies. Eigen and Winkler's games are played by means of cellular automata (CA) models, and they call them 'games' to emphasise, first, their heuristic nature, second, the interplay that you can see in them between local action and global structure, and third, between the fixed 'rules of the game' and the creativity that can take place within their frame. In what follows we suggest instead of CA, our FACS models, and we employ the notion 'game' to emphasise, in addition to the above three properties of Eigen and Winkler's games, the role of the human free agent in the process, and the usefulness of the game in the context of urban decision making and planning.

Like Eigen and Winkler's games of life, our urban games too are computerized, mathematical, simulation models of various forms and complexity. In this respect they are not new and follow the simulation models of the quantitative geography of the Sixties, as well as the self-organization models of Chap. 3. And yet they are different: unlike the simulation models of the past and those of Chap. 3, their aim is not to best-fit a simulated scenario to a real one – our view is that in a self-organizing city of the postmodern age this is somewhat problematic. These models were designed as heuristic and hermeneutic games. 'Heuristic' in the sense that their essence is to be means to learn and explore the properties of cities. And 'hermeneutic', in the sense that they include a strong element of free phenomenological interpretation, and in this respect they are not very different, we would claim, from hermeneutics as practiced in social theory.

The usual methodology of social theory and its SMH cities, as discussed in Chap. 2, is hermeneutics: a free phenomenological interpretation by means of a discourse, normally rhetorical or textual. The product of this process is an ever growing and accumulating body of knowledge, mainly oral and textual. This is different from the usual methodology employed in the self-organizing cities of Chap. 3. In the latter the almost standard methodological tool is a simulation model, which is designed to mimic as close as possible (i.e. best-fit) empirical data. The basic, often implicit, assumption is that empirical analysis genuinely describes reality. As a consequence, the model is assumed to describe reality and to have the capability of predicting its future.

On the face of it, there is an unbridgeable gap between the two 'cities' – the one of social theory and the one of self-organization. Our view is that the gap results not so much from inherent contradictions, as from different developments and traditions of scientific disciplines. First, as already noted, theories such as self-organization have found in the domain of matter properties hitherto assigned to the human domain, and they thus come much closer to the discourse of social theory then the mechanistic theories which preceded them. Second, we submit that every simulation model is characterized by what might be termed an *interpretation gap*. Every model has in it an element of free, intuitive, phenomenological, interpretation of reality, or in short an element of hermeneutics. From this perspective, city models are first and foremost a method of representation of ideas about the dynamics and structure of the phenomenon we term city. For example, the cities of Chap. 3 are constructed on the basis of an idea that the city is a self-organizing system. And this idea is supported by our general experience and knowledge of cities and urbanism as accumulated in a long process of hermeneutic discourse among students of cities and urbanism.

Self-organizing cities have a strong hermeneutic element not only at their very start – in their phenomenological foundations, but also at their end products which usually come in a mathematical, geometrical and a map-like scenario evolving on the computer screen. Such evolving map-like scenario has no explicit, self-evident results or answers in it. It is a visualization of some sort, which still requires an interpretation. You as a scientist must say what you see in this simulated 'Rorschach shape', you must provide a good imaginative interpretation. And whether you like it or not, what you see, that is to say, your interpretation, is strongly influenced by your cultural and social environment, as well as by your personality. (And in this respect you are very similar to our virtual free agents which, in line with the virtual memory we've installed inside them, move about, and take decisions, in the cellular space that in our imagination symbolizes a city.)

So, is there no difference at all between the hermeneutics of social theory and that of self-organization theory? Well, there is, and it is in the nature of the language used. In social theory it is usually (but not exclusively) verbal and textual; in self-organization it is usually mathematics, geometry and simulated shapes. Compared to the hermeneutics of social theory, those of self-organization rely heavily on the visual. The visual is the most prominent and direct medium to represent reality – in our case – the city. The (simulated) city of self-organization is very visual.

5. City: The Greens and the Blues*

5.1 Introduction

In this chapter we make a start in our general attempt, in Part II, to explore the conjunction between a city and self-organization in relation to issues of social–spatial urban segregation of national, ethnic and other socio–cultural groups. More specifically, our aims in this chapter are the following:

1. To further examine cellular automata (CA) as a means to study processes of self-organization in a city, especially with respect to the issues concerning social and cultural phenomena in the city. So far most explicit applications of self-organization to the urban dynamics were formulated in terms of differential equations. We follow here the preliminary suggestions made by Tobler (1979) regarding the geographical potential of Cell Space models in general and of Couclelis (1985, 1988) and Phipps (1989) regarding the potential of cellular automata.
2. To highlight a systemic situation which is central to a city but has not as yet been studied, namely, the existence of local regions of stability and instabilities within an otherwise stable urban system. We discuss this issue in relation to socio–spatial residential segregation. (For a different type of discussion on stability and instability in a city, especially with respect to diffusion, see Zhang 1989, 1990.)
3. To explicitly consider the location of individuals in space. In most applications of self-organization to a city context no attention is paid to the actual location and movement of individuals in space. They enter the city from the 'outside' and are located 'somewhere' in the urban space. This is so since most applications were statistical, regional and global, considering the city's population and space as a whole. As we shall see below, one of the advantages of using cellular automata is the ability to simultaneously and explicitly consider the actual location and location decision of the individual, and the spatial structure of the city as a whole.
4. To be more specific and thus more realistic with respect to human behavior. The various applications of self-organization to the socio–urban domain have assumed an identity between wants, motivations and actual

* With Itzhak Benenson and Itzhak Omer

behavior. This assumption which might be appropriate to, say, ecological systems in nature, is inappropriate to the human systems. In the latter, one must make a distinction between wants, motivations and inclinations, on the one hand, and actual behavior in space, on the other.

5. To be more specific and realistic with respect to the variability of wants, motives and intentions in the socio–human domain. Most applications of the theory of self-organization to the socio–human domain have assumed homogeneous motivation on the part of individuals i.e., the assumption of economic man. We prefer not to assume any fixed, predetermined human wants and motives, but to allow for their variability and to treat them heuristically. One aim of our model is to examine how various configurations of wants and preferences are related to actual behavior and to the various socio–spatial properties of cities.

We start the discussion below by introducing City – a heuristic model we have designed in line with the aims and principles declared above and translated into the formalism of the CA model. We then discuss the properties of City and 'play' on it three games corresponding to various configurations of individuals and their socio–spatial preferences. As already noted, our usage of the terms 'game' and 'play' is borrowed from the *Laws of the Game* by Eigen and Winkler (1983). The games enable us to elucidate the dynamics of stability and instability in a self-organizing city. We close the chapter by considering the city's socio–spatial reproduction process and by introducing the associated notion of the *captivity principle.*

5.2 City: A Heuristic Model

In order to stay within the CA formalism, we consider City's territory as a rectangular lattice of cells. Each of these cells might be seen as a house or a place. Individuals, say households, occupy or leave the various places and thus participate in generating the migration dynamics of City and its socio–spatial structure. Individuals decide to stay in, or leave, the places they occupy on the basis of their residential preferences regarding the properties of the individuals in the neighboring places. For example, a Blue individual who prefers Blue neighbors has a high probability to occupy a certain place as long as all the individuals in the nearest places are also Blue. As the number of non-Blue neighbors increases, also the probability increases that this individual will leave the place and migrate.

The vacant places in City are subject to a competition between individual 'place hunters' who also participate in forming the migration dynamics of City and its socio–spatial structure. The latter group is composed of immigrants from the outside and veteran residents of City, who for reasons described above, have decided to migrate. Individual place hunters decide to occupy a vacant place on the basis of their preferences regarding the properties of the individuals in their neighboring places. For example, the probability that a Green individual who prefers to live among Greens will occupy a certain vacant place is high if the neighboring places are occupied by Greens, and vice versa. This decision and migration process of residents and place hunters takes place simultaneously all over the city's space and at every (machine) time-step.

At each time step, leaving and occupying places proceed in three consecutive sub-steps: first leaving, next occupying by internal migrations, then occupying by immigrations. At the first sub-step, based on the properties of the individual and the constitution of his/her neighborhood, we calculate the probabilities 'to leave' each of the occupied places; individuals then leave or stay in the various places according to these probabilities. The resultant migration wave creates a new socio–spatial order according to which, in the second sub-step, we calculate the probabilities 'to occupy' each of the empty places. Individual place hunters choose empty places in light of these probabilities. The same process is repeated in the third sub-step with respect to external migration.

To conclude the description we have to define the number of individuals of each type who are looking for places and the fate of those who leave the places. Below we consider a simple case in which, first, at each time-step a constant number of individuals enter the system; second, the probability that an entering individual is of a certain type is constant. Third, an individual who leaves a place and does not succeed to occupy a new one at the next time step, leaves the system altogether.

Based on the above we can estimate the probability that a given place on the lattice will change its state; for example, that a place occupied by a Green individual at time t, will be occupied by a Blue individual, or will remain empty, at time t+1. We estimate these probabilities by means of Monte-Carlo simulations.

It must be noted that in a formal CA model there are no individuals nor immigrants who take decisions, but only cells which can be in various states according to the rules of the specific CA model. We thus have to translate the informal model as described above into the formalism of CA.

5.3 City: A Formal Description of the Cellular-Automata Model

As discussed in Chap. 4, CA models form a sub-group within 'Cell Space' models (Albin 1975, Couclelis 1988). Their aim is to describe the dynamics and the various stages in the evolution of a spatially distributed system. Cell space models have been applied in a variety of domains and to many problems and all of them are based on autonomous CA of minimal configuration, described by (4.1). We begin with the generalization of (4.1), given by (4.2).

The formulation of ours accounts for the dependence of the dynamics of the Cellular Automata A on its global properties and on the external parameters:

$$S^H(t+1) = F_t\left(S^H(t), S^{U(H)}(t), S^{\mathcal{A}}(t), E(t)\right).$$

Let us recall, that $S^H(t)$ is the current state of cell H, which belongs to $S = \{S_1, S_2, \ldots, S_k\}$, $U(H)$ is a neighborhood of cell H, and $S^{U(H)}$ denotes the neighborhood's configuration (the set of states of the neighboring cells). By $S^{\mathcal{A}}(t)$ we denote the global characteristics of $\mathcal{A}$, and $E(t)$ designates the external parameters. F_t gives the transformation rule according to which the new state of H is selected out of S. Below City is considered as a rectangular $I \times J$ lattice of cells H_{ij}, $i \in [1, I]$, $j \in [1, J]$.

We have designed City as a probabilistic CA model (Mikhailov and Loskutov 1991) in which the transformation rule F_t gives for each cell H the probabilities of transition from its current state $S^H(t)$ at a time-step t to each one of the k possible states $\{S_1, S_2, \ldots, S_k\}$ at time step $t+1$.

The City transformation rules F_t depend on time and are a superposition of transformation rules for three sub-steps, that we denote below as t, $t+1/3$ and $t+2/3$. The rules F_t and $F_{t+1/3}$ correspond to internal migrations, while $F_{t+2/3}$ correspond to external migrations. These three sets of rules are repeated in turn.

To implement the heuristic model we distinguish in City CA between five possible states of the cell:

$$\begin{array}{ll} \text{Em} - \text{Empty}, & \\ \text{GS} - (\text{Green, Segregative}), & \text{GN} - (\text{Green, Neutral}), \\ \text{BS} - (\text{Blue, Segregative}), & \text{BN} - (\text{Blue, Neutral}). \end{array} \tag{5.1}$$

For convenience let us call the states GS, GN, BS and BN as 'occupied'.

Here and further a cell's neighborhood $U(H)$ includes all cells in a 5×5 square with H at its center. With the boundary effects taken into consideration it gives

$$\begin{aligned} U(H_{ij}) = \{H_{kl} \mid & \max(1, i-2) \le k \le \min(I, i+2), \\ & \max(1, j-2) \le l \le \min(J, j+2)\} \end{aligned} \tag{5.2}$$

A salient feature of our model is that when formulating transformation rules at sub-steps t, $t+1/3$ and $t+2/3$ it accounts for the current global characteristics of A. That is, F_t depends on the fractions of cells of A that are in each one of the occupied states. $F_{t+1/3}$ depends on the fractions of cells in A that have changed their states from occupied to empty between sub-steps t and $t+1/3$. $F_{t+2/3}$ depends on the parameters of the immigration flow. The transformation rules for each sub-step are as follows:

5.3.1 Internal Migrations

5.3.1.1 Leaving the Places: Rules F_t

The transition probabilities at the first sub-step are as follows:

State of a cell at $t + 1/3$

State of a cell at time-step T		Em	GS	GN	BS	BN
	Em	1	0	0	0	0
	GS	p_{GS}	$1 - p_{\text{GS}}$	0	0	0
	GN	p_{GN}	0	$1 - p_{\text{GN}}$	0	0
	BS	p_{BS}	0	0	$1 - p_{\text{BS}}$	0
	BN	p_{BN}	0	0	0	$1 - p_{\text{BN}}$

To define the values of p_i, $i =$ GS, GN, BS, BN, in the table above (the 'repel probabilities'), consider cell H whose state is one of occupied ones, say GS. In the text we say that the decision of an individual to leave a place depends on the fraction of repelling individuals in the neighborhood of that place. In terms of CA we denote the states that 'repel' GS as $R(\text{GS})$. According to our model the repelling states are as follows:

$$\begin{aligned} R(\text{GS}) &= \{\text{BS}, \text{BN}\}, \\ R(\text{GN}) &= \text{O}, \\ R(\text{BS}) &= \{\text{GS}, \text{GN}\}, \\ R(\text{BN}) &= \text{O}. \end{aligned} \tag{5.3}$$

For cell H in state S^H the values of the repel probabilities depend on the fraction $v^H(t)$ of cells within neighborhood $U(H)$, whose states belong to $R(S^H)$:

$$v^H(t) = \sum_G \{1 \mid G \in U(H),\ S^G \in R(S^H)\} / N^H(t), \tag{5.4}$$

where

$$N^H(t) = \sum_G \{1 \mid G \in U(H),\ S^H \neq Em\} \tag{5.5}$$

is the number of cells in the occupied states in $U(H)$.

Based on the fractions given by (5.4) we calculate the probabilities for an occupied cell H (say, in state GS) to change its state to an empty one as follows:

$$p_{\text{GS}} = \begin{cases} 0 & \text{if } \alpha_{\text{GS}} + \beta_{\text{GS}} \exp(\gamma_{\text{GS}} v^H(t)) < 0 \\ 1 & \text{if } \alpha_{\text{GS}} + \beta_{\text{GS}} \exp(\gamma_{\text{GS}} v^H(t)) > 1 \\ \alpha_{\text{GS}} + \beta_{\text{GS}} \exp(\gamma_{\text{GS}} v^H(t)) & \text{otherwise.} \end{cases} \tag{5.6}$$

Below we do not differentiate between repel functions for the Green and Blue segregatives. That is $\alpha_{\text{GS}} = \alpha_{\text{BS}} = \alpha_{\text{S}}$, $\beta_{\text{GS}} = \beta_{\text{BS}} = \beta_{\text{S}}$, $\gamma_{\text{GS}} = \gamma_{\text{BS}} = \gamma_{\text{S}}$. For the Neutrals we suppose $\beta_{\text{GN}} = \beta_{\text{BN}} = 0$, $\gamma_{\text{GN}} = \gamma_{\text{BN}} = 0$, $\alpha_{\text{GN}} = \alpha_{\text{BN}} = \text{const}$. The chosen sets of coefficients α, β, γ give the 'repel' curves, presented in Fig. 5.1.

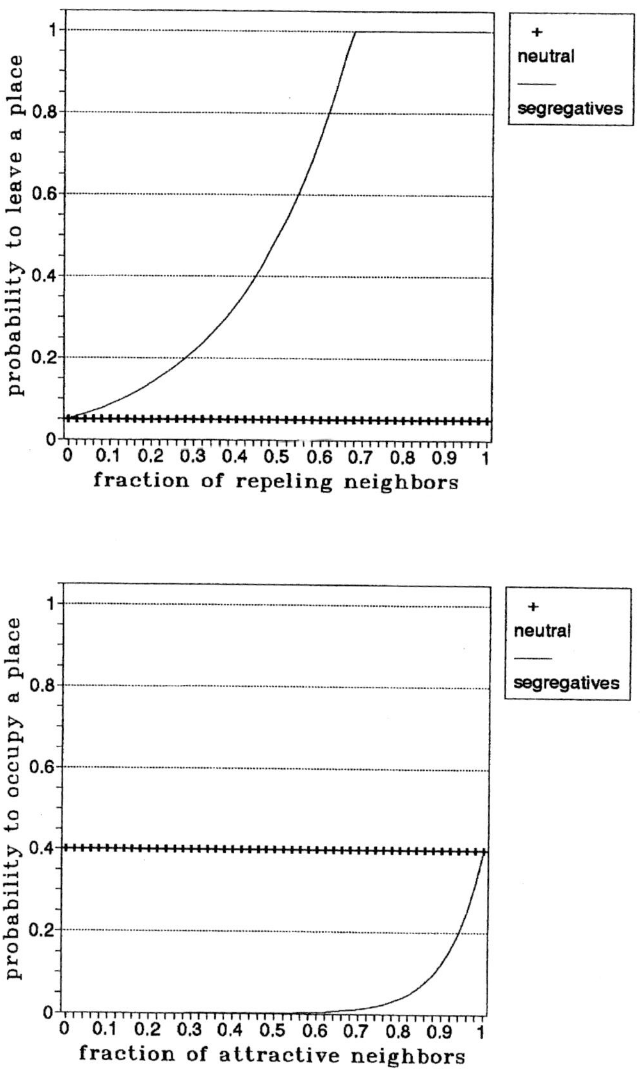

Fig. 5.1. *Top*: the probability, for neutrals and Segregatives, to leave a place. *Bottom*: the probability, for neutrals and Segregatives, to occupy a place

5.3.1.2 Occupying the Places: Rules $F_{t+1/3}$

The transformation rules $F_{t+1/3}$ depend on several global characteristics of $\mathcal{A}(t+1/3)$. First, a cell H is 'active' if at least three cells in $U(H)$ are in an occupied state. The 'active' part $\mathcal{A}_{\mathrm{a}}(t+1/3)$ of cellular automata $\mathcal{A}$ is obtained as

$$\mathcal{A}_{\mathrm{a}}(t+1/3) = \{H \mid N^H(t+1/3) > 3\}. \tag{5.7}$$

We denote the number of cells in $\mathcal{A}_a(t)$ as $N_a(t)$ and the overall number of cells in an occupied state, as $N_{Oc}(t)$.

We then denote the numbers of cells that have changed their state from an occupied one at t, to Em at $t+1/3$, as $C_i(t+1/3)$, $i=$ GS, GN, BS, BN. Qualitatively, $C_i(t)$ are the numbers of cells in occupied state i that are set free after the first sub-step (Sect. 5.3.1.1).

We now define the probabilities for a cell to change its state from empty to occupied. Structurally, the matrix of transition probabilities for the second sub-step is:

		State of a cell at time-step $t+2/3$				
		Em	GS	GN	BS	BN
State of a cell at time step $t+1/3$	Em	q_{Em}	q_{GS}	q_{GN}	q_{BS}	q_{BN}
	GS	0	1	0	0	0
	GN	0	0	1	0	0
	BS	0	0	0	1	0
	BN	0	0	0	0	1

In the text we say that the decision of an individual to occupy cell H depends on the fraction of attractive individuals in $U(H)$. In terms of CA, this implies that a cell within $U(H)$ can 'attract' an empty cell H to change its state. We denote the states that attract a given cell, say GS, as $T(\mathrm{GS})$. According to our definitions, $T(\mathrm{GS})$ and $T(\mathrm{BS})$ are as follows:

$$T(\mathrm{GS}) = \{\mathrm{GS}, \mathrm{GN}\},$$
$$T(\mathrm{BS}) = \{\mathrm{BS}, \mathrm{BN}\}.$$

An empty cell changes its state to a neutral GN or BN no matter what the neighborhood configuration is.

Let us denote the fractions of the cells in the neighborhood of an empty cell H that 'attract' the states GS and BS as $w^{\mathrm{GS}}(t)$ and $w^{\mathrm{BS}}(t)$. For an empty cell H the fraction of cells within the neighborhood $U(H)$, whose states belong to, say, $T(\mathrm{GS})$ equals

$$w^{\mathrm{GS}}(t) = \sum_G \{1 \mid G \in U(H),\ S^G \in T(\mathrm{GS})\} / N^H(t), \tag{5.8}$$

and the attractivity for an empty cell to change its state to a GS-state is calculated as:

$$a_{\mathrm{GS}} = \begin{cases} 0 & \text{if } \delta_{\mathrm{GS}} + \varepsilon_{\mathrm{GS}} \exp(\lambda_{\mathrm{GS}} w^{\mathrm{GS}}(t)) < 0 \\ 1 & \text{if } \delta_{\mathrm{GS}} + \varepsilon_{\mathrm{GS}} \exp(\lambda_{\mathrm{GS}} w^{\mathrm{GS}}(t)) > 1 \\ \delta_{\mathrm{GS}} + \varepsilon_{\mathrm{GS}} \exp(\lambda_{\mathrm{GS}} w^{\mathrm{GS}}(t)) & \text{otherwise} \end{cases} \tag{5.9}$$

As above, coefficients δ, ε and λ in (5.9) depend on the state we consider. Below we do not differentiate between attractivity functions for the Blue and Green segregative individuals. That is $\delta_{\mathrm{GS}} = \delta_{\mathrm{BS}} = \delta_{\mathrm{S}}$, $\varepsilon_{\mathrm{GS}} = \varepsilon_{\mathrm{BS}} = \varepsilon_{\mathrm{S}}$, $\lambda_{\mathrm{GS}} = \lambda_{\mathrm{BS}} = \lambda_{\mathrm{S}}$. Coefficients δ_{S}, ε_{S}, λ_{S} give the 'repel' curves, presented in Fig. 5.1. For Neutrals $\varepsilon_{\mathrm{N}} = 0$, $\lambda_{\mathrm{N}} = 0$, $\delta_{\mathrm{S}} = \mathrm{const}$.

Based on attractivities, the transition probabilities of the second sub-step are calculated as

$$\begin{aligned} q_{\mathrm{Em}} &= 1 - a_{\mathrm{GS}}d_{\mathrm{GS}} - a_{\mathrm{GN}}d_{\mathrm{GN}} - a_{\mathrm{BS}}d_{\mathrm{BS}} - a_{\mathrm{BN}}d_{\mathrm{BN}}, \\ q_{\mathrm{GS}} &= a_{\mathrm{GS}}d_{\mathrm{GS}}, \\ q_{\mathrm{GN}} &= a_{\mathrm{GN}}d_{\mathrm{GN}}, \\ q_{\mathrm{BS}} &= a_{\mathrm{BS}}d_{\mathrm{BS}}, \\ q_{\mathrm{BN}} &= a_{\mathrm{BN}}d_{\mathrm{BN}}, \end{aligned} \tag{5.10}$$

where a_i is the 'attractivity' of an empty cell for a state GS, GN, BS or BN, respectively, and

$$d_i = C_i(t)/(N_{\mathrm{a}}(t) - N_{\mathrm{Oc}}(t)). \tag{5.11}$$

is the fraction of the empty cells among those cells that could become occupied.

Formulas (5.10) are based on the postulate that one individual only can try to occupy a certain place during one time step.

5.3.2 Occupation by Immigrants: Rules $F_{t+2/3}$

The formulas for the transition probabilities that reflect the processes of leaving and occupation ($p_{\mathrm{GS}} - p_{\mathrm{BN}}$ and $q_{\mathrm{Em}} - q_{\mathrm{BN}}$ above) are the same as at the second sub-step. The difference is in the formula for d_i. In the text the immigration is determined by the number of immigrants of each type M_i and the overall number of immigrants $M = M_{\mathrm{GS}} + M_{\mathrm{GN}} + M_{\mathrm{BS}} + M_{\mathrm{BN}}$. In CA terms it gives:

$$d_i = \begin{cases} M_i/(N_{\mathrm{a}}(t+1/3) - N_{\mathrm{Oc}}(t+1/3)) & \text{if } N_{\mathrm{a}}(t+1/3) - N_{\mathrm{Oc}}(t+1/3) > M \\ M_i/M & \text{otherwise.} \end{cases} \tag{5.12}$$

We use Monte-Carlo simulations to run the model. To ensure the contiguity of City we allow the cells of the active part of the CA only to change the state from the empty to an occupied one at the sub-steps $t + 1/3$ and $t + 2/3$.

5.4 The Properties of the Model

Most explicit applications of self-organization to urban and regional modeling tend to structure their work in terms of differential equations and dynamic paths that bifurcate so that a certain steady state is suddenly replaced by multiple stable equilibria, or behave in a chaotic manner (Dendrinos and Sonis 1990). We prefer to use CA models for several reasons. First, though until very recently not common among students of self-organization in urban and regional modeling, CA models have been used quite extensively to simulate processes of self-organization in many domains (for details see Sect. 3.6). Second, in certain situations the use of differential equations might make computation very tedious. In such situations, especially when one is interested mainly in qualitative results, analysis by means of CA might be sufficient (see Sect. 3.6).

Third, as also noted in Sect. 3.6, the discrete spatial structure of real cities, makes CA models with their discrete cellular structure a natural tool to represent cities. The same applies to the fact that in both cities and CA models, the properties of a given local spatial unit (a building, parcel of land or a CA cell) are determined, to a large extent, in relation to their immediate neighbors. This is apparently the reason for the growing wave of urban and regional CA models that we see in the last few years. City, the specific CA model we present here is part of this wave. Its specific properties are as follows:

5.4.1 Self-Organization

As defined above, City is a typical system in self-organization. It is an open and complex system; open, in the sense that it is part of its environment through a flow of immigrants from the environment inward into City and from it outward to the environment. This flow and the fact that the properties of every place (i.e. cell) in the system are a derivation from the properties of the neighboring places and thus change with every move in City, keep City in a far from equilibrium situation. It is a complex system in the sense that it involves interaction among a large number of individuals, each with their own 'personal' properties, and a large number of places whose properties are ever changing with every new iteration. As our aim is to model a real city as a self-organizing system, we do not determine the internal rules, which lead from local interaction among individuals and places to the global pattern and structure of the system. Given various configurations of initial conditions, our aim is to study such processes. As formulated above and as we shall see below, City's global properties are not the sum of its parts and it exhibits phenomena of non-linearity.

5.4.2 A Probabilistic CA Model

From the description above it is clear that our model can be considered prob abilistic cellular automata as defined, for example, by Demongeot, Goles and Tchuente (1985). This is not a very frequent application of the CA models, however. In most applications the transition rules are deterministic and only rarely stochastic (For a general discussion see Wolfram 1984). Our model (as the models of Phipps 1989 and Atlan 1985) belongs to this latter sub-group.

5.4.3 A Genuinely Spatial Model

CA models are, by their very nature, spatial models. Global spatial patterns are developed as a consequence of local spatial relations among the cells. This is also the case in our model which explicitly considers space: individuals take decisions to occupy or leave a place on the basis of the properties of their neighbors, and the quality of a certain place is derived from the properties of its immediate neighbors; the global pattern of the city as a whole is generated out of these processes of spatial interaction.

5.4.4 A Heuristic Model

Apart from the laws of the CA game, that is to say, a simultaneous interaction among spatially distributed cells, no fixed assumption is made as to how individuals should behave or make migration decisions, or how the state of cells is determined and so on. All such issues or the starting conditions of a certain City-game are topics we examine and learn by means of the model. For example, a typical City-game might start with the following question: 'given a certain distribution of individuals with their specific composition of preferences and a certain socio–spatial configuration of residential segregation, what scenarios can result'?

5.4.5 Between the Local and the Global

By their very nature, CA models allow for a local determination of global properties. For example, a typical CA city-game that starts with local rules of interaction among individuals, and with a random urban landscape, might typically develop into a global and segregated urban landscape – a genuine bottom-up process. But, note the meaning of 'a random' landscape – it is a property of the system as a whole and in this respect one might say that we have here a global determination of local events, or a top-down process. In a similar way one can say that the categorization of the individuals in the model into Blue and Green is also a property of the global system as a whole; and so are the various rules of the game and other initial conditions. Our position is that the local versus global determination issue is not the relevant question, since we are dealing with a system in self-organization; such a system is not causally determined, not globally nor locally, but it self-organizes itself in response to global and/or local stimuli.

5.5 Three Heuristic Games

In the following we use our heuristic model City as means to examine and investigate two aspects of a city as a system in self-organization: the gap between intentions versus behavior and the interplay between stability and instability. For this purpose we assume the following initial conditions: two groups of individuals, Greens and Blues, each of which is subdivided according to the preferences of its members into Segregatives and Neutrals.

Segregatives prefer to spatially reside among neighbors of their own kind – Greens among Green neighbors and Blues among Blue neighbors. When hunting a new place to live, or when evaluating their actual location in the city, they are thus attracted by their own kind and repelled by the other kind. Neutrals are indifferent as to their neighbors.

We have postulated that for the Segregatives (types GS and BS) the probability to leave a place monotonically increases with the increase in the fraction of the individuals of the other social groups; while the probability to occupy a place monotonically increases with the increase in the fraction of the individuals of the same (social) group. We assume that repel and attractivity curves for Blue and Green Segregatives are identical (Fig. 5.1), and, according to the definition, no one repels the neutral individuals, thus, the attractivity of a place for them does not depend on the neighbors of the place. Below we assume that the rate of individuals entering the system at each time-step is relatively high – 200, that is, 8% of the overall number of cells in the system (2500).

Throughout the discussions below we use the Dissimilarity Index for a 5×5 square net (Duncan and Duncan 1955) as a measure for segregation. Actually, in all runs we have used also the Gini and Entropy coefficients and found that in all the situations examined the three indices behaved synchronously. For a comparative discussion regarding the three indices, see James and Taeuber (1985), Massey and Denton (1988). Each of the games starts with a relatively small domain of randomly distributed individuals of all types at the center of the lattice – Fig. 5.2, $t = 0$.

5.5.1 Game 1: Segregative Blues Versus Segregative Greens

The first game is between types GS and BS. We assume that their proportions are equal among the newcomers. As we see from Fig. 5.2, the urban process (i.e. the game) starts with fast changes in the spatial pattern. Then, gradually, the pattern becomes completely segregative, the number of free places is balanced by the number of newcomers and the spatial pattern changes very slowly (from $t \sim 25$ and further). At this stable state of City the changes are mainly due to *slow* changes in the position of the boundaries between the homogeneous areas of Green and Blue individuals.

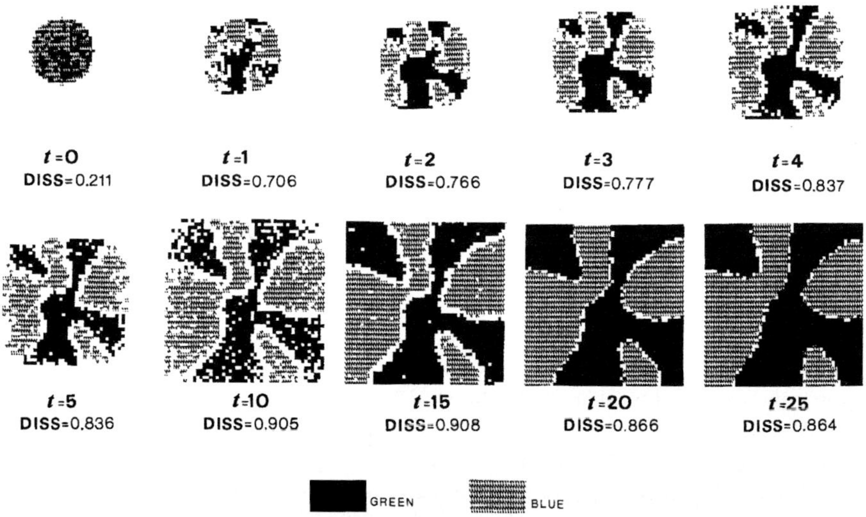

Fig. 5.2. *Game 1* (segregative Greens versus segregative Blues): The time evolution of a segregated city

5.5.2 Game 2: Neutral and Segregative Greens Versus Segregative Blues

What happens if the fraction of neutrals among the Green immigrants is growing in a step-wise manner (say, from 33% to 67% and 100%) while all the Blue immigrants remain segregatives? As we see from Figs. 5.3, 5.4 col. 1, in spite of the presence of Neutral individuals among the Greens, the global pattern continues to be segregative – very similar in principle to that obtained in game 1. This is so even when all the Greens are neutrals (DISS = 0.876), with the implication that one segregative group is sufficient to reach a global systemic segregation. In Figs. 5.3, 5.4 (col. 1, lines 2–4), it can further be seen, that an increase in the rate of Neutrals among the Greens does not entail an increase in the fraction of Greens penetrating into the region occupied by the purely segregative Blues, with the implication that the system remains segregative. An interesting outcome of the latter is that with the introduction of neutral Greens, it is the Green area (that is, the area of the non-segregatives) which is becoming the most segregative – not a single Blue lives in the Green neighborhood. On the other hand, the Blue neighborhood – that of the segregatives, is becoming a mixed neighborhood.

On the face of it the above is a contradictory result: we intuitively tend to assume that a culturally, or ethnically homogeneous neighborhood indicates that its inhabitants are segregative too, and vice versa. Our model shows a different relationship. Empirical evidence as well as a second, somewhat dialectical, thought, suggests that the results of the model make a lot of sense: the Neutrals are prepared to live everywhere in the city, even among

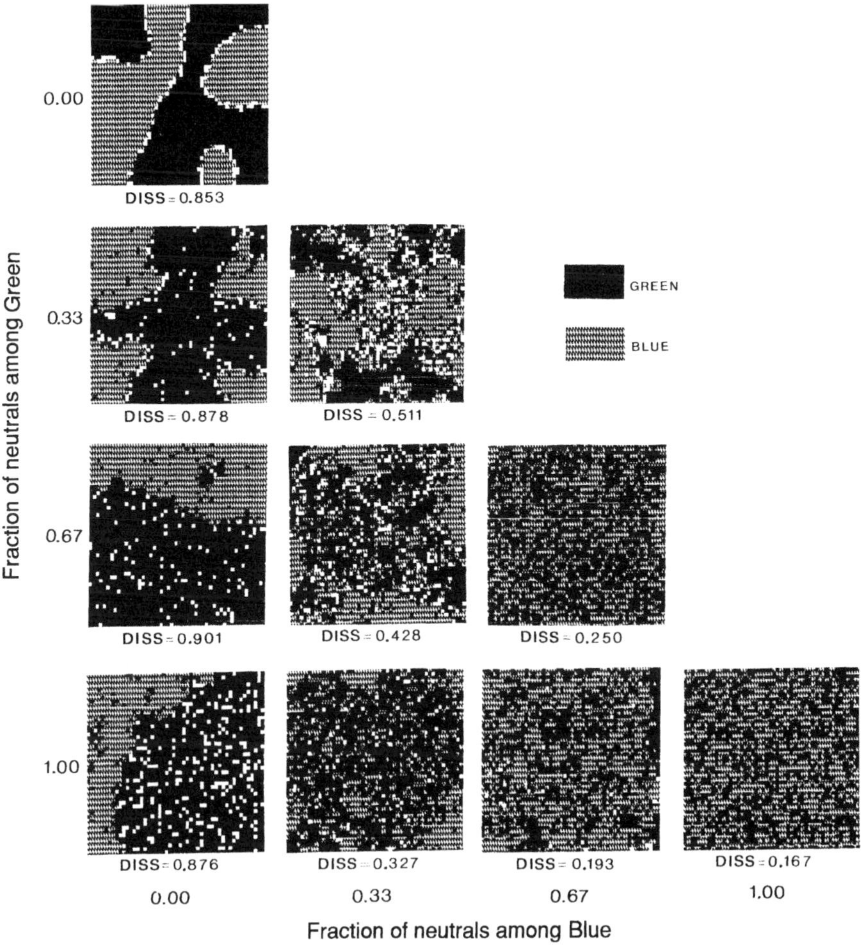

Fig. 5.3. Spatial distribution of Greens and Blues with increasing proportions of neutrals and segregatives in both groups. DISS is the dissimilarity index. *Game 2* is represented by col. 1, and *Game 3* by the whole figure

another cultural group the members of which might be segregatives. Segregatives, per contra, do not even attempt to search a location outside their own kind.

5.5.3 Game 3: Neutral and Segregative Blues Versus Neutral and Segregative Greens

Consider now a game between neutral and segregative Blues and Greens. As shown in Figs. 5.3, 5.4, a step-wise increase (by fractions of 33%) in the proportions of neutral individual newcomers in both groups entails a

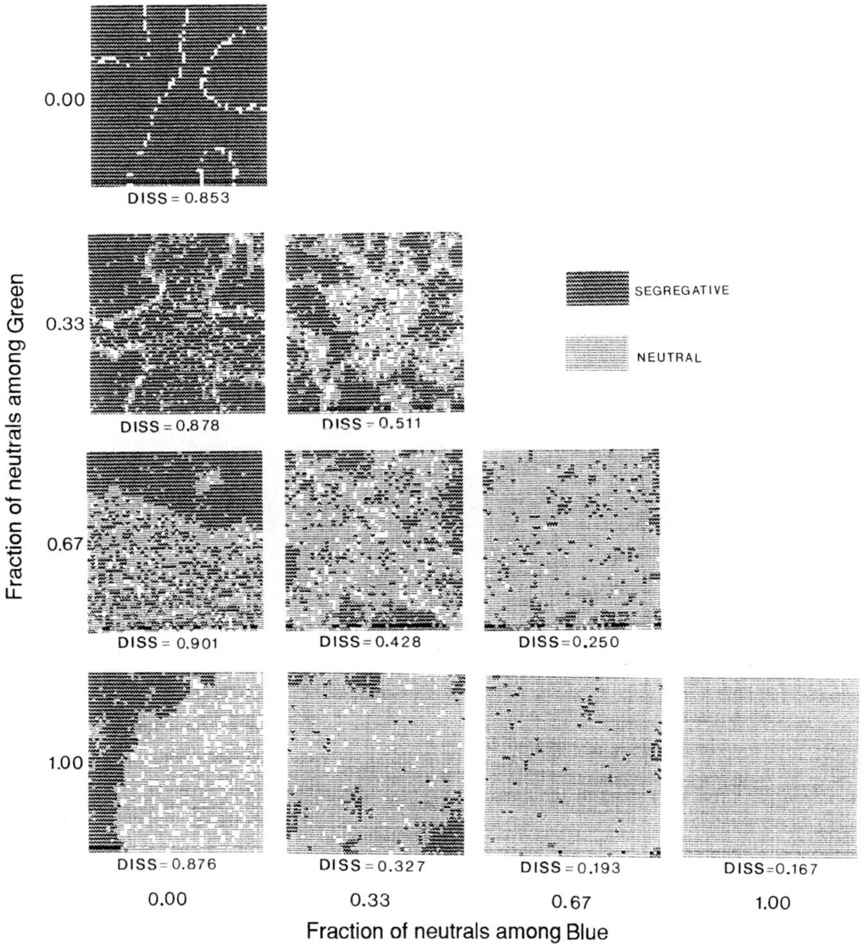

Fig. 5.4. The spatial distribution of Neutrals and segregatives when their proportion among the Blues and Greens is increasing as in Fig. 5.3. DISS is the dissimilarity index. *Game 2* is represented by col. 1, and *Game 3* by the whole figure

decrease in the level of segregation, until for 100% Neutral individuals in both groups we obtain a random distribution. When we examine this process by means of our segregation index (Figs. 5.3, 5.4) we see that the process is essentially non-linear. Similar results emerge also from a visual inspection of the changes in Figs. 5.3, 5.4: 33% Neutrals in both groups produce a visually segregative pattern, but 67% do not. In fact, when we have increased the fraction of Neutrals by 5% steps we have found that the partition between a segregative to homogeneous pattern is between 50% and 60% Neutrals in both groups. An important feature of the segregated spatial distributions of Figs. 5.3, 5.4 is the observed concentration of Neutral individuals close to the boundaries between the clusters. This phenomenon is pronounced if the fraction of Neutral individuals among at least one group does not exceed 30–35%.

5.6 Intention Versus Behavior in a City

In the above CA simulations the actors are defined in terms of their probability of locating close to their own kind. This formulation enables us to isolate the impact of the preference for socially or ethnically homogeneous neighborhoods from other complex location factors operating in the real world, and trace the implications of such preferences on residential patterns (and as we shall illustrate below, also on stability and instability of the urban space). As our model enables us to separate intentions from behaviors, it can also show whether, to what extent, and where in the city, people behave in line with, or counter to, their intentions. In other words, 'how many and what proportion of the city's population live in a state of socio–spatial cognitive dissonance?' Separating intention from behavior and studying the symmetry and asymmetry between them thus enabled us to identify regions of cognitive dissonance. As we shall see below (and in Chap. 7), this dissonance keeps the system in a far from equilibrium condition and is directly related to the stability–instability surfaces of the city we discuss in the next section below.

In our model and analyses below we thus make no assumption as to what is the genuine human motivation. In principle we allow for as many motivations and intentions with respect to residential location as there are individuals, and the name of the game is to see what socio–spatial individual migratory behaviors, systemic patterns and structures, result from a certain configuration of motivations, preferences or wants. In addition, we are interested to find the proportion of individuals who behave according to, or counter, their wants and the resultant spatial patterns.

The above theorization, which makes a distinction, and studies the interplay, between actual behavior and intention comes close, as we've just seen, to Festinger's notion of cognitive dissonance. It is also connected to the question of the relations between the individual and the system that stands at the center of social theory, social geography and the theory of self-organizing systems. In Chap. 7 we reiterate this issue and make it a basis for the study of socio–spatial morphogenesis. That is to say, the emergence of a new socio–spatial, cultural entity, out of the dynamics of the city as a self-organizing system.

5.6.1 Intention Versus Behavior in Games 1, 2 and 3

As defined above, Blue and Green segregatives prefer to live among their own kind while Neutrals are indifferent. Thus, for each of the segregatives,

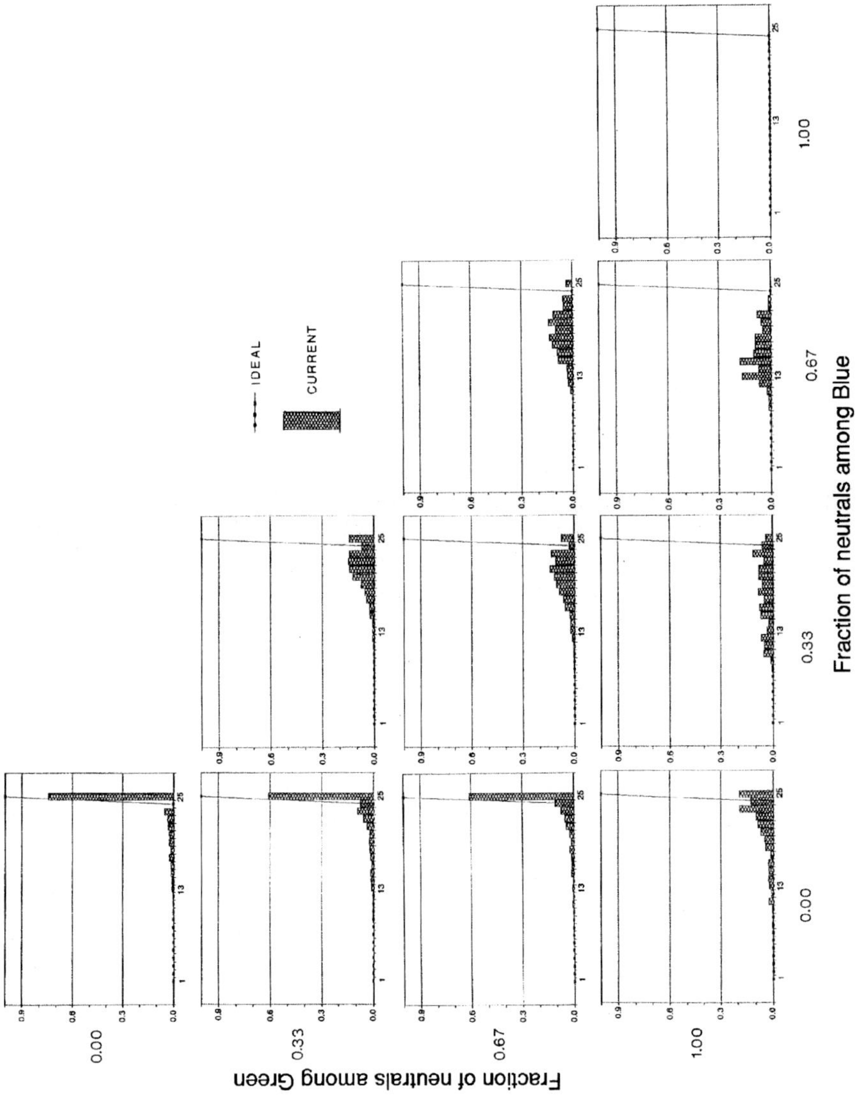

Fig. 5.5. a. Ideal versus current (or actual) socio-spatial distribution of neutrals corresponding to the game presented in Fig. 5.3, 5.4

the ideal socio–spatial distribution is such that 100% of their neighbors are of their own kind. Neutrals are indifferent as to the various distributions of Greens and Blues. Consider a random distribution of Greens and Blues as 'ideal' for Neutrals.

In order to evaluate the relation between wants and actual behavior we compare, in Figs. 5.5a, 5.5b, ideal socio–spatial distributions as defined above with those obtained in the various CA games. In Game 1, the majority of the

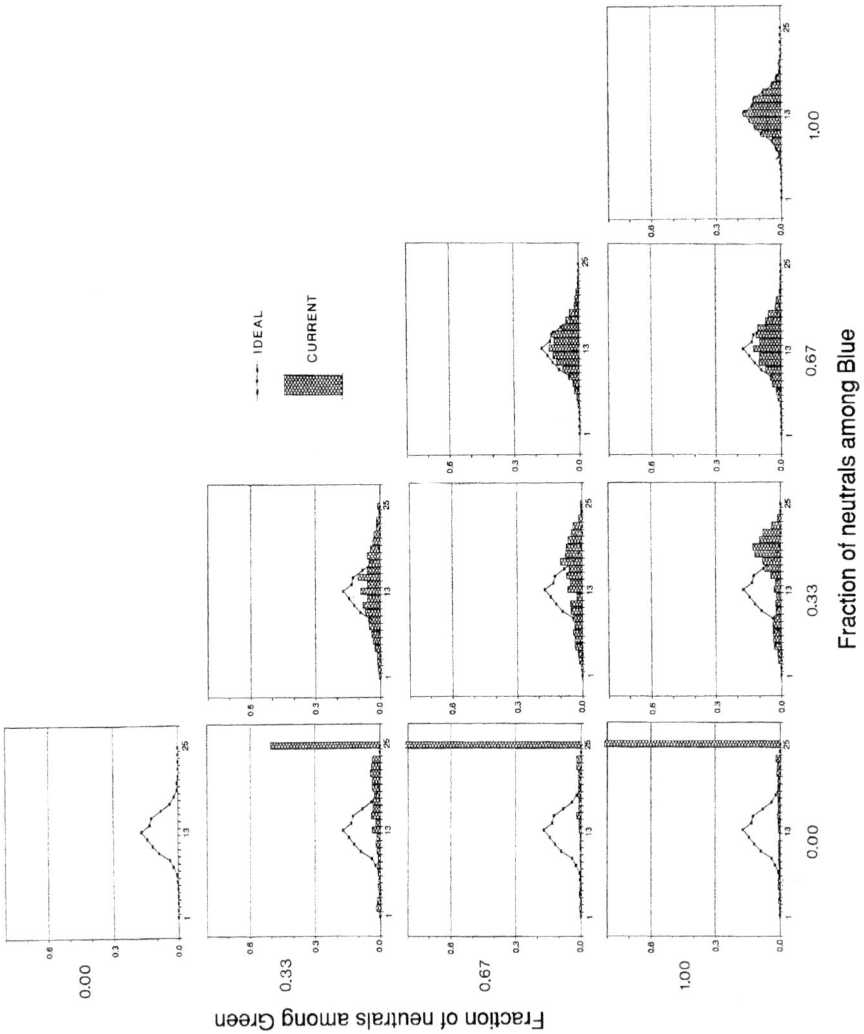

Fig. 5.5. b. Ideal versus current (or actual) socio–spatial distributions of segregatives corresponding to the game presented in Figs. 5.3, 5.4

individuals from the two groups behave according to their preferences; a minority of about 25% behaves counter their wants in varying degrees (Fig. 5.5a, column 1, top). Members of the latter group are spatially distributed along the boundaries between the Green and Blue areas.

In the specific realization of Game 2, the effect of a step-wise increase of neutral Greens (33%, 67%, 100%) did not alter significantly the picture obtained in Game 1. While a distinguished concentration of individuals who behave counter their intentions can be seen on the boundaries (Fig. 5.3), one can also observe an increase of a counter-preferences behavior within the Blue areas. An interesting feature of these results is that almost all the Neutrals do not behave randomly, but exhibit a socio–spatial segregation behavior (column 1 in Figs. 5.5a, 5.5b).

Once we introduce a step-wise increase of neutrals within the Blues also, say 33%, 67% and 100%, (columns 2, 3, 4 in Figs. 5.5a, 5.5b), the picture changes significantly in a non-linear fashion: the spatial distribution of the Neutrals and Segregatives is approaching randomness, with the consequence that most Segregatives behave counter their preferences – much beyond the ratios obtained when the proportion of Neutrals was the same, but of Greens only.

Such a comparison shows, first, that the gap between intention and behavior is typical to all three games, that is to say, some individuals always behave differently than their preferences. Second, the gap is relatively small in the results from Game 1, slightly larger in Game 2, and very large in Game 3. That is to say, with the increase of Blue and Green Neutrals in the system, or more generally, with increasing complexity of population composition.

The implications from the above are fourfold. Firstly, they illustrate the property of non-linearity (in this case between intentions and behavior) which is typical of the dynamics of cities and of systems in self-organization in general. Secondly, in our CA games the property of non-linearity is a property of the spatiality of the relations between the various individuals. For example, in all three games the difference between intentions and behavior is specifically high on the boundaries between the homogeneous Green and Blue areas. Thirdly, with respect to the notions of economic person and to Sonis' social person, we see that the finding that in certain circumstances people behave economically or 'like others', does not necessarily reflect human motivation, inclination or intentions. Rather, in certain socio–spatial situations people are forced into certain behaviors despite their intentions – in the language of Haken's synergetics we would say that they are slaved by the system's order parameter. Fourthly, as we shall see in the next section, the gap between intentions and behavior is closely tied with phenomena of stability and instability in a city.

5.7 Stability and Instability in the City

Self-organization is a characteristic, as noted, of open and complex systems in far from equilibrium conditions. Such systems are always in motion and consequently stability and instability refer to stable versus unstable forms of motion. Put within a city context this viewpoint implies that the city, and/or various parts in it, is always in motion, stable or unstable. This view, in fact, reflects the formal mathematical approach in which 'stability' is meant as 'structural stability'.

5.7.1 Predictability Versus Unpredictability

Stability and instability refer, among other things, to the level of predictability. A system in motion as above is stable in the sense that from a certain point in space–time the direction of its motion is predictable; it is unstable, in the sense that from that space–time point the direction of its motion is unpredictable or chaotic. For example, a city which is growing in population and expanding in size in predictable quantities, forms and directions, as in Fig. 5.2, can be regarded as stable; if, on the other hand, the changes in population, form and directions are unpredictable, we would say that it is unstable. In the same token it must be said, first, that a system in self-organization is, in principle, unpredictable; second, and as a consequence of the above, one can discuss predictability and unpredictability only with reference to the past. For example, it can be said that 'within, or for, a certain space–time period the system behaved in a predictable/unpredictable way', or alternatively, that 'judging from the past, it is probable that the system will behave in a certain way'.

5.7.2 Local Versus Global Stability and Instability

Global instability refers to the behavior of the system as a whole – in our case to the evolution of the city as a whole. A typical phenomenon in cities is that while the city as a whole persistently evolves, local areas within the city might exhibit unstable behavior. The literature on urban processes in plentiful with such examples (though not as here, with respect to the issue of local–global systemic behavior), and some of the most typical cases are the *zone of transition*, in classical urban ecology (Berry and Kasarda 1977), the phenomenon of invasion–succession in processes of ethnic segregation (Burgess 1925, Cressey 1938, McKenzie 1929), as well as the more recent phenomenon of *gentrification* which characterizes (post)modern, Western cities (Smith and Williams 1986). All such cases are associated with local socio–spatial instabilities, though, as will be illustrated below in Chaps. 7, 8, in certain circumstances these local events might develop into a global one with a consequent global change.

5.7.3 Stability Versus Instability in Games 1, 2 and 3

In the above CA games, we say that cells in states GS or GN are in a Green (G) state, while cells in states BG or BN are in a Blue (B) state. From the model we can estimate numerically the unconditional probabilities for each cell to be in states B, G or Em at the time step t. Let $r_{\mathrm{B}}^{H}(t)$ and $r_{\mathrm{G}}^{H}(t)$ be the probabilities that a cell H be occupied at a time step t by a Blue or Green individual respectively. Based on probabilities $r_{\mathrm{B}}^{H}(t+1)$ and $r_{\mathrm{G}}^{H}(t+1)$ we define the state of a place H as 'predictable' if one value only from $r_{\mathrm{B}}^{H}(t+1)$ and $r_{\mathrm{G}}^{H}(t+1)$ is greater than 0. If both $r_{\mathrm{B}}^{H}(t+1)$ and $r_{\mathrm{G}}^{H}(t+1)$ are greater than

0, then the state of the place is not predictable – the greater the minimum of $r_{\mathrm{B}}^{H}(t+1)$ and $r_{\mathrm{G}}^{H}(t+1)$ the higher the 'unpredictability' of the future state of that cell. We call the surface of minimum values of the probabilities $r_{\mathrm{B}}^{H}(t)$ and $r_{\mathrm{G}}^{H}(t)$ stability and instability surface (SIS), and the sum of these minimal values over the whole territory an instability index (INS). We start the discussion by considering the SIS obtained in Games 1–3 above.

5.7.3.1 Game 1

Figure 5.7.3.1, column 1, top, is the SIS obtained in Game 1. As can be seen, there are large stable areas cut by belts of low level instability. A comparison with Figs. 5.3 and 5.4 reveals that these belts are actually the boundaries between homogeneous and stable Blue and Green areas. This situation, by which unstable zones are formed on the boundaries between stable areas, is similar as noted to the zone of transition empirically observed in connection with the urban ecology of cities (see above). Thus, in urban scenarios such as in Game 1, changes occur rarely, global processes such as boundary movements are slow, and zones of low-level instability are formed on the boundaries between stable-homogeneous territories. An interesting feature of the above results is that the locally unstable boundary zones exhibit global spatio–temporal stability.

5.7.3.2 Game 2

Figure 5.7.3.1, column 1 shows, that the introduction of neutral Greens increases the degree of instability on the boundaries between the Blue and Green areas, and interestingly, also within the Blue areas. The reasons being that in their search for housing, neutral Greens penetrate into the Blue areas, while the segregative Blues do not. The rate of penetration of these invaders is low and the global segregative pattern remains stable – very similar to that of the previous game.

As already noted, this finding highlights the dialectics of the urban space: a property of the Green guys affects, not the nature of the Green areas, but that of the Blue areas. Once we've 'revealed' this property, we realized that it typifies many real-life situations such as 'neutral' blacks penetrating white segregative neighborhoods, neutral religious residences entering secular communities (i.e. in Jerusalem, see Shilhav 1983), or lower middle-class families entering working class zones (e.g. Camden Town, Islington or Chalk Farm in London, or the Florentine area in Tel-Aviv). In many of these cases these 'invasions' entailed local instabilities which in certain circumstances ended up with a global systemic change.

5.7.3.3 Game 3

Figure 5.7.3.1, columns 2–4 show that the introduction of Neutrals into both Blues and Greens influences instability in a non-linear and non-monotonic

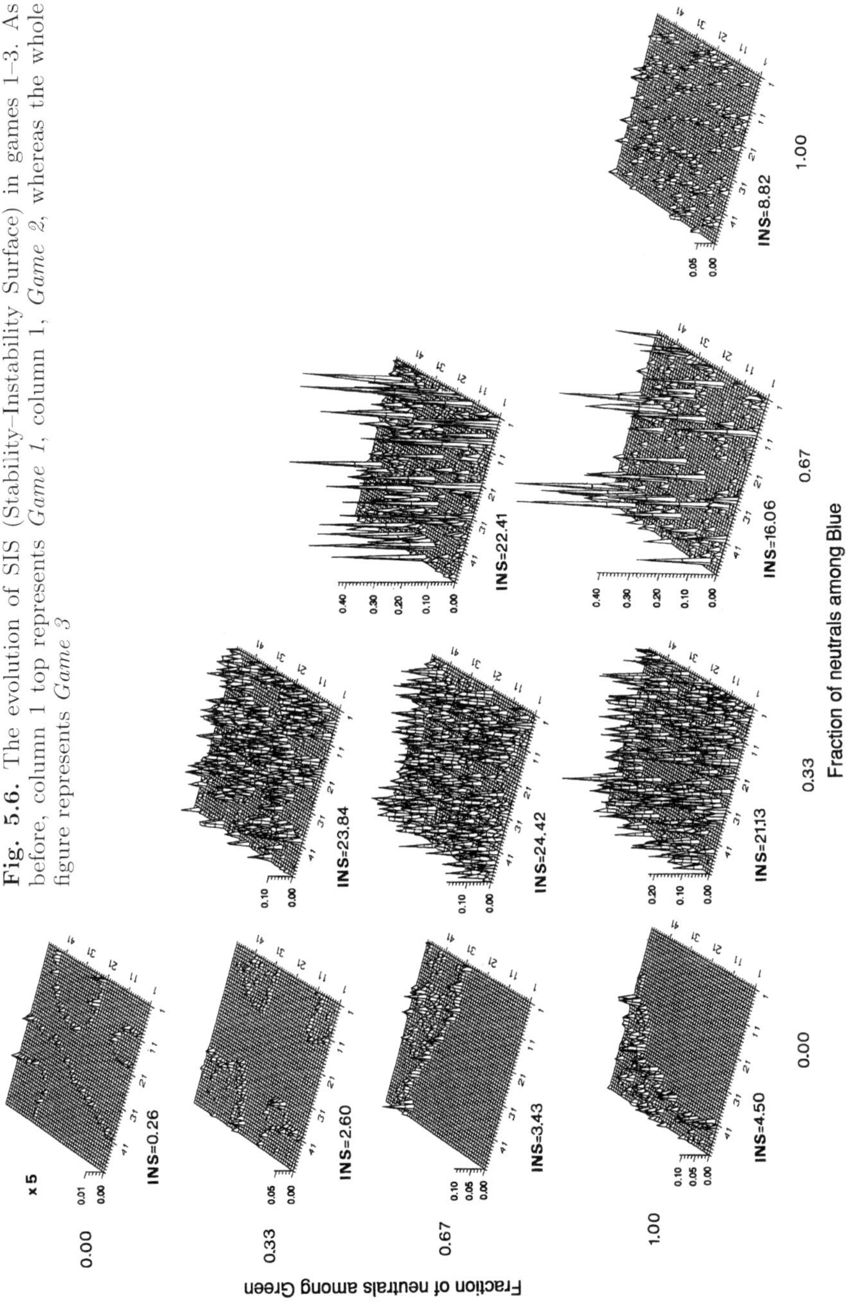

Fig. 5.6. The evolution of SIS (Stability–Instability Surface) in games 1–3. As before, column 1 top represents *Game 1*, column 1, *Game 2*, whereas the whole figure represents *Game 3*

fashion. This can be seen and more so in Fig. 5.7. First, the level of instability increases much beyond the proportion of Neutrals in the system as a whole: 33%+33% Blue and Green Neutrals create more instability than 67% Blues or Greens. This is a consequence of the dialectics of space noted above. Neutral Greens penetrate Blue areas and vice versa, with the consequence that none is stable any more. This trend continues up to 50% Neutrals in both color groups. As the proportions of Neutrals further increases, instability decreases, until with 100% Neutrals in both groups it reaches its lowest level. The fact that some level of instability still persists at this latter situation results from the non-zero probabilities for Neutrals to leave the places they occupy and from the existence of empty places in the system.

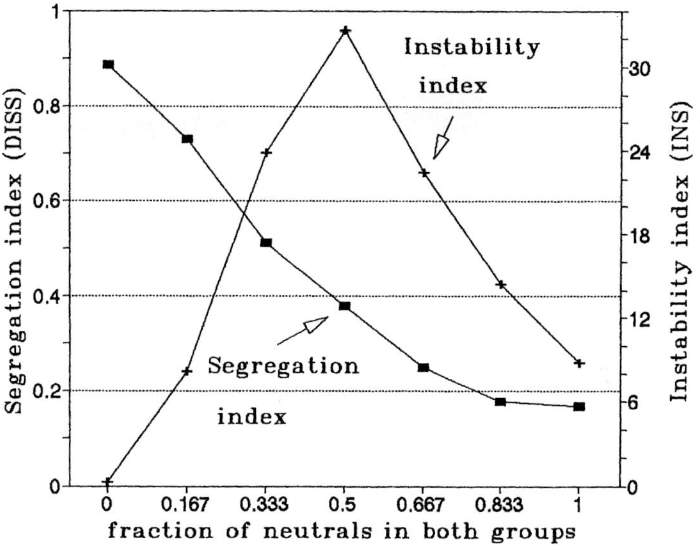

Fig. 5.7. Segregation (DISS) and Instability (INS) indices as a function of increasing rates of Neutrals (when the rest are Segregatives) in both Blues and Greens. Note that DISS decreases in a non-linear fashion, and that INS first increases and then decreases

5.8 The Captivity Principle

In a self-organizing system that is in a far from equilibrium condition and is thus always in motion, stability refers to a process by which the system generates reproducible relations among its components. This is the process of reproduction, which implies that the system is capable of withstanding internal and external, as well as local and global, instabilities or disturbances. How does the system generate reproducible relations among its components?

As we have seen above, the theory of self-organization suggests that open, complex and far from equilibrium systems exhibit a distinct pattern of development: relatively long periods of stable steady state followed by abrupt, short periods of instability and chaos. According to Haken's (1983a, 1983b) synergetics approach, while in steady state, the system is governed by a certain (or a few) collective variables he terms order parameter(s), while in instability or chaos, several order parameters coexist. The very act of self-organization, according to this view, starts at this stage: several order parameters coexist and compete until one 'wins' in the sense that it enslaves competing orders to its motion. This process is termed by Haken the slaving principle and as he noted on several occasions, it is central to his synergetics approach to self-organization.

With reference to the question posed above we would thus suggest that the system generates reproducible relations among its parts by slaving their motion. Slaving is the classical principle in Haken's synergetics approach to self-organization, but not the only one; as he notes, in several circumstances the competition among order parameters ends up with a cooperation. The latter case might be termed the *cooperation principle.*

In Chap. 4 above (Figs. 4.3–4.5) we have presented Portugali's (1993) three forms of systemic change in self-organizing socio–spatial systems: stratigraphic, furcative and hermeneutic. The first and third refer to situations where a certain order parameter wins by enslaving competing order parameters; in stratigraphic change by eliminating them, while in hermeneutic change by re-interpreting, and thus re-defining, them. In furcative change the competition among order parameters ends up either with a cooperation (in which case, quite often, one order parameter dominates the others), or by "imprisoning". In the latter case a certain order parameter wins by keeping competing order parameters spatially captive. The latter scenario can be termed *the captivity principle* and it refers to cases were the system achieves reproducible steady-state relations by keeping the disturbing forces captive within a certain space–time niche.

Within a city context the process of slaving is typical to cases of invasion–repelling, invasion–succession, as well as to the more recent process of gentrification. In all these cases a foreign element invades a certain area in the city and creates a local disturbance; in the first case the result is that the foreign element is being pushed out, in the second it is being followed by other invaders of its kind with the consequence that the veteran elements are being pushed out; a similar process typifies the third case. The common denominator to all such examples is that a certain socio–spatial order enslaves competing ones by eliminating them. Note that Games 1,2 and most of 3 as played above do not refer to this form of slaving. This is so since slaving by elimination would imply either that one group is giving up its identity (say, the Greens stop being Green), or else all individuals in the game are Neutrals. This is actually the case noted above with respect to Fig. 5.7.3.1, column 4 (and the fact that some low-level instability can still be observed is a consequence of the existing empty places).

The captivity principle is typical, among other things, of a case where local instabilities and disturbances are not being eliminated, but imprisoned or captured within a certain spatial niche in the city. A well known example can be the zone of transition in Burgess' (1925) ecological theory of city morphology, while another example is the slums of modern western cities. The common denominator here is the situation by which society or 'the city', maintains reproducible stable relations and development by socio-spatially imprisoning local unstable elements which threaten its global stability. Other examples, at a smaller spatial scale, are the prison and the madhouse, which according to Foucault's (1967) *Madness and Civilization*, are among the social institutions within which society keeps its unstable elements spatially captive, in order to allow a smooth process of social reproduction.

To further investigate the notion of slaving by captivity we turn once again to our heuristic game city, and follow, in Fig. 5.8, the space–time evolution of SIS in a situation were 33% of the Greens are Neutrals, while the rest of the Greens and all the Blues, are segregatives. The game starts, as can be seen, at the center of the board, with a very unstable surface (reflecting the initial random distribution). With the inflow of new immigrants and the resultant increase in population and density, the system exhibits a process of self-organization by which, first, instability drops and fluctuates between INS = 7.57 and 10.52; then it drops once again and from $t = 14$ onwards INS fluctuates between 2.17 and 3.03.

The whole description is a typical process of self-organization by means of captivity. From $t = 14$ onward one can observe, first, a steady state and second, that instability is spatially captured in the boundary belts and spatially confined to the Blue areas. Similar results can be observed when we refer back to Fig. 5.7.3.1: the high-level instability at the initial stages of the game is gradually being imprisoned until a steady state situation is reached. In the latter, instability remains spatially captive in the boundary belts, thus enabling the smooth reproduction of a stable segregative city. Changing the proportions of Segregatives and Neutrals among the Green guys, as in Game 2, entails an increase in the overall level of instability in the city, and a change in its spatial distribution. First, as before, instability is captured in the boundary belts, though belts areas are wider and the levels of instability are higher. Second, we see a phenomenon of spatially distributed instability, but only within the Blue areas. The global pattern of the city remains intact, and the system reproduces itself as a stable segregative city by capturing its instabilities, partly concentrated in the boundary belts and partly distributed in dots within the Blue areas. Changing the proportions of neutrals and segregatives as in game 3, increases the level of instabilities along the boundaries

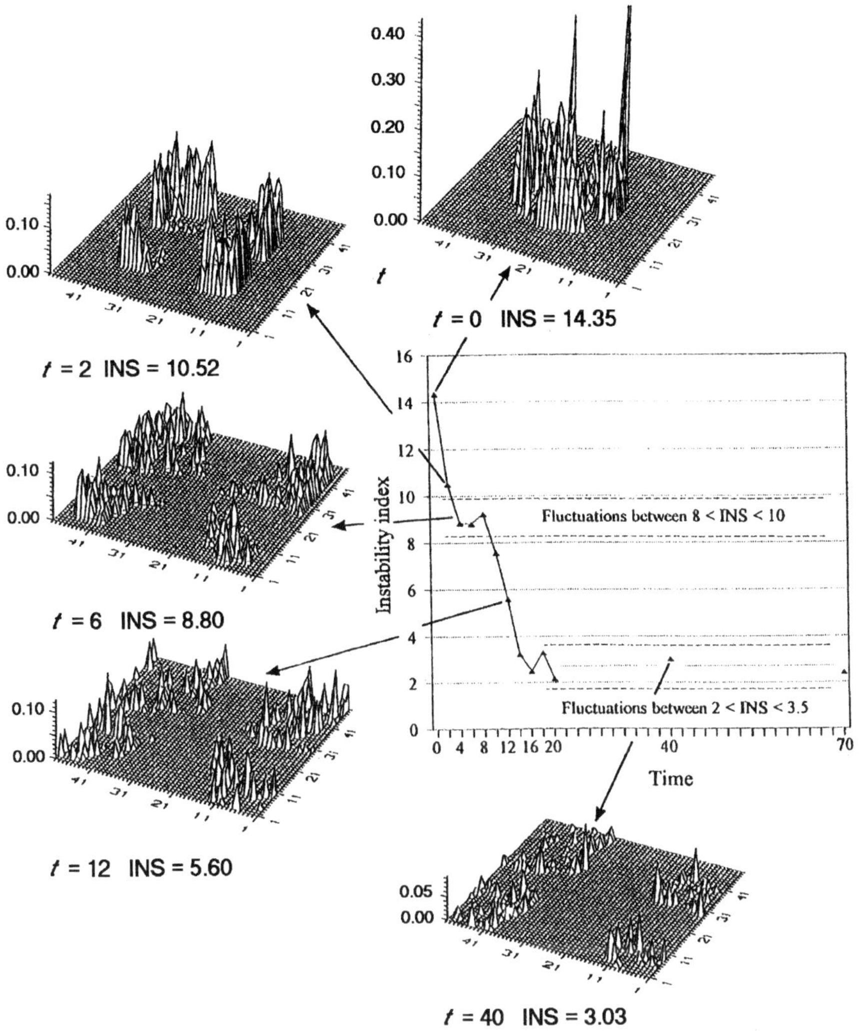

Fig. 5.8. Time evolution of INS and consecutive stages of SIS in the development of a city with 33% neutral Greens when the rest of the Greens and all the Blues are segregatives

and the extent of its distribution among the Green and Blue areas; in fact it brings the system into a bifurcation point – an issue we'll discuss in Chap. 7 below. It is interesting to note the interrelations between the various SISs and the gap between wants and behavior discussed above. Unstable areas are, in fact, areas many of whose residents behave counter their inclination. As can be seen from Fig. 5.4 the unstable areas are populated mainly by Neutrals and unsatisfieds.

As can be seen, the notion of the captivity principle introduced above adds a new dimension to Haken's synergetic approach to self-organization. The more general and formal properties of this new notion are currently under investigation and will be published in the near future (Haken and Portugali, in preparation).

5.9 Conclusions

In this chapter we have developed and used City as a heuristic computer game with which to examine several phenomena of the city as a system in self-organization. We have examined phenomena of socio–spatial segregation, the gap between intentions and actual behavior, stability and instability in a city and the new notion of captivity principle. These phenomena are actually the various facets of self-organization in a city, but not the process itself. The latter, which includes phenomena of phase transition, bifurcation and systemic change, has yet to be studied. This is the task of Chaps. 6–10 that follow.

6. International Migration and the Internal Structure of Cities*

6.1 Introduction

Our aim in this chapter is twofold: Firstly, to further develop our model City and transform it from an implicit to an explicit and genuine FACS model. As noted above, City was essentially a CA model where the free-agent properties of individuals appeared only implicitly. In City-1, the model we develop below, these properties form an explicit aspect of the model's formalism. Secondly, we wish to study the relations between international migration and the internal structure of cities – a general urban phenomenon typical to cities in immigration absorbing countries. In this chapter we shall develop this issue with special reference to the recent migration waves to Israel from ex-USSR countries.

6.2 City-1: An Heuristic Immigration Absorption Model

6.2.1 A Formal Description of City-1

City-1 is built upon, and thus an extension of, our previous model City as developed in Chap. 5. The latter is a cellular automata model that evolves by means of an interaction between cells and their nearest neighboring cells. The properties of the occupant of the cell H are translated in City into the state S^H of the cell and the general dynamics of this cellular automata $\mathcal{A}$ were described above as follows:

$$S^H(t+1) = F_t\big(S^H(t), S^{U(H)}(t), S^{\mathcal{A}}(t), E(t)\big),$$

where $S^H(t)$ is the current state of cell H, which belongs to a finite set of states $S = \{S_1, S_2, \ldots, S_k\}$, $U(H)$ is a neighborhood of cell H and $S^{U(H)}$ denotes the neighborhood's configuration (the set of states of the neighboring cells). By $S^{\mathcal{A}}(t)$ we mean the global characteristics of $\mathcal{A}$, and vector $E(t)$ designates the external parameters. The new state of H at time-step $t+1$ is selected out of S according to the 'transformation rule' F_t.

* With Itzhak Benenson

Two problems emerge from the analysis of our City CA model in Chap. 5. First, the interpretation of migrations in CA terms demands a roundabout technique with resulting transformation rules, which are far too complex. Second, a CA framework is internally limited in its ability to account for the changing characteristics of the cells' occupants (e.g. an individual's economic status). In order to overcome these difficulties we transform the single-layer cellular automata model (City) into a two-layer, agent-based, FACS model of residential dynamics. The new model (City-1) is built of an infrastructure layer which is a cellular automata that represents houses; and a superstructure layer which explicitly represents the movement of the individual free agents. The latter enter the city, occupy houses and reside there, change their residential location, or leave the City when the conditions are unsatisfactory. The general structure of the model (we still denote it as $\mathcal{A}$) can be written as follows:

$$
\begin{aligned}
S^H(t+1) &= F_t\big(S^H(t), A^H(t), S^{U(H)}(t), A^{U(H)}(t), S^{\mathcal{A}}(t)\big),\\
A^H(t+1) &= G_t\big(S^H(t), A^H(t), P^A(t), S^{U(H)}(t), A^{U(H)}(t), S^{\mathcal{A}}(t), E(t)\big),\\
P^A(t+1) &= L_t\big(S^H(t), A^H(t), P^A(t), S^{U(H)}(t), A^{U(H)}(t), S^{\mathcal{A}}(t), E(t)\big),
\end{aligned}
$$

where $S^H(t)$, as above, is a state of a cell H, $A^H(t)$ is a state of a human agent A occupying H, $P^A(t)$ denotes the agent's location, $A^{U(H)}(t)$ is a set of states of the human agents in $U(H)$, G_t defines the state of an agent occupying cell H and L_t defines the location of A at the next time step.

The transition from CA to FACS-based modeling does not change the representation of the city's territory: as before, the latter is a rectangular $I \times J$ lattice of cells H_{ij}, where $i \in [1, I]$, $j \in [1, J]$), each of which represents a house. In City-1 the state of each house is represented by its *value*. Individuals (or families) occupy or leave the houses and thus participate in generating the migration dynamics of the city and its socio–spatial structure. One individual/family only can occupy a house.

Each individual in City-1 is characterized by two economic properties (*status* and *tendency*) and a socio–cultural *origin*. Status refers to the individual's current economic ability, while tendency to his/her potential to improve status. In City-1 individuals might have one of the following two origins: *Veterans* and *Olim* (Hebrew for Jewish new immigrants; singular *Ole*). Note that the latter division is similar, in principle, to the Greens and Blues in the previous chapter.

Depending on status and tendency, each individual can either buy or rent a house. As a consequence, each house in the model has two forms of occupancy: rented or privately owned. The interrelation between the CA layer of the houses and the layer of the free agents is bi-directional. First, an individual influences the value of the house when occupying it. The value of an occupied house is a function of the status of the occupant, of the value of the neighboring vacant houses and the status of the occupants in the neighboring occupied houses. The value of a vacant house is determined by the values of its neighboring houses and the status of the occupants in the neighboring occupied houses. Second, when buying or renting a house, the individual has to pay for it. This payment is proportional to the value of the house and it affects the status of the paying individual.

At each sequential time-step every individual takes a series of decisions as described in Fig. 6.2.1. A resident of the city decides whether to stay in, or leave his/her present house, and in the latter situation, whether to rent or buy a new house in the city, to altogether leave the system, or just to leave his house in the city. In the latter case the resident becomes 'homeless' and enters the *queue* for houses. A new immigrant (veteran or Ole) enters the city by joining the queue and then decides whether to buy/rent a house, further wait in the queue for housing, or altogether leave the queue (and thus the city). Individuals take such decisions by comparing their own properties (status, tendency) to the properties of their houses, their vacant neighboring houses, and their nearest neighbors.

The individual's decisions have probabilistic and deterministic components. They are probabilistic in the sense that the larger the gap between the properties of the individual and his/her neighbors, the higher is the probability that he/she will decide to leave the house, and vice versa with respect to occupying a place. They are deterministic in the sense that the individual can buy or rent the house when his/her status is sufficiently high compared to the value of a house, and has to leave the city when he/she cannot afford paying the rent, or the mortgage, of a house in the city. The result of the latter is a *queue* of individuals who want to occupy a house within the city, but cannot afford doing so.

Consider individual A with status $S_A \in [0,1]$, tendency $T_A \in [0,1]$, and origin $O_A \in \{\text{Veteran}, \text{Ole}\}$. When that individual occupies a house/cell H whose value is $V^H \in [0,1]$, we denote that occupation form as $F_A \in \{\text{Owner}, \text{Renter}\}$. The payment of an individual occupying house H is $W^H \in [0,1]$. S_A, F_A, V^H and W^H depend on time. We shall mark it explicitly, when necessary.

As in the previous chapter, we consider as neighbors individuals in the 5×5 square around the house/cell.

In the following we consider in some details the sequence of decisions each individual agent takes at every time-step. This sequence is shown in the flow chart of Fig. 6.2.1. The numbers of the titles below correspond to box numbers in the flow chart.

1. To move house?

The probability R_A that an individual A will leave his/her house H depends on the occupant's origin, the fraction of his/her neighbors whose origin differs from the occupant's one and the difference between the occupant's status and the status of his/her neighbors:

$$R_A = R(O_A, |S_A - S_{\text{neighbors}}|, 1 - O_{\text{neighbors}}), \tag{6.1}$$

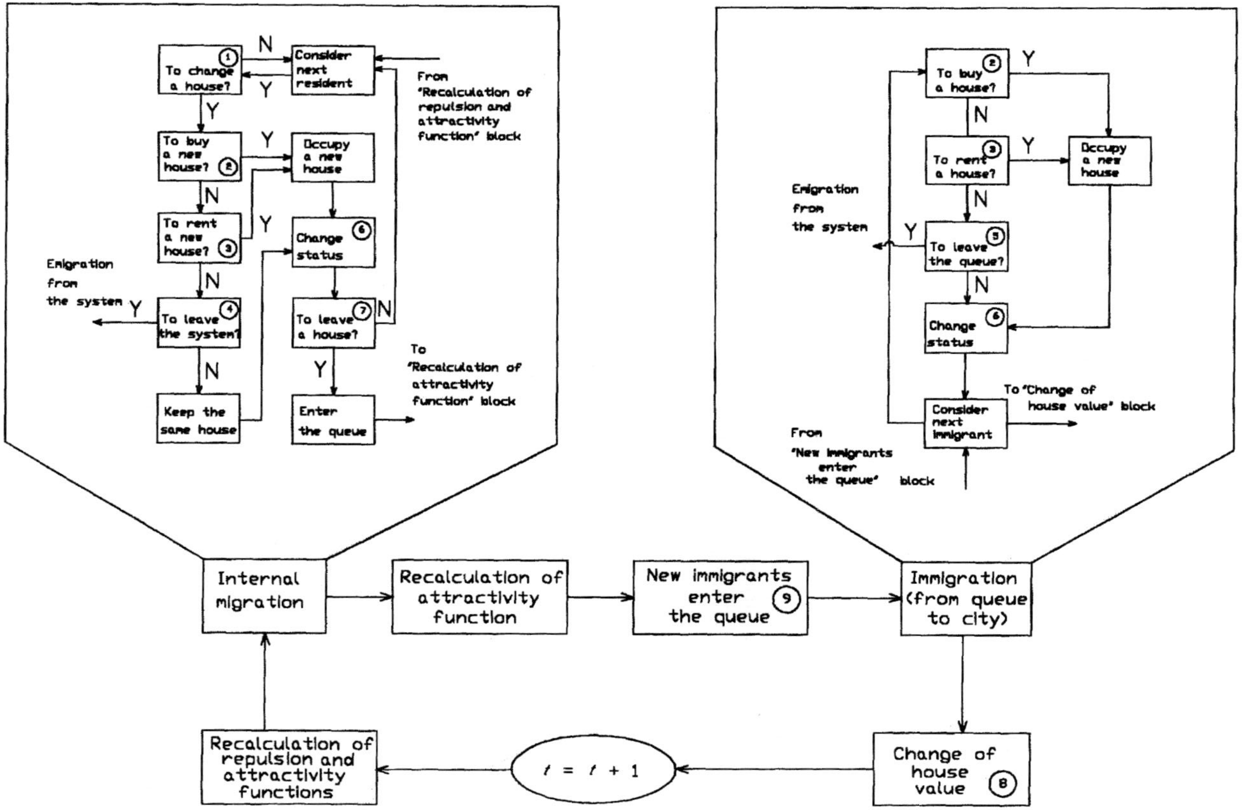

Fig. 6.1. The series of decisions taken by every individual in City-1 at each sequential time-step

where

$$S_{\text{neighbors}} = \sum_{A} \{S_A \mid A \text{ occupies house } G \in U(H), G \neq H\}/N$$

is the average status of A's neighbors,

$$O_{\text{neighbors}} = \sum_{B} \{O_B \mid B \text{ occupies house } G \in U(H), G \neq H\}/N$$

is the fraction of the individuals of the same origin as O_P in the $U(H)$ and

$$N = \sum_{B} \{1 \mid B \text{ occupies house } G \in U(H), G \neq H\}$$

is the number of B's neighbors.

The specific realization of equation (6.1) we use below is similar to (5.6) of Chap. 5 but accounts for both differences in status and origin:

$$\begin{aligned} R_A = {} & \alpha_1(O_A) + \beta_1(O_A)\exp(\gamma_1(O_A)|S_P - S_{\text{neighbors}}|) \\ & + \alpha_2(O_A) + \beta_2(O_A)\exp(\gamma_2(O_A)(1 - O_{\text{neighbors}})), \end{aligned} \tag{6.2}$$

We choose α_i, β_i and γ_i so that R_A increases monotonically with the increase of $|S_A - S_{\text{neighbors}}|$ and $1 - O_{\text{neighbors}}$ (Fig. 6.2).

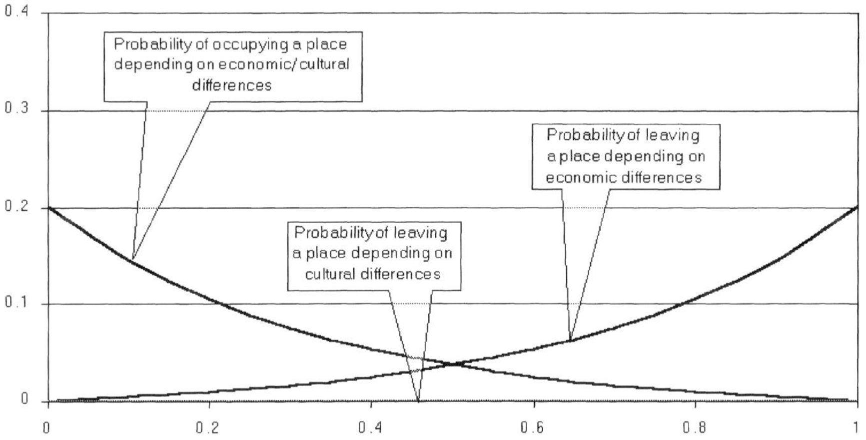

Fig. 6.2. Cultural and economic components of repellency and attractivity functions

2. To buy?
The probability of buying a house H is defined by the balance between status S_A and tendency T_A of the individual, on the one hand, and the value V^H of the house, on the other. This deterministic condition is applied to queuing individuals who want to enter the city as well as to residents who want to change their houses. An individual A can buy a house H of value V^H if his/her status and tendency are sufficiently high relative to the value of the house. That is:

$$S_A > \delta_0 + \delta_1 V^H, \tag{6.3}$$
$$T_A > \tau_0 + \tau_1 V^H, \tag{6.4}$$

where δ_0, δ_1, τ_0, τ_1 are constants.

An individual A chooses a house from the set of vacant houses satisfying (6.3) and (6.4). Potentially, each vacant house can be bought by agent A. The 'attractivity' Q_A^H of a house H for an individual A is estimated as follows:

$$Q_A^H = Q(O_A, O_{\text{neighbors}}, S_{\text{neighbors}}). \tag{6.5}$$

Below we choose for Q_A^H an analytical expression, which is similar to (5.9) from Chap. 5:

$$\begin{aligned} Q_A^H &= \alpha_3(O_A) + \beta_3(O_A)\exp(\gamma_3(O_A)S_{\text{neighbors}}) \\ &\quad + \alpha_4(O_A) + \beta_4(O_A)\exp(\gamma_4(O_A)O_{\text{neighbors}}) \end{aligned} \tag{6.6}$$

with α_i, β_i and γ_i chosen in such a way that Q_A^H monotonically decreases with an increase in $S_{\text{neighbors}}$ and $O_{\text{neighbors}}$ (Fig. 6.2).

We interpret 'attractivity' as the probability of occupying (buy or rent) a vacant 'house' at cell H when it is the only possible choice. An individual occupies one of the attractive houses or fails to occupy any of the houses in the following way. Let us denote as Λ the set of the houses, satisfying (6.3) and (6.4). As H_{00} we denote the 'negative' result of the residential choice, corresponding to failure of the attempt to occupy.

For an individual A we define the probability p_{ij} of choosing a house $H_{ij} \in \Lambda$ for residence in the following way:

$$p_{ij} = \frac{s_{ij}}{\sum_{ij}\{s_{ij} \mid H_{ij} \in \Lambda\}} \tag{6.7}$$

where

$$s_{ij} = \frac{Q^{H_{ij}} s_{00}}{1 - Q^{H_{ij}}}, \quad s_{00} = \prod_{ij} \{(1 - Q^{H_{ij}}) \mid H_{ij} \in \Lambda,\ H_{ij} \neq H_{00}\} \tag{6.8}$$

(index A for attractivity Q_A^H is omitted).

3. To rent?
An individual who failed to buy a house, will try to rent one. Similar to (6.3) and (6.4) above, here too the individual chooses from the set of houses satisfying:

$$T_A > \tau_2 + \tau_3 V^H \tag{6.9}$$

where τ_2 and τ_3 are constants.

Each vacant house, satisfying (6.9), can be rented. The 'attractivity' of the vacant houses is estimated according to (6.6). The probability of renting one of the houses H, satisfying (6.9), is calculated according to (6.7).

4. To leave the system?
A resident who could not buy a new house nor rent one, stays at his/her current house in the city or leaves the system. It is supposed that the probability of leaving the system L_A is the same for all the individuals, that is

$$L_A = L = \text{const.} \tag{6.10}$$

A new immigrant who failed to buy or rent a house either returns to the queue or else leaves the system.

5. To leave the queue?
An individual entering the queue is trying to occupy a house at each time-steps. If he/she didn't succeed in occupying a house during some pre-determined time-step $t_{\text{threshold}}$, he/she leaves the system. Below

$$t_{\text{threshold}} = \text{const.} \tag{6.11}$$

6. Change of status.
Simultaneously with the above activities regarding spatial location, the individual's status changes at every time-step. The status, at the next time-step, of individual A located at house H, depends on his/her current status, tendency and payment:

$$S_A(t+1) = \begin{cases} \max\{0, S_A(t) + T_A(t) - W^H(F_A, t)\} \text{ if} \\ T_A(t) - W^H(F_A, t) \leq 0 \\ \min\{1, S_A(t) + T_A(t) - W^H(F_A, t)\} \text{ if} \\ T_A(t) - W^H(F_A, t) > 0 \end{cases} \tag{6.12}$$

We define payment $W^H(F_A, t)$ as the right parts of (6.4) and (6.9):

$$\begin{aligned} W^H(F_A, t) = \tau_0 + \tau_1 V^H(t) \;\; & \text{for } F_A = \text{Owner} \\ \tau_2 + \tau_3 V^H(t) \;\; & \text{for } F_A = \text{Renter}. \end{aligned} \tag{6.13}$$

Queuing individuals have no payment and consequently the changes in their status are determined as follows:

$$S_A(t+1) = \min\{1, S_A(t) + T_A(t)\}. \tag{6.14}$$

7. To leave the city?
This is a deterministic rule. An individual will leave the city's territory and enter the queue if his/her status decreases below a certain threshold level $S_{\text{threshold}}$:

$$S_A(t) < S_{\text{threshold}} = \text{const.} \tag{6.15}$$

8. Change of value.
The value of a vacant house H at the next time-step is defined as the average of the status of individuals occupying houses and the value of the vacant houses in the house's neighborhood $U(H)$.That is

$$V^H(t+1) = \tag{6.16}$$

$$D\left(S_{\text{neighbors}} + S_A(t) + \sum_G \{V^G(t) \mid G \in U(H), G \text{ unoccupied}\}\right)/25$$

where $D < 1$ is a decrement of value, and the second expression in the parenthesis gives the values of vacant houses in $U(H)$.

Thus, at each time-step every individual takes up to five decisions. For a new immigrant entering the city they are 'to buy a house', 'to rent a house', 'to wait in the queue', 'to leave the system'. For a resident they are 'to change a house', 'to buy a new house', 'to rent a new house', 'to leave the system', 'to enter the queue'. It is important to note that each individual in the system takes this sequence of decisions simultaneously and at each time-step.

6.2.2 Some Properties of City-1

In City-1 we have tried to introduce some additional features of urban dynamics which seem essential to our understanding of cities as self-organizing environments. In particular the feedback social and economic interrelations between the local cell/house to its neighbors and between the individual decision-maker to his/her local neighbors and neighborhood. Such interrelationships that are commonplace in the literature of urban geography have not, as yet, been incorporated in full into urban modeling (Bertuglia et al.1987, Dendrinos and Sonis 1990, Fischer et al. 1990, White and Engelen 1993).

More specifically, City-1 elaborates on City in several important respects. As already noted, unlike City that was a CA model City-1 is a two-layer agent based model and in this respect more general. As a consequence, it has several new features: First, in City individuals differ only with respect to their cultural group affiliation, in City-1 in addition to this they differ also with respect to their economic status and tendency. Second, in City we translate individuals' behavior into the states of the cells. In City-1 they can explicitly occupy cells (i.e. houses), either rent or buy them. Third, in City there is no explicit entity 'land value'; in City-1 land value is an explicit entity which exists in complex feedback and circular causation interrelationships to the status and tendency of individuals, as well as to their group affiliation. Fourth, another new feature of City-1 is the queue in which 'homeless' individuals can wait for an opportunity to occupy a house.

The above properties make City-1 more complex and realistic than City and thus more suitable to study the specific processes we would like to address here: the relations between global international migration and local intra-national and intra-urban migrations. A central aspect of international migration has to do with structural differences between the origin and destination countries. Quite often, a person moving from a rich to a poor country improves his/her status relative to the one in the country of origin (i.e. colonialism). On the other hand, when moving from a poor to a rich country, the person often loses his or her relative status, despite the fact that this might be associated with an absolute economic gain. This situation is typical of many 'guest workers', for example.

The decline/gain of status is specifically distinguished during the first period after arrival to the destination country; with time, some immigrants might regain their previous status, while the status of others might deteriorate even further. This depends, of course, on the personality of the immigrant (ambition, drive, initiative, etc.), but also on the demand in the new country for his/her specific qualifications. If such a demand exists, the relatively low-status starting point of the immigrant in the new country is only temporary and there is a high possibility that he/she will realize their potential position in the new country in a relatively short time. This latter scenario characterizes many immigrants from the ex-USSR, especially professionals and academics. The story is very different with immigrants whose qualifications are not in demand in the new country. As the fate of many Oriental Jews in Israel indicates, the status loss upon arrival in the new country might become a permanent situation. The above distinctions are captured in our model by means of the relations between the notions of status and tendency.

6.3 Selected Results

As noted above, our aim in constructing City-1 was to suggest an improved model which simulates the city as a self-organizing system, and on which one can study an issue which is currently at the focus of interest in Israeli society: the relations between international migration to the country and the internal structure of its cities. The actual, full-scale, analysis of these relations will not be discussed here. Such a discussion requires an independent study, which would examine a large number of initial conditions and their associated urban scenarios, in comparison with empirical data. We leave such a discussion for a subsequent work and in the following we thus examine selected results that illustrate the properties of the model we have developed and presented above.

6.3.1 Stability of the Socio–spatial Structure of the City in Face of Massive Waves of International Migration

One of the more interesting results that emerges from the many model runs we have examined concerns the stability of the socio–spatial structure of the city and its associated space economy in face of massive waves of international migration. In several of our tests we have run the model, first with veteran population and migration only and then, once the structure of the veteran city was shaped, we started to input increasing rates of Olim among the immigrants who want to enter the city: starting with 1% veterans versus 0.33% Olim, and ending with 1% veterans versus 1% Olim, when the initial status of Olim is, on the average, four times lower than veterans, while their tendency is four times higher. The results have been that up to a certain threshold (which in these runs was between 0.33% and 0.67% Olim immigrants into the city), no change in the space economy of the city (status map and land-value map) could be visually observed (Fig. 6.3). The only change visually observed was the residential spatial segregation of Olim around the central area, which was predominantly occupied by veterans (Fig. 6.3). However, once we have compared this Olim–veterans residential spatial segregation map with the status spatial segregation map, it became clear that the segregation between veterans and the Olim is a seeming segregation; in effect it reflects the relatively initial low status of Olim new immigrants. The latter, when entering the city, usually settle in the low-status areas and do not penetrate the very high-status veteran areas. Within the low-status areas of the city, no veterans–Olim residential spatial segregation could be observed. However, once we have crossed the above threshold of Olim among the immigrants to the city, the socio–economic spatial structure of the city has changed. This can be observed in Fig. 6.3. Note that since the spatial distribution of tendency in Fig. 6.3 it visually random, it is likely that had we let the process continue, the result would be an economic spatial segregation irrespective of Olim–veterans distinction.

It is interesting to note that the above scenarios conform to the empirical results from studies on the first and second residential locations of ex-USSR Olim in the Israeli settlement system. These studies show that the recent massive wave of new immigrants from ex-USSR countries has not revolutionized the Israeli urban structure (as many predicted it would), but instead 'adapted' itself to the previous veteran structure. This is so with respect to the Israeli settlement system as a whole as well as to the internal structure of its towns and cities.

In several of the games we have tried to 'break' the (economically determined) stable veteran city structure. This was done in several ways. First, as shown above, by increasing the rate of the Olim among the immigrants to the city beyond a certain threshold (as in Fig. 6.3 above). Second, by introducing a qualitative change in the behavior of individuals in one (or two) of the groups with respect to residential location, and third, by changing the

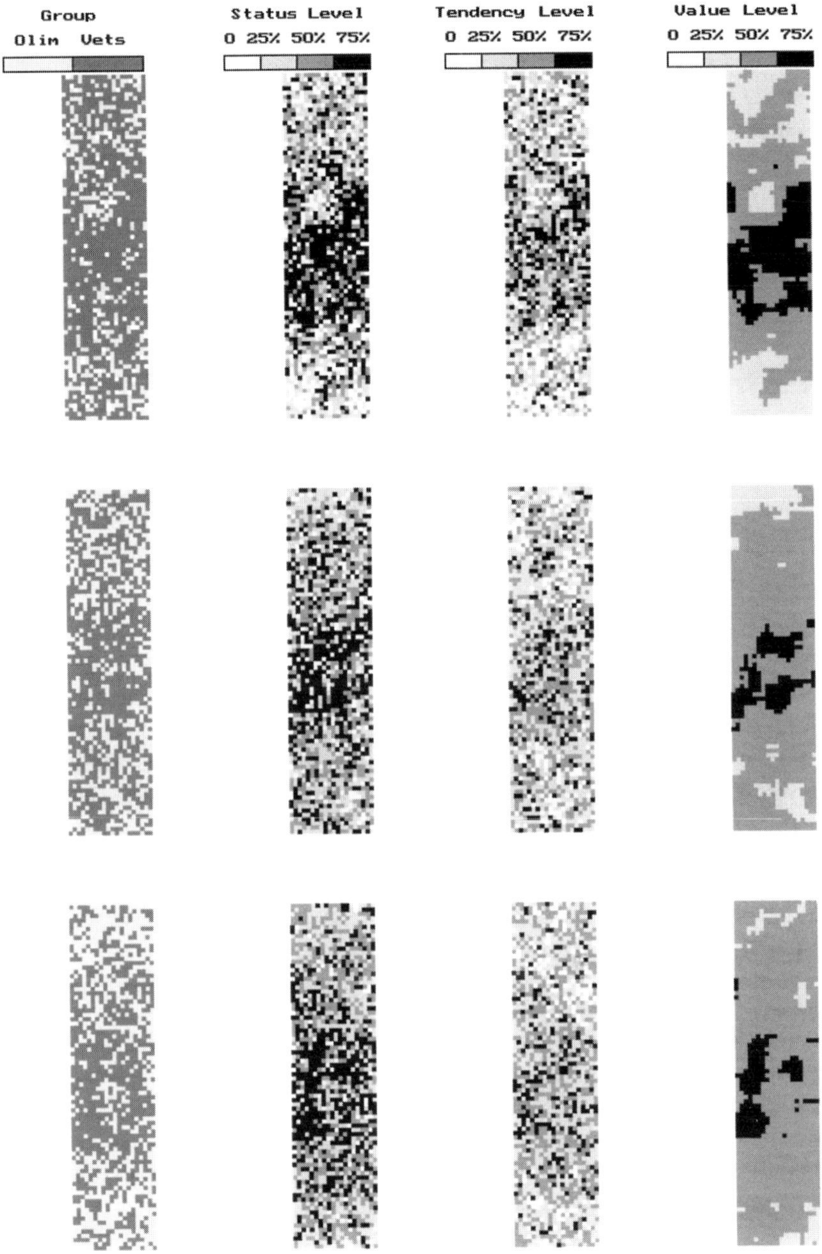

Fig. 6.3. City-1 spatial structures at $t = 80$ when veterans' immigration rate is 1%, while Olim immigration rates are 0.33% (row 1), 0.67% (row 2) and 1.00% (row 3)

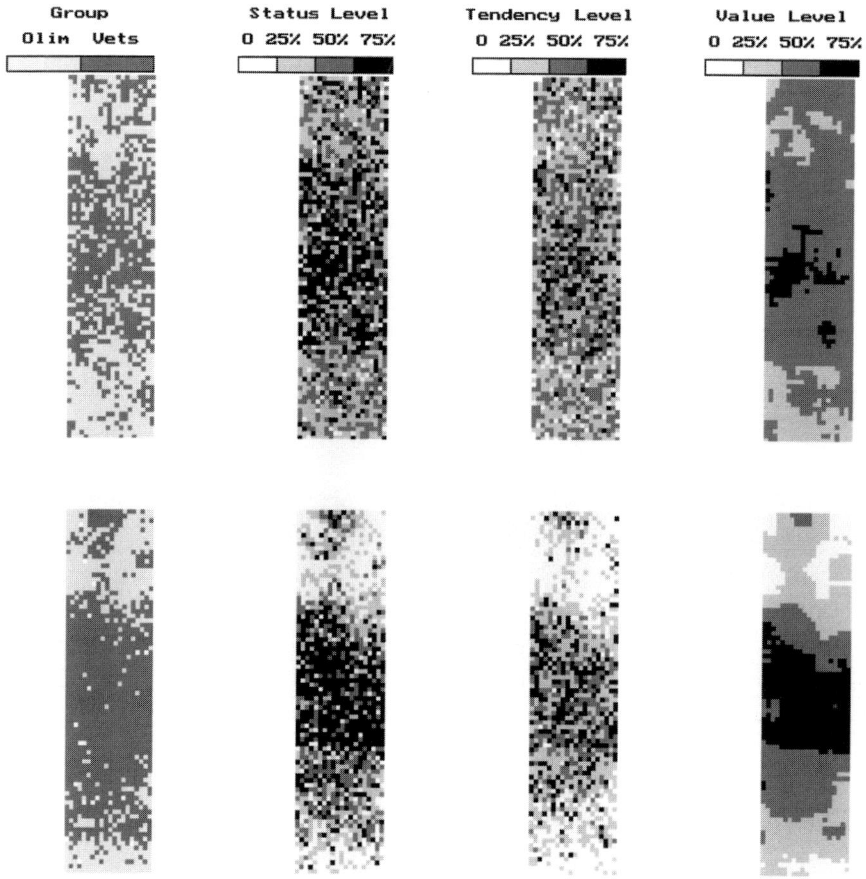

Fig. 6.4. City-1 spatial structure when veterans are antagonistic to Olim. *Upper row*: Mean tendency of Olim is higher than that of veterans. *Lower row*: Mean tendency of Olim is equal to the mean tendency of the veterans

tendency of the Olim relative to the veterans. To examine the second way on City-1 we postulated that the veterans are antagonistic to the Olim and prefer not to locate or live near them. In the model this implied making the repel function of veterans dependent not only on the status of neighbors, as before, but also on the fraction of Olim neighbors (see (6.2) and (6.6) above), while retaining the previous composition of 1% Olim and 1% veterans immigrants to the city (as in Fig. 6.3, row 3). The results have been a visually observed residential spatial segregation between veterans and Olim all over the city (Fig. 6.4, upper row), but no changes in the spatial distribution of status and tendency in the city, nor major morphological changes (compared to Fig. 6.3) in the spatial distribution of land values (Fig. 6.3, row 1).

To examine the third way we've played on City-1 a game, which differed from the previous game in that the tendency of the Olim immigrants was the same as the tendency of the veterans. The results, as can be seen in Fig. 6.4, lower row, have been a visually observed change in all domains relative to the previous situation (Fig. 6.4, upper row): spatial residential segregation of veterans versus Olim is accompanied now by spatially segregated status and tendency surfaces, as well as by a spatially polarized land value surface. The fact that the three maps describing the 'existing situation' (Olim–veterans, status, land value) are similar to the fourth map which indicates the tendency to change the existing situation, suggests that the system entered a stable steady state (i.e. a stable path of evolution).

These results are important as they indicate how subtle social and cultural spatial relations, which in the city often exist 'under the surface', are related to the city's structure and space-economy. They are also important with respect to the reality of the Israeli urban space. They indicate, first, that the rates of flow of new immigrants to Israel and its cities have not as yet crossed the threshold beyond which the Israeli settlement system would be subject to a major qualitative change. Second, that culturally, no veterans–Olim antagonism, strong enough to affect status and land-value spatial distribution in cities, has yet developed in the domain of residential spatial behavior. This latter point conforms with what is known from empirical studies: The absorption of the immigrants from ex-USSR countries, which form the majority of the immigrants to Israel in recent years, is relatively smooth in the sense that it was not followed by spatial antagonism in the housing market. This is not the case, however, with 'Ethiopian' immigrants which differ from 'Russian' Olim in two important respects: First, both their status and tendency were lower than the average among veterans; second, and possibly as a consequence of the above, there have been several reports in the media about veterans' residential antagonism against Ethiopian Olim. This latter reality is well captured by the above games and indeed the 'Ethiopians' in current Israel are spatially segregated – very much in line with the patterns observed in Fig. 6.4, lower row. However, given the relatively small number of 'Ethiopians' in the recent migration waves, it is unlikely that this phenomenon will have a substantial impact on the urban structure of Israel.

The above discussion is based, as noted, on visual inspection of the various patterns as they evolved on the screen. In addition to the changing patterns, various statistics related to these patterns can be considered. Some of these, which describe changes in the time evolution of (a) the number of inhabitants, (b) queuing individuals and (c) the residential (dissimilarity) segregation index, can be seen in Fig. 6.5. The nature of these measures is such that they refer to the city as a whole and not to local variations on its various surfaces. Comparing the patterns with these measures adds further insight into the relations between local–spatial and global–temporal, changes in the city. It can be seen, for example, that the introduction of spatial antagonism to City-1 has not changed its global characteristics relative to the previous situation: all measures remain the same. On the other hand, the change of tendency – equalizing the tendency of Olim with that of the veterans – had a major effect on the overall global structure of the city; this can be seen in all measures presented in Fig. 6.5, including the segregation index which does not decrease below 0.6. It is interesting to note that similar global characteristics were obtained in a game where all immigrants to the city (2% each iteration) were veterans (Fig. 6.5).

6.3.2 Competing Order Parameters in a Self-Organizing City

As elaborated above, the evolution of self-organizing systems follows a distinct routinized path: long periods of steady state, interfered by relatively short periods of strong fluctuations and chaos. According to Haken (1983a, 1985b) while in steady state, the system is governed by a certain order parameter; while in fluctuations and chaos, several order-states compete, until one wins, enslaves the system, brings it to a new steady state and thus becomes the new order parameter. In this section we use City-1 as means to study cultural–spatial segregation in light of the interrelation between two order parameters which play a central role in shaping the socio–spatial structure of cities: the cultural order parameter (COP) and the economic order parameter (EOP).

For this purpose we've played on City-1 three scenarios, that describe the dynamics of the city, and the interplay between the EOP and COP, when

(i) The EOP is dominant in a certain way;
(ii) The COP is dominant in a certain way; and
(iii) Both the COP and EOP act together.

In the following we describe each scenario in sequence and examine their evolution by means of three devices: segregation indices (Fig. 6.6), three moments ($T = 10$, $T = 100$, $T = 300$) of the evolving spatial–cultural segregation (Fig. 6.7), and the same three moments of the evolving land value surface (Fig. 6.8). Note that Fig. 6.8 can be regarded as a surrogate for the evolving spatial economic segregation of individuals in the city. In all scenarios the initial mean value and STD of the status of veterans is four times higher than that of Olim; mean tendency of Olim is four times higher than that of veterans; STD of Olim tendency is four times lower than that of veterans.

6.3.2.1 When the Economic Order Parameter is Dominant

The EOP is dominant when we assume two forms of economic antagonism: First, the probability for an individual to leave his house increases with the growth of the difference between the individual's and his neighbors' economic status. Second, the attractivity of a house increases with the increase of the economic improvement of its environment (Fig. 6.2). As can be seen in

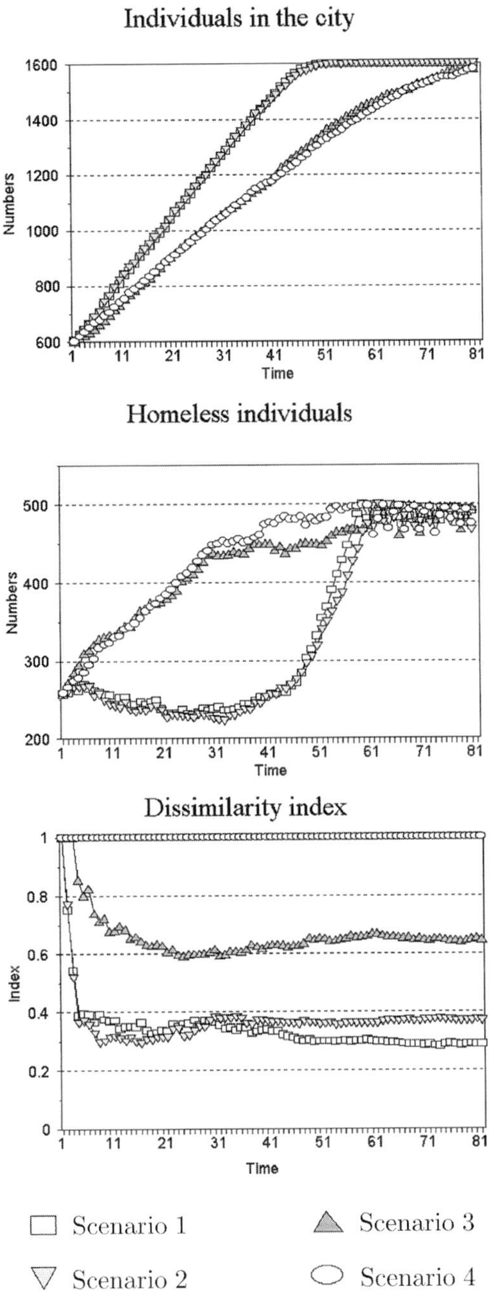

Fig. 6.5. Time evolution of the global parameters of City-1. *Top*: individuals in the city. *Middle*: 'homeless' individuals. *Bottom*: dissimilarity indices for the various scenarios. Scenario 1 – Immigration rates of Olim and veterans are 1%; no antagonism; mean tendency of Olim is four times higher than that of veterans. Scenario 2 – Same as scenario 1, except that veterans are antagonistic to Olim. Scenario 3 – Same as scenario 2, except that mean tendency of Olim is equal to that of veterans. Scenario 4 – Immigration rate of veterans is 2% and no Olim immigrants

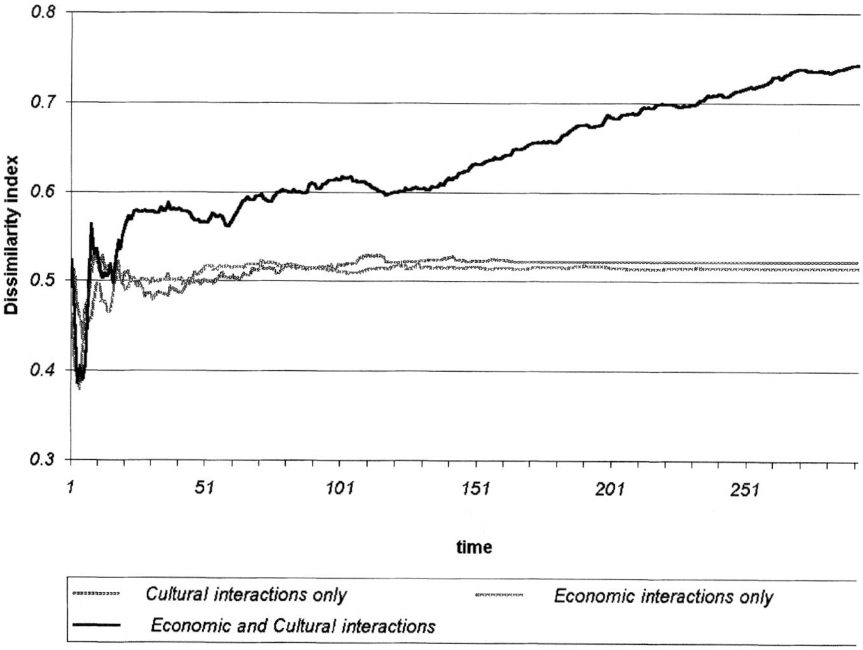

Fig. 6.6. Dynamics of cultural segregation for the three scenarios studied

Figs. 6.6, 6.7, in this case the city evolves with no cultural–spatial segregation. The city's evolution is thus dominated by the EOP as some form of economic segregation can be observed (Fig. 6.8). It is important to note, however, that despite the fact that this scenario is governed by economic forces only, economic segregation is not very distinguished.

6.3.2.2 When the Cultural Order Parameter is Dominant

Here there is no economic antagonism and the COP is dominant in the sense that we assume a 'mild' cultural antagonism between veterans and Olim. The latter shows up, first, in that the individual's decision to leave a house does not depend on the characteristics of its neighbors, and second, in that when choosing a new house the individual prefers neighbors of his/her own kind. 'Prefers' implies that the attractivity of a fully friendly environment is 0.45 compared to 0.1 of a fully strange one (Fig. 6.2). In this case the evolution of the city and the interplay between EOP and COP is more complex than in the previous scenario. At the beginning ($0 < T < 100$), the COP enslaves the EOP and dominates the city's evolution: the city is segregated culturally (Figs. 6.6, 6.7) and economically (Fig. 6.8). Then gradually the EOP and the COP neutralize each other, until at $T = 300$ no cultural segregation (Fig. 6.7),

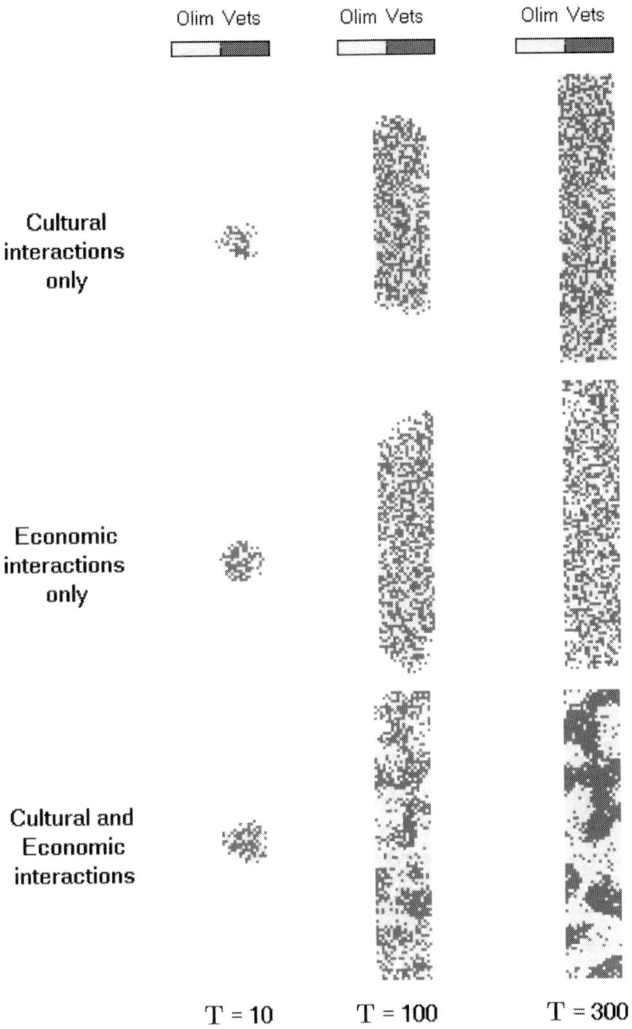

Fig. 6.7. Three moments ($T = 10$, $T = 100$, $T = 300$) in the evolving spatial–cultural segregation in the city

nor economic segregation (Fig. 6.8) can be observed. Apparently this is a result of the higher economic tendency of Olim relative to the veterans.

6.3.2.3 When the Cultural and Economic Order Parameters Interact

Here we assume the existence of both economic and cultural antagonism as above, and the repel and attractivity functions are thus the sum of both antagonisms as defined above (Fig. 6.2). This scenario is rather interesting.

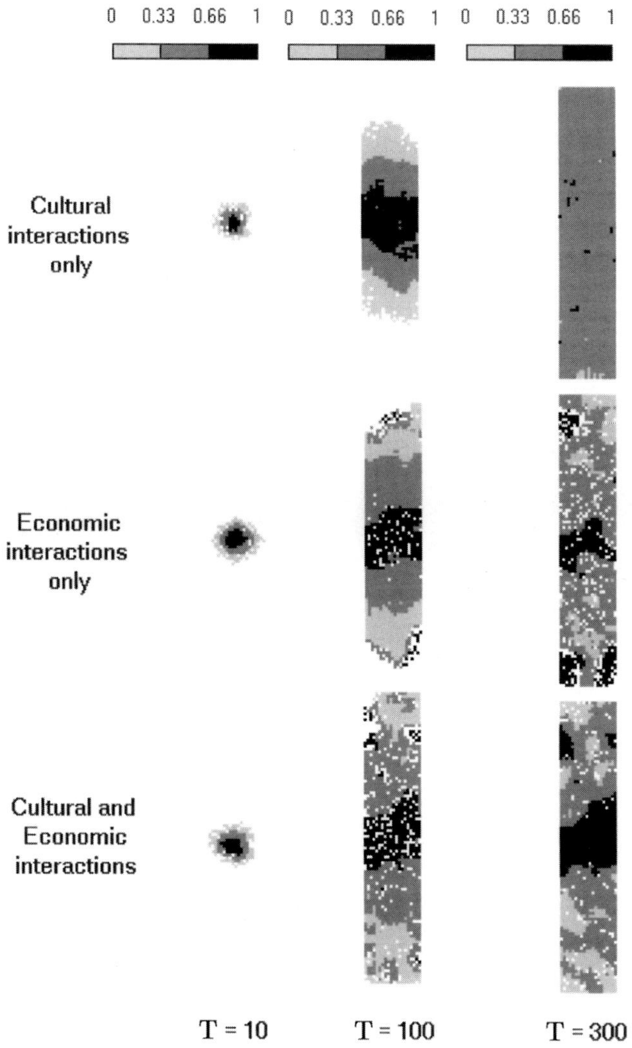

Fig. 6.8. Three moments ($T = 10$, $T = 100$, $T = 300$) in the evolving land-value surface of the city

At first, the evolution of the city is governed fully by the COP – a strong trend towards cultural segregation can be observed in Fig. 6.6. Then segregation decreases but does not disappear, and in fact it can be observed all the way to $T = 300$. The interesting part is the role played here by the EOP. Indeed, as can be seen in Fig. 6.6, it acts to moderate cultural–spatial segregation, but at the same time it does not eliminate it, but actually acts as a catalyst and reinforces it: the COP enslaves the EOP and the city becomes culturally (Fig. 6.7) and economically (Fig. 6.8) segregated in line with the COP.

6.4 Concluding Notes

We would like to conclude this discussion by emphasizing a point concerning the position of our model within the field of self-organization in regional analysis. It can be said that it stands somewhere in between a very abstract theorization of cities as self-organizing systems, and more operational models, which were built from the start as empirical tools. This midway position is the property of agent-based models in general which are, on the one hand, par-excellence theoretical devices, but on the other, are intuitively related to the physical structure of cities: cells as 'houses' and individuals which take residential location decisions in light of the properties of the cells/houses and the neighbors around them. As we shall demonstrate below, in Chap. 12, this intuitive connection between the theoretical model and the actual structure of cities makes cell space models such as ours a rather convenient planning tool. It enables us to give planners a highly theoretical and abstract device on which, and with which, he or she can examine real situations in a general manner.

7. Spatial Cognitive Dissonance and Socio–spatial Emergence in a Self-Organizing City*

7.1 Introduction

Modern, or rather postmodern, cities are characterized by spatial, social and cultural pluralism: the city can be described as a spatial mosaic of coexisting cultural and social groups (see Chap. 2, Sect. 2.13). Some of these groups, such as Little-Italy or China Town, are the product of 'old' ethnic groups who emigrated to the city as already established cultural groups, while others, and this is the more recent phenomenon, are the dialectical product of the city itself (Yuppies, Gays, Lesbians). If in the two previous chapters our aim was to study the way already established cultural groups participate in constructing the city's cultural and socio–spatial mosaic, our aim here is to examine the way a new cultural, or social, spatial entity emerges out of the city dynamics. The general question here concerns, of course, morphogenesis: the way a new spatial socio–cultural entity is born.

To answer this question we refer back to one of the findings in our discussion in Chap. 5: the gap we've observed emerging between individuals' intentions and wishes, on the one hand, and their actual behavior, on the other. We suggest that such a gap might lead to a spatial cognitive dissonance on the part of individual residents in the city, and that in certain cases that we investigate, this dissonance leads individuals to adopt a new socio–spatial cultural identity. We start the discussion in this chapter by elaborating on the various facets of the relations between intention and behavior, and then construct City-2 as our heuristic model, on which we 'play' and thus study spatial cognitive dissonance and socio–spatial emergence in a self-organizing city.

City-2 can be described as a two layer model composed of a migration sub-model describing the inter- and intra-city migration movements, superimposed on a Cellular Automata sub-model describing the dynamics of the urban landscape itself. As such it elaborates on, and continues, our two previous heuristic models City and City-1. However, it differs from these two models in the sense that in both City and City-1 the socio–cultural properties of individuals (Greens versus Blues and Olim versus veterans) and their socio–spatial intentions (segregatives or neutrals) were determined externally,

* With Itzhak Benenson and Itzhak Omer

outside the models. In City-2 – the model we construct and use in this chapter – we allow for these properties to be determined internally, inside the model. Such a model will enable us to examine the very dynamics of (a) the tendencies, values and intention of individuals, (b) the way a new socio–spatial group with its own cultural identity is born, and (c) the implications thereof to residential spatial segregation in the city.

It is important to note that the questions of human intentionality and socio–spatial emergence are central to social theory. In this context this chapter, in fact, suggests and adds our own theory or view on that matter. It differs from the usual discourse in social theory (and in social geography) in the means employed for that purpose: in addition to, and in place of, the 'usual' hermeneutic discourse which is typical of social theory, we'll make an extensive use of City-2 which is the specific heuristic–hermeneutic model we construct below.

7.2 Intention Versus Behavior

Our point of departure in this chapter is a phenomenon which emerged out of the various games played on our first model City – a gap between individuals' intentions and values, on the one hand, and their actual behavior and action, on the other. Separating intentions from behavior goes counter mainstream regional science and urban and regional modeling which are currently dominated by a mixture of Economism and Behaviorism. According to the latter,

> researchers interested in a science of behavior ought to restrict themselves strictly to public methods of observation .... [and] ought to focus exclusively on behavior: researchers ought assiduously to eschew ... such concepts as plans, desires or intentions (Gardner 1987, p. 11).

By Economism we mean the assumption of the rational economic person (see discussion below). The result is that in regional science, intention and behavior are usually treated as causally related: intention is the cause of behavior, or alternatively, behavior is some product of the optimization of intentions. In the model we propose below intention and behavior are two relatively independent entities which might affect, complement, negate, or compete with, each other. By departing in this respect from the approach of mainstream regional science, we in fact follow disciplines such as psychology or cognitive sciences (the past strongholds of behaviorism) which since the mid 1950s have turned their backs to behaviorism and today regard "the theoretical claims of behaviorism (though not its various applied achievements) ... [as] largely of historical interest" (Gardner, ibid, p. 110). We also follow classical social theory, and what we consider as some of the most important implications of the theory of self-organization. These theoretical bodies provide the conceptual background to the model and theory we develop in this chapter. In the following we present and discuss each of them in turn.

7.2.1 Classical Social Theory

The interrelation between the individual's intentions and value system, his/her actual behavior, and society, forms a central theme in social theory and philosophy. The notion of ideology, for example, is directly related to the tension created between a person's value system and the person's actual behavior and action (Larrain 1982). Such a tension often leads to what Hegel and later Marx have called 'ideological false-consciousness' which obscures people's vision from their real conditions of existence. A central controversy in social theory is between Marxists who claim that a person's value system (including intentions) is dialectically determined by his/her conditions of existence (i.e. actual behavior), and liberal humanists who consider human action as an outcome of human intentionality. Giddens' (1984) recent theory of *structuration* aims to synthesize the two views: on the one hand the individual is a free agent whose intentions determine his/her actions; on the other hand, the individual is acting in a relatively autonomous social structure with its own rules and thus the individual's actions and behaviors have 'unintended consequences'. (In Chap. 10 we further discuss this concept.)

In the last two decades, mainstream social geography was strongly influenced by social theory and by Giddens' structuration, and is thus very critical of behaviorism and its application to geography and regional science (Thrift 1983). Arguing from the perspective of social theory, it accuses behaviorists of blurring the dialectical relations between the human agency and his/her socio–spatial structure. That is, the refusal of behaviorism to consider the subjectivity of the individual with his/her wants and intentions, and to study the ways socio–spatial structures such as cities, determines individuals' intentions and ideas (for a discussion and bibliography see Jackson and Smith 1984, Chap. 3; Gregory and Urry 1985). The model we present below thus provides a basis to reintegrate social geography with regional science.

7.2.2 Self-Organization Theory

One of the cutting edges of regional sciences of the last few years is the theory of self-organization. Now, the very idea of self-organization with its property of nonlinear relations implies, almost by definition, a gap between intentional causes and behavioral effects, as well as various forms of unintended consequences. Furthermore, one of the logical consequences of applying the theory of self-organization to the domain of behavior has been to discredit behaviorism. This is implicit in Prigogine's theory especially where he criticizes

> the particularly unfortunate use of concepts [such as optimization, since while] it is obvious that the management of human society ... tends to some aspects of behavior ... , to consider optimization as the key ... is to risk confusing causes with effects (Prigogine and Stengers 1984, p. 207).

This is even more prominent and explicit in Haken's synergetics approach to self-organization in which the order parameter (e.g., an emerging spatial pattern) enslaves the individuals' behaviors. Within the domain of human behavior this was beautifully illustrated in a series of experimental and theoretical studies designed by Kelso who has used Haken's synergetics approach to self-organization as his framework (Kelso 1984, 1990, Haken, Kelso and Bunz 1985; Haken 1990). In Kelso's experiments intention and behavior are found to be methodologically and scientifically separated though dialectically related. Interesting discussions and results were also obtained by Stadler and Kurse (1990) with respect to cognitive processes in general. Separating intention from behavior in the model we develop below, thus follows directly from our view of the city as a self-organizing system.

7.2.3 Cognitive Dissonance

Unlike social theory and Giddens' structuration theory whose main concern is society, Festinger's (1957) theory of cognitive dissonance focuses directly and exclusively on the cognitive processes of the individual. The idea is straightforward: a cognitive gap or dissonance between an individual's intentions and his/her actual behavior and action, is cognitively unbearable – it creates a cognitive tension which eventually will have to be resolved either by a change of behavior and action, or by a change of intentions and value system. Festinger's theory was examined in various laboratory experiments as well as in real life observations and is now generally accepted. More recently the theory got further support by Gazzaniga's (1985, p. 80) studies on the modular structure of the brain: "the new brain science [adds to Festinger's theory] the knowledge that [cognitive dissonance is related to the fact that the brain] is organized in ... relatively independent modules that are capable of initiating disparate behavior in the first place". Cognitive dissonance as formulated by Festinger and elaborated by Gazzaniga provides the starting point to the notion of spatial cognitive dissonance which stands at the center of our discussion below.

So far there have been only a few attempts to consider the behavioral implications of self-organization theories in urban and regional modeling, and only one attempt to our knowledge (Timmermans 1990) to take notice of the post-behaviorism approaches which since the 1950s are prevalent in behavioral sciences such as psychology or cognitive studies (for a collection of papers on these and related issues see Fischer, Nijkamp and Papageorgiou 1990; see also Desbarts 1983). With respect to the first issue Fischer et al. (1990) have examined (among other things) the 'master equation approach' and the 'ecological approach' in urban and regional modeling and made the distinction (further elaborated by Sonis 1991) between two behavioral models: the economic person (*homo oeconomicus*) versus the social person (*homo socialis*). The first

> is a totally egoistic, rational omniscient creature who is supposed to accomplish rational free choice ... on the basis of the utility maximization principle', whereas the second's behavior is based on co-interaction [e.g. among individuals] and information flows through contacts with the 'nearpeers' ... and through mass media... (Sonis 1991, p. 36).

Intuitively, one could say that the social person behaves 'like others' – the more people behave in a certain way, more people will follow – up to a certain limit, with the implication that as a whole it takes a logistic form. People thus might change their behavior as a consequence of the various socio–spatial interactions in the system.

Since Sonis' social person behaves like others, the specific original intentions of that individual are of no significance and by implication no intention–behavior tension can arise and persist; the model thus conforms to strict behaviorism. Sonis' discussion of the social person remains on an aggregate 'meso level'. As such it leaves open questions such as 'how many, or where in the city, people change their behavior in the process of self-organization?', 'do they change their behavior voluntarily?', 'what happens to those who do not?'. As noted, strict behaviorism commands that 'researchers ought assiduously to eschew' such questions. We say that obeying this command diffuses the achievement of Sonis' social person as it leaves the homo socialis in a no better position than the homo oeconomicus: both concepts attempt to predetermine human nature; for the economic person utility maximization is the prime human motivation, for the social person – being like others. We suggest following the implications of self-organization and the synergetics studies on behavior and make a preliminary step toward an explicit consideration of the individual with his/her subjective beliefs and intentions, in relation to the global system – the city as a whole. Such a theorization is close in nature to Timmermans' (1990) recommendation that students of regional science should take notice of the 'variety seeking choice behavior' models. Elaborating on this notion he brings 'various motivation theories [which] offer insight into why individuals may seek variety'. Among these is Festinger's theory of cognitive dissonance noted above, which like other theories of this kind, makes a distinction, and studies the interplay, between actual behavior and intention. Timmermans suggests the 'variety seeking choice behavior' as an addition to the usual utility maximization approach. We suggest that separating intention from behavior enables us to address one of the problems which stands at the center of social theory, social geography and the theory of self-organizing systems: the relations between the individual and the system.

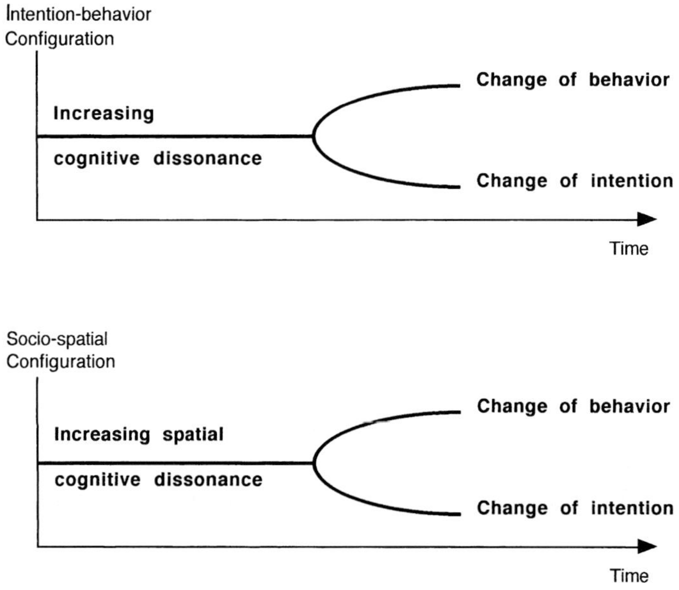

Fig. 7.1. Cognitive dissonance formulated in terms of a bifurcation diagram. *Upper*: general. *Lower*: spatial cognitive dissonance

In line with our perception in this chapter that the city is a self-organizing system, and with several previous attempts to consider individual cognition and behavior in terms of self-organization, it is interesting to note that Festinger's theory can easily be described by a bifurcation diagram as in Fig. 7.1. That is to say, from the perspective of the individual, a situation of cognitive dissonance drives the individual into a cognitive bifurcation point: to change behavior, or to change intentions and value system.

A typical case within a city would be that of an individual living in a neighborhood where he or she does not want to live. This frustrating situation can be resolved either by a change of wants, or else by migration. A recent empirical examination of the intention–behavior gap, within a city context, seems to support this view. It suggests that with time, people's reaction to the intention–behavior gap becomes polarized: people living for a long period of time (18+ years) in a neighborhood of another cultural group (Jews among Arabs and vice versa), become either integrative (i.e. change of intentions), or extremely segregative in their value judgment of their actual situation. This is illustrated in Fig. 7.2.

As noted above, this dissonance between intention and actual behavior was also obtained as a by-product from the various simulation games played on our first model City, in Chap. 5. We have used City as means to examine socio–cultural spatial segregation as resulting from two groups of individuals (Greens and Blues) divided into various configurations according to their

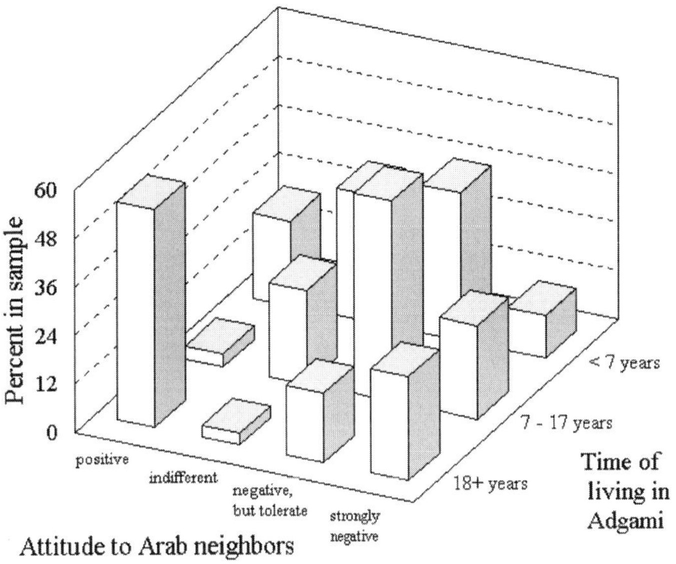

Fig. 7.2. The changing reaction in time of Jewish inhabitants living in Adgami (the Arab community of Tel Aviv-Yaffo) to their Arab neighbors. Four attitude groups were found: positive attitude about living among Arab neighbors; people indifferent as to their neighbors; negative attitude – prefer to live among Jewish neighbors; strongly negative – strong objection to living among Arabs

intentions: Blue or Green Segregatives who want to spatially reside among their own kind, and Blue or Green Neutrals who are indifferent as to their location in the city. In the various games played it was found, first, that in all the simulations there were always individuals who actually behaved (i.e. were located in space) counter their intentions (e.g. a segregative Green located among Blue neighbors). Such people live in what we shall define as 'spatial cognitive dissonance'. Second, in the majority of the simulation games these individuals were spatially segregated in the city, thus creating a relatively homogeneous area of intense cognitive dissonance.

7.3 The Model

7.3.1 Preliminaries

The model City-2 is structurally similar to City-1 described in Chap. 6 above. It explicitly accounts for the human properties of the actors, and it has two layers: One is a population layer representing and describing the migration, interaction, properties and activities of individuals. This population layer is superimposed on a CA layer which represents and describes the city's physical infrastructure. City-2 is specifically designed as a heuristic tool to examine, first, the circumstances by which spatial cognitive dissonance in the city may arise and lead to changes in individuals' intentions, cultural affiliation and identity. Second the implications thereof to residential spatial segregation in the city. As a result, City-2 is governed now by means of the following system of equations:

$$
\begin{aligned}
S^H(t+1) &= F_t(S^H(t), A^H(t), S^{U(H)}(t), A^{U(H)}(t), S^A(t)), \qquad (7.1)\\
A^H_{\mathrm{e}}(t+1) &= G_{\mathrm{e}t}(S^H(t), A^H(t), P^H(t), S^{U(H)}(t), A^{U(H)}(t), S^A(t), E(t)),\\
A^H_{\mathrm{c}}(t+1) &= G_{\mathrm{c}t}(S^H(t), A^H(t), P^H(t), S^{U(H)}(t), A^{U(H)}(t), S^A(t), E(t)),\\
P^A(t+1) &= L_t(S^H(t), A^H(t), P^A(t), S^{U(H)}(t), A^{U(H)}(t), S^A(t), E(t)),
\end{aligned}
$$

where $S^H(t)$ is a state of a cell H, $A^H(t)$ describes the state of an individual occupying H and consists of two evolving components: economic – $A^H_{\mathrm{e}}(t)$ (as in City-1) and cultural – $A^H_{\mathrm{c}}(t)$, that is $A^H(t) = (A^H_{\mathrm{e}}(t), A^H_{\mathrm{c}}(t))$. $P^A(t)$ denotes the agent's location, $G_{\mathrm{e}t}$ and $G_{\mathrm{c}t}$ define the state of an agent occupying cell H and L_t defines the location of A at the next time step.

As in City-1 so in City-2, the immigrants who enter the city, and the inhabitants living in the city, interact with each other and the system of cells/houses. These interactions result in (a) intra- and inter-city migration dynamics, (b) changes in the properties of the cells, (c) changes of the economic and cultural properties of individuals. The latter form the focus of attention in this chapter.

The population of City-2 is composed of individuals, either inhabitants in the city, immigrants who try to enter it, internal immigrants who try to change location within it, or emigrants who leave the city. Each individual has a specific intention to spatially live among individuals who have cultural identity similar to his/her own. There are two types of cultural identity: *original identity* with which that individual starts the game, and *current identity* referring to the individual's identity in any specific iteration of the game.

At the start of the game original and current identities are equal and there are only two polar types of individuals in City-2: Blue individuals whose intention is to live among Blue neighbors and Green individuals whose intention is to live among Green neighbors. All individuals at this stage are segregatives in the sense that their intention is to spatially locate among neighbors of their own kind. This applies also to all new immigrants when they first come to the city.

The original identities with their implied intentions might change in the course of the game to become current identities and intentions. That is to say, individuals might become neutrals (i.e. indifferent as to the Blueness or Greenness of their neighbors), Blue might change towards Green and vice versa, and a totally new socio–spatial identity might emerge. Formally, individual A has an original cultural identity with implied intentions $O_A \in \{\text{Blue}, \text{Green}\}$, (compare to Blues and Greens in Chap. 5 and Veterans and Olim in Chap. 6) and a new current cultural identity $C_A \in [0, 1]$, where we set original Green identity as 1 and Blue as 0.

As in City-1, each individual has a changing economic status. We characterize an individual by a current economic status $S_A \in [0, 1]$ and the rate of status' growth – economic tendency – $T_A \in [0, 1]$. C_A and S_A depend on time and we mark it explicitly when needed.

The infrastructure of City-2 is a 60×50 rectangular lattice of cells/houses. Each house can be either vacant or occupied. An occupied house obtains a cultural property defined by the current cultural identity of its inhabitants. At the beginning of the process it is Blue if occupied by a Blue individual and Green if by a Green. The neighborhood $U(H)$ of house H is a standard 5×5 square with H in the center.

Every house has an economic value $V^H \in [0, 1]$, calculated as an average of the economic status of its resident, its neighbors and the value of unoccupied houses in $U(H)$. We denote the payment for occupying house H as $W^H \in [0, 1]$. Additionally, in City-2 every house, vacant or occupied has a *cultural image*, calculated as an average of the cultural properties of its neighbors and its inhabitant.

New immigrants who want to enter the city, join the queue, from which, in their turn, they 'try' to find a house in the city at the next time step. Inhabitants who decided to change their home in the city but were not successful doing so, remain in their houses.

A typical city simulation game starts with a small area occupied by individuals whose original cultural intention is randomly assigned as Blue or Green. At each time-step a constant number of individuals enter the city. Their initial cultural identities and implied intentions are also either Blue or Green. The initial status and tendency of the immigrants are assigned randomly and independently according to given normal distributions with

$$m = m(O_A), s = s(O_A); \tag{7.2}$$

that is, the mean and standard deviation of initial economic status and tendency could differ for individuals of different origin.

7.3.2 Spatial Cognitive Dissonance and Its Implications

As discussed above, spatial cognitive dissonance arises when an individual actually behaves (i.e. is located in a certain house) counter his/her intentions.

Let us define cultural cognitive dissonance for an individual A occupying house H as

$$\Delta C_A = C_{\text{neighbors}} - C_A \tag{7.3}$$

and his "economic" dissonance as

$$\Delta S_A = S_{\text{neighbors}} - S_A \tag{7.4}$$

where

$$C_{\text{neighbors}} = \sum_B \{C_B \mid B \text{ occupies } G \in U(H), G \neq H\}/N,$$

$$S_{\text{neighbors}} = \sum_B \{S_B \mid B \text{ occupies } G \in U(H), G \neq H\}/N,$$

and $N = \sum_B \{1 \mid B \text{ occupies } G \in U(H), G \neq H\}$ is the number of A's neighbors.

$C_{\text{neighbors}}$ and $S_{\text{neighbors}}$ are the mean intention and mean status of the neighbors. Below we use the distribution of $C_{\text{neighbors}}$ as a cultural image of the city.

The individuals in the queue try to locate in a house that satisfies their intentions. That is to say, the lower the values of $|\Delta C_A|$ and $|\Delta S_A|$ for some cell/house H are, the higher is the chance for an individual A to locate at H.

Consider, for example, a Blue individual who is segregative by intention and is located in a house whose neighbors (in the 5×5 square around it) are all Green. This individual will try to resolve this tension between intention and behavior by leaving his/her house and buying or renting another house in the city, or, by leaving the city and joining the queue, or by altogether leaving the system. This involves the sequence of decisions (Fig. 7.3), which differs from the decisions of the individual in City-1 in one important respect – the ability to change cultural identity.

The formal rules that govern the individual's decision in City-2 are close to those for City-1 regarding the economic component. The cultural component is accounted now in a different manner; thus we follow, in short, the formulas that describe the stages of the individual's decision in Fig. 7.3.

1. To leave the house?
The expression for the probability for an individual A to leave house H is similar to the one we use in City-1 simulations. The difference is in explicit usage of the cultural and economic dissonance:

$$R_A = \alpha_1 + \beta_1 \exp(\gamma_1 |\Delta C_A|) + \beta_2 \exp(\gamma_2 |\Delta S_A|). \tag{7.5}$$

As usual, we suppose that R_A increases monotonically with the increase of $|\Delta C_A|$ and $|\Delta S_A|$ – see Fig. 7.4.

2. To occupy a new house?
This condition is applied to queuing individuals who want to enter the city as well as to residents who want to change their houses. First, according to economic limitations identical to those of City-1, an individual A can occupy house H of value V^H if his/her status and tendency are sufficiently high relative to the value of the house. An individual A chooses a house from the set of vacant houses satisfying the conditions:

$$\begin{aligned} S_A &> \delta_0 + \delta_1 V^H \\ T_A &> \tau_0 + \tau_1 V^H \end{aligned} \tag{7.6}$$

where δ_0, δ_1, τ_0, τ_1 are constants.

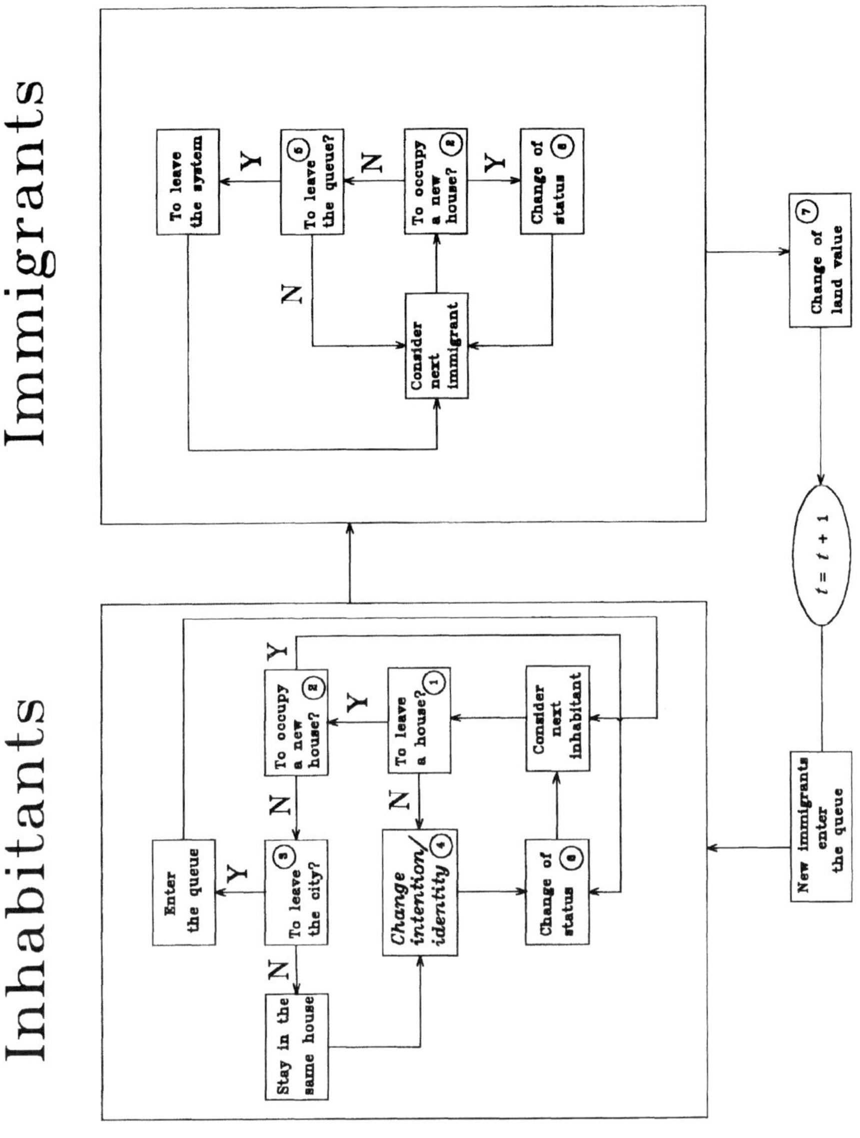

Fig. 7.3. The flow chart of City-2

The 'attractivity' of a house H for an individual A is estimated as follows:

$$Q_A^H = \alpha_3 + \beta_3 \exp(\gamma_3 |\Delta C_A|) + \beta_4 \exp(\gamma_4 |\Delta S_A|). \tag{7.7}$$

We suppose that Q_A^H monotonically increases with a decrease in $|\Delta C_A|$ and $|\Delta S_A|$, and interpret 'attractivity' as the probability to occupy a vacant house H when it is the only possible choice (Fig. 7.4). The process of the choice between several attractive houses is described in Chap. 6.

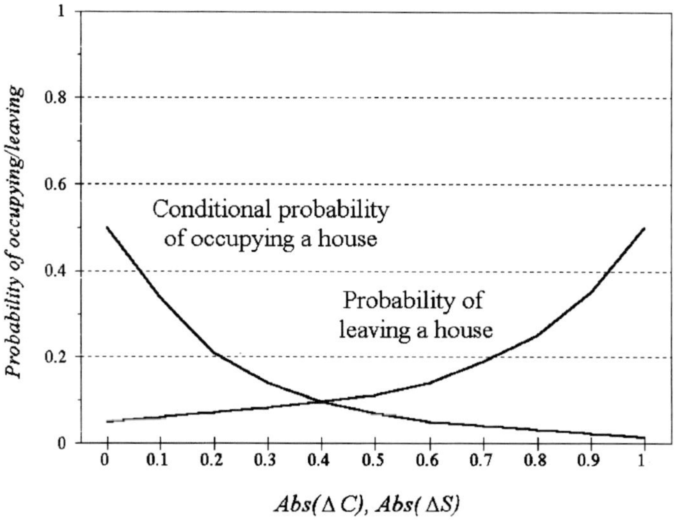

Fig. 7.4. Repulsion and attractivity functions in City-2

3. To leave the city?
An inhabitant, who tried to move to a new house and didn't succeed, will either continue to occupy the current house (with its high cognitive dissonance), or will leave the city and join the queue. We suggest that the probability to leave the city for an individual during an unsuccessful search is constant:

$$L = \text{const.} \tag{7.8}$$

An additional reason to leave the city is an economic one. If an individual's status at a current time step is below some threshold level $S_{\text{threshold}}$,

$$S_A(t) < S_{\text{threshold}} = \text{const}, \tag{7.9}$$

he/she leaves the city and joins the queue.

4. The change of intentions
If, for whatever reason, the above attempts to resolve cognitive dissonance fail, the individual stays at his current cell/house and the tension between intention and behavior persists, the result would be a change in that individual's intention: the individual will become either more segregative, or, more neutral. For a segregative Blue individual that would mean becoming 'more Blue' than before in his/her intentions, i.e. more segregative, or 'less Blue', and more neutral, in the sense that he/she will become more indifferent as to his/her neighbors. The change of cultural intention for individual A located at house H is calculated in the following way:

$$C_A(t+1) = C_A(t) + m\Delta C_A(t), \tag{7.10}$$

where m is a common sensitivity of the model individuals to a cognitive dissonance.

7.3.3 In the Queue

As noted, the queue is composed of new immigrants who have not as yet found a house in the city. (In reality you can think of them as living temporarily in hotels, friends' places, and so on.) At each time-step every individual in the queue examines his/her situation and takes a sequence of decisions: 'to buy a house', 'to rent a house', 'to wait in the queue', 'to leave the queue' (Fig. 7.3). The first two decisions are calculated as defined above. The last two decisions are calculated in the following way:

5. To wait in, or leave the queue?
An individual entering the queue is trying to occupy a house at each time-step. If he/she didn't succeed in occupying a house during some predetermined time-step $T_{\text{threshold}}$ he/she leaves the system.

A new immigrant who failed to occupy a house and didn't leave the system returns to the queue.

7.3.4 Further Structural Changes in the City

The above sequences of decisions entail the above cultural structural changes associated with cognitive dissonance and in addition some structural changes associated with the economic properties of the individuals and the cultural and economic properties of the housing stock itself (i.e. the landscape of cells/houses). The following structural changes take place:

6. Change of status
Simultaneously with the above activities regarding spatial location, the individual's economic status changes at every time-step. The status, at the next time-step, of individual P located at house H, depends on his/her current status, tendency and payment in the following manner

$$S_A(t+1) = \begin{cases} \max\{0, S_A(t)+T_A(t)-W^H(t)\} & \text{if } T_A(t)-W^H(t) < 0 \\ \min\{1, S_A(t)+T_A(t)-W^H(t)\} & \text{if } T_A(t)-WH(t) > 0 \end{cases} \tag{7.11}$$

where index t denotes the number of time-step and $S_A(t)$, $T_A(t)$ are as defined above.

We define the payment $W^H(t)$ for the house H as the right parts of the second of (7.6):

$$W^H(t) = \tau_0 + \tau_1 V^H(t). \tag{7.12}$$

Queuing individuals have no payment and consequently the changes in their status are determined as follows:

$$S_A(t+1) = \min\{1, S_A(t) + T_A(t)\}.$$

7. Change of value
The value of a vacant house H at the next time-step is defined as the average of the status of individuals occupying houses or the value of the vacant houses in the neighborhood $U(H)$. That is

$$V^H(t+1) = \tag{7.13}$$
$$D\big(S_{\text{neighbors}}(t) + S_A(t) + \sum_G V^G(t) \mid G \in U(H), G \text{ unoccupied}\}\big)/25$$

where D – const < 1 is a decrement of value.

Queuing individuals do not change their cultural affiliation.

8. Change of cultural image of houses
The cultural image of the house H occupied by an individual A at time step t is equal to the value of $C_{\text{neighbors}}(t)$. It changes according to the changes of the cultural intentions $C_A(t)$ of the individuals in the occupied cells/houses of $U(H)$.

7.3.5 Socio–spatial Emergence

Spatial cognitive dissonance can result in changes of the individual's intentions and location preferences. An originally Blue or Green segregative might gradually change his/her cultural intention and become a Blue or a Green neutral. A neutral is defined below as an individual with $1/3 \leq C_A \leq 2/3$, a Blue segregative as an individual with $0 \leq C_A < 1/3$, and a Green segregative as an individuals with $2/3 < C_A \leq 1$. For neutral individuals $|\Delta C_A| \leq 2/3$ for all possible neighborhoods, thus the repulsion function (7.5) weakly depends on $|\Delta C_A|$. This can be seen in Fig. 7.4.

As defined above, from the locational point of view the level of segregativeness–neutrality forms one of the cultural ties between the individual and his/her original cultural group. Being neutral thus implies weaker cultural ties. Now, a neutral individual might find himself in, or in between, the following three spatial situations: among neighbors of his original cultural group, among neighbors of the other cultural group, and among neutrals of the two groups. Whereas the first two situations will not affect the cultural composition of the city (Green or Blue individual), the third situation might do exactly this – it might give rise to a new cultural identity: Blue or Green neutrals, spatially located among neutral neighbors (in the 5×5 cells around them), for a sufficient period of time, might acquire a new cultural identity and become Red. In the present model this will happen if, at the same time, most neutrals in the city are in a similar locational situation. More formally, a neutral individual will acquire a new cultural identity and become *Red*, if two conditions hold simultaneously: first, the individual neutral must have a majority of neutrals in the 5×5 neighborhood around his/her house; second, the majority of the neutral individuals in the city as a whole must be in a similar situation.

We use the Lieberson (1981) isolation index LS_X in order to estimate the level of segregation in the city. In terms of Lieberson's index the above condition means that LS_X for neutrals is sufficiently high. In the discussion below we fix LS_X at 0.6 with the implication that the Red cultural entity will emerge when LS_X for neutrals exceeds this value, that is to say, in neighborhoods where at least 60% of the neighbors are also neutrals.

7.3.6 City-2 as a Heuristic–Hermeneutic Model

In the introduction we have suggested that City-2 be visualized as a heuristic–hermeneutic model. Heuristic, in the sense that its aim is not to achieve the "best fit" with reality, but to serve as a means to discover various socio–spatial properties of the city as a self-organizing system. This discovery is done, at least partly, by means of hermeneutics, that is, 'free' phenomenological interpretation which is 'evidenced by the senses' (The Concise Oxford Dictionary 1956, p. 803). On hermeneutics as a method and philosophy see Levi (1986); Bleicher (1980). From this perspective City-2 is a hermeneutic model in three respects: First, the model itself, as constructed above, is a phenomenological interpretation of reality. Second, City-2 aims to visualize the city as a self-organizing system and it thus exhibits phenomena of non-linearity and randomness. As a consequence, each run, or game, of City-2 produces a different socio–spatial scenario and configuration. That is to say, a different 'text', or interpretation of the possibilities enfolded in a given situation. Third, these scenarios, or socio–spatial configurations, provide the 'text' for us to interpret – an interpretation which is, at least in part, phenomenological and evidenced by the senses (e.g. by observing various socio-spatial configurations as they evolve on the screen). As could be seen above and as will be further seen below, the theory of synergetics plays a central role in this process of interpretation. As we elaborate below, we employ here what in Chap. 13 we define as the 'conceptual–hermeneutic approach' to synergetics, which does not use its mathematical formalism, but rather its conceptual–hermeneutic framework.

7.3.7 The Interpretation Screen

Every simulation game on City-2 implies a recursive process by which at each time-step the city changes in line with the rules of the game as defined in the above model, and in line with the various parameters as set at the start of the game. In order to fix the parameters in City-2, as in City-1 we work with a 'starter screen' on which, at the start of the game, the player can determine, or change some parameters. On the 'interpretation screen' the player can observe various aspects of the city as it evolves. The interpretation screen we have used in this study is illustrated in Fig. 7.5.

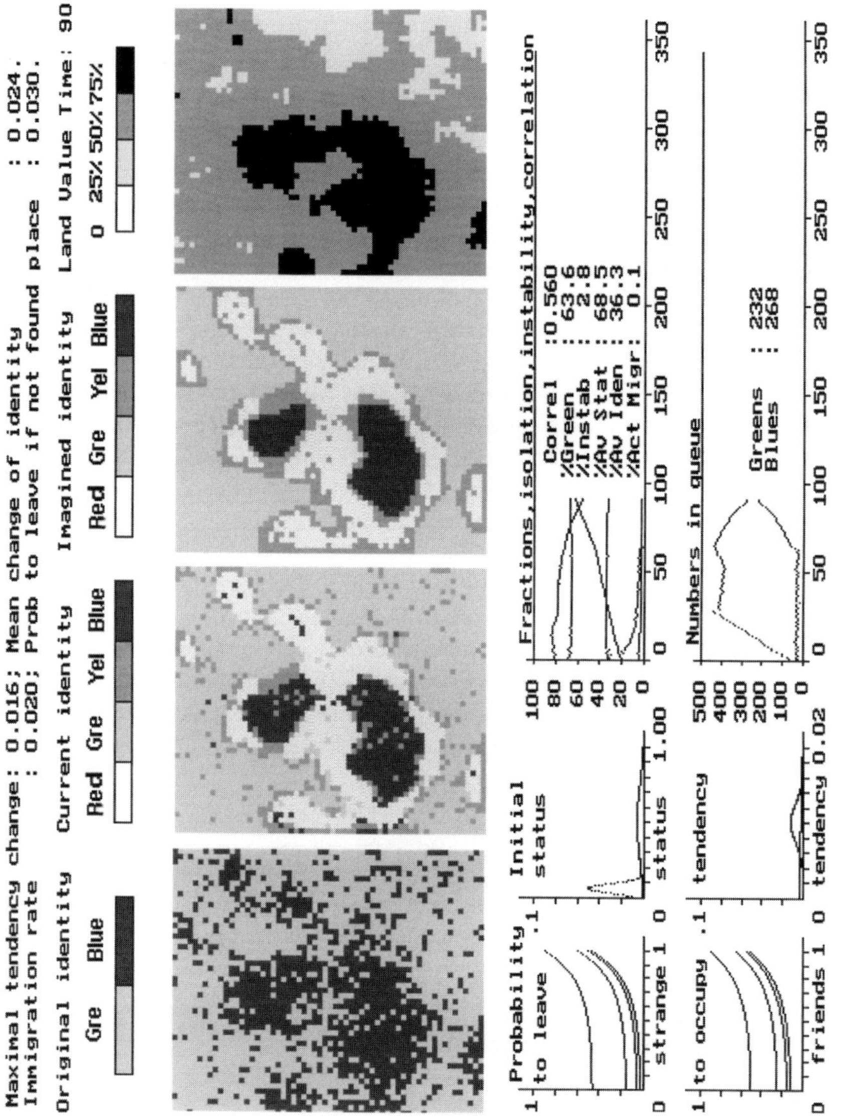

Fig. 7.5. The interpretation screen of City-2. Green and Blue are two cultural groups

For the purpose of this study we present on the interpretation screen four maps. From left to right they are original cultural identity of the inhabitants, current identity (of the inhabitants), cultural image (of the houses/cells) and economic value (of houses/cells). It is important to note that the above maps represent our choice as to what is specifically relevant and important to the issues discussed in this chapter.

7.4 Results and Interpretations

City-2 was so designed that the forces responsible for its dynamics are partly socio-cultural and partly economic; in fact City-2 was specifically designed to illustrate the conjunction between the two. As was just noted, we leave the full discussion on this conjunction for a subsequent study, and present, in this section, results which bear directly on spatial cognitive dissonance as resulting from the gap between intention and behavior, its relation to the emergence of a new socio–spatial entity, and the implications to residential segregation in the city. To study the above phenomena we've examined on the interpretation screen quite a number of scenarios, or simulation games, of the city's evolution. In the following we'll discuss scenarios which differ in two parameters: in the probability of individuals leaving the city [L in (7.8)], and, in the mean sensitivity m of individuals to a situation of spatial cognitive dissonance [see (7.10)]. We present and interpret here two series of runs: one with very low probability to leave the city ($L = 0.01$), and the other with high probability ($L = 1$). In each of the two series we consider six different values of sensitivity m, varying from 0.01, 0.02, ... to 0.06, and denoted as $L1, L2, \ldots, L6$, for the low probability of leaving scenarios, and as $H1, H2, \ldots, H6$, for the high probability of leaving scenarios. We run each scenario for 100 iterations and fix the threshold for transforming a neutral Blue or Green into Red, i.e. the Lieberson index LS_X, at 0.6.

Consider Figs. 7.6, 7.7, which represent several snapshots from the evolution of scenarios $H3$ and $L3$, respectively. In both figures the second map from the left shows the changes in intentions and cultural identities as they actually evolve in space and time. To enable identifying the neutrals as they evolve, we have colored them Yellow. Comparing the maps in the two figures it can be seen how, as a consequence of the urban dynamics, in both scenarios the intentions of a growing number of Blue and Green individuals are gradually being transformed from segregatives to neutrals. However, while in both scenarios neutrals are created, it is only in Fig. 7.7 that some of them change their cultural identity and become Reds. That is to say, only in scenario $L3$, where the level of internal migration is low, neutrals become spatially exposed to other neutrals to such an extent that their Lieberson index LS_X crosses the threshold of 0.6 and a new socio–cultural entity – the Red – emerges in a process of self-organization.

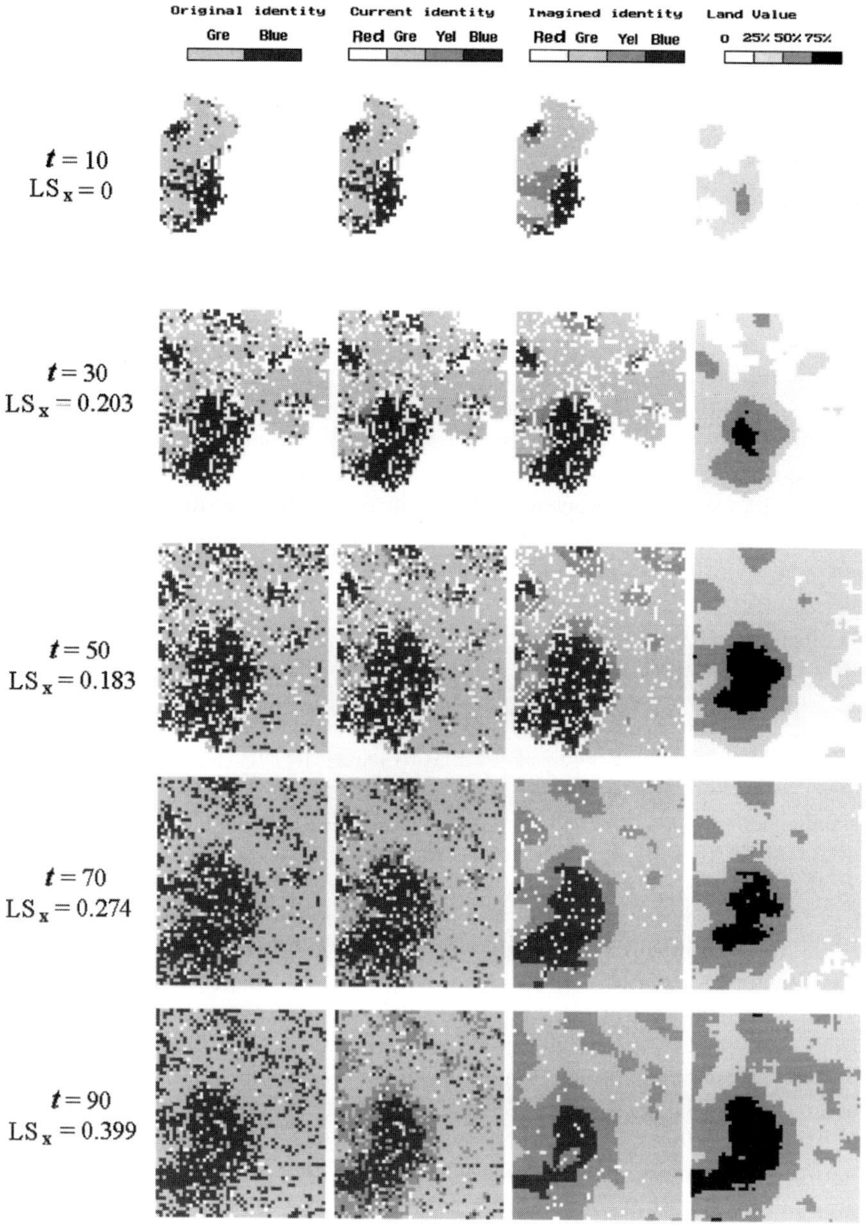

Fig. 7.6. Several snapshots from the evolution of scenario $H3$. Green and Blue are two cultural groups; Yellow symbolizes Green or Blue neutrals (see text)

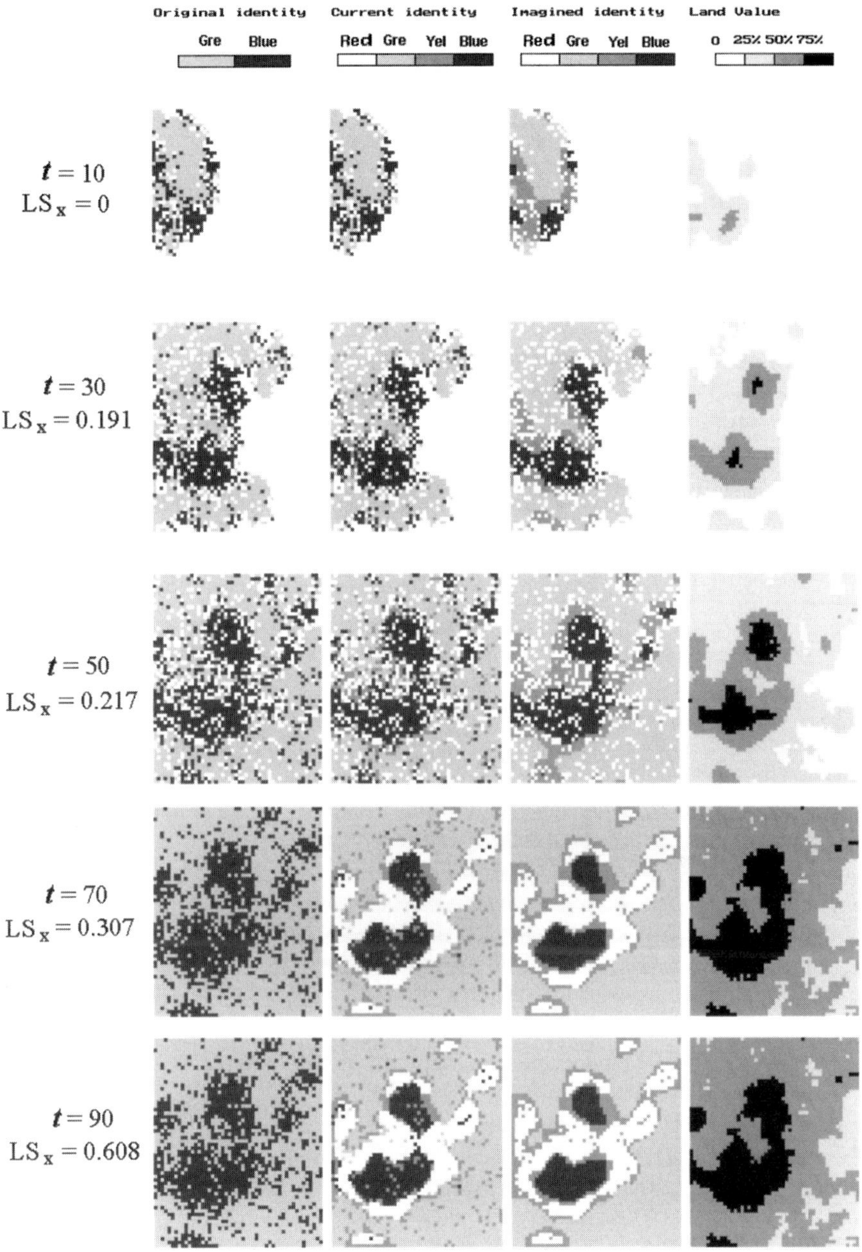

Fig. 7.7. Several snapshots from the evolution of scenario $L3$. Green, Blue and Red are three cultural groups; Yellow symbolizes Green or Blue neutrals (see text)

7.4.1 Spatial Dialectics

This phenomenon by which socio–cultural spatial segregation leads to spatial cognitive dissonance and then to the emergence of a new socio-spatial identity has been termed by Portugali *spatial dialectics*. In a study on *Implicate Relations: Society and Space in the Israeli-Palestinian Conflict*, Portugali (1993) has shown that spatial dialectics was, and still is, an important force behind the emergence and space–time diffusion of nationalism as a social order parameter, that it played an important role in the creation of the Palestinian national identity, as well as in the emergence of the Israeli Arabs as a distinct socio-spatial entity. Similar results, on an intra-urban scale, are currently emerging from an ongoing study in Adgami – the Arab community of Tel-Aviv-Jaffo: local Arab inhabitants of Adgami tend to strongly dissociate themselves from other Palestinian Arabs living in their neighborhood, despite the fact that they are all Palestinian Arabs.

A most interesting aspect of the above process of spatial dialectics and the emergence of the Reds as a distinct socio–spatial entity concerns the space–time morphology of this change. First, observing the evolving maps in Fig. 7.7 it can be seen that the appearance of the Reds is abrupt – suddenly at about $t = 80$ they can be seen in the city. Observing the process from $t = 0$ to $t = 80$ we know that this dramatic appearance was preceded by a relatively slow and long process during which, as a consequence of spatial cognitive dissonance, Blue and Green segregatives have been transformed into neutrals (i.e. Yellows), thus paving the way to the visual bifurcation that followed. From the graph in Fig. 7.8 we learn that the process was not linear: there was one jump at about $t = 10$ when neutrals first appear, and another one (apparently the one associated with the emergence of the Reds) at about $t = 60$.

The nonlinear and bifurcative form of the process is even more distinguished in Fig. 7.9 describing the time evolution of the Lieberson indices associated with scenarios $L1$, $L6$, $H1$ and $H6$. Again we see that the $L1$, $H1$ scenarios do not take off, but the $L6$, $H6$ scenarios do – two jumps or a combined sequence of two S-shape curves can be observed: the first curve starts at about $t = 5$ when neutrals first appear, speeds up to $t = 30$ from which it slows down and doesn't change much until $t = 60$. Then a second S-shape process begins, speeds up, crosses the threshold of LS_X, just to stabilize and become flat again from $t = 70$ onward. Whereas the first half of the process is associated with the emergence of neutrals, the second half describes the emergence of the Reds.

Further insight into the relations between L, m and LS_X, can be gained from Figs. 7.10, 7.11. From Fig. 7.10 it can be seen that, within the limit of 100 iterations, the H-trajectories reach the LS_X threshold later than their corresponding L-trajectories. In Fig. 7.11 we can see the possible implications of different configurations of L and m, given that LS_X is 0.6. As can be seen, there is a domain where neutrals are created but do not self-organize and

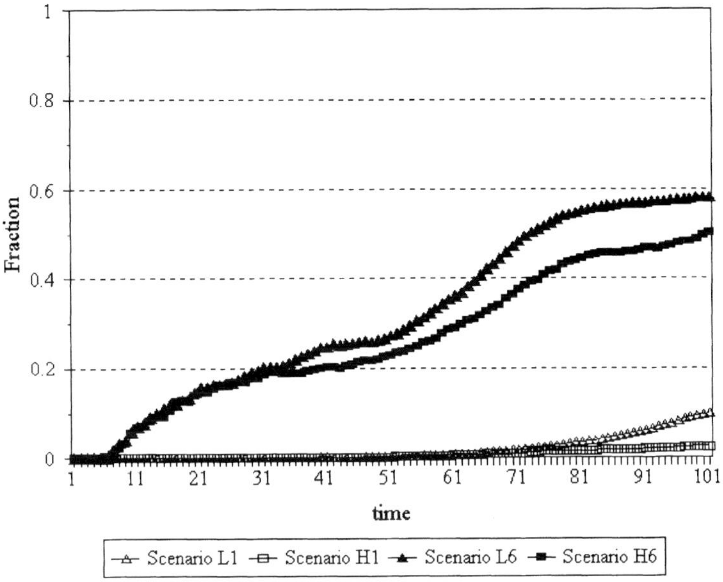

Fig. 7.8. Growing fraction of neutrals in selected scenarios

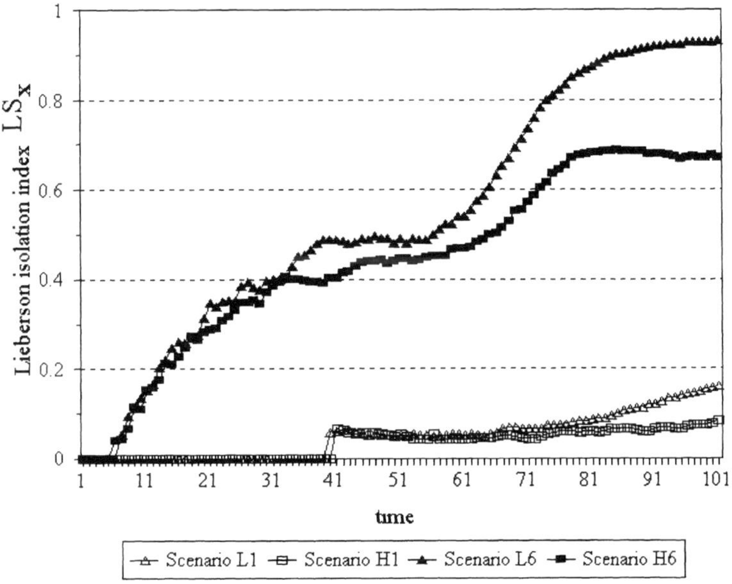

Fig. 7.9. The time evolution of the Leiberson index for selected scenarios

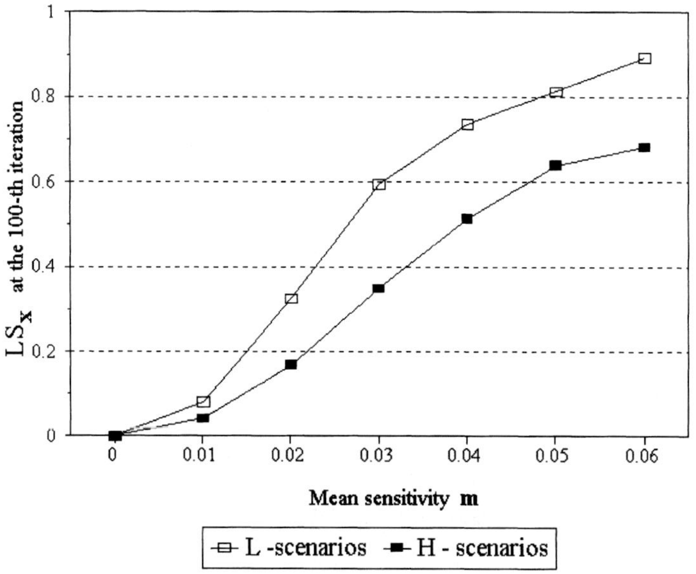

Fig. 7.10. Varying Lieberson values for changing mean sensitivity, at the 100th iteration

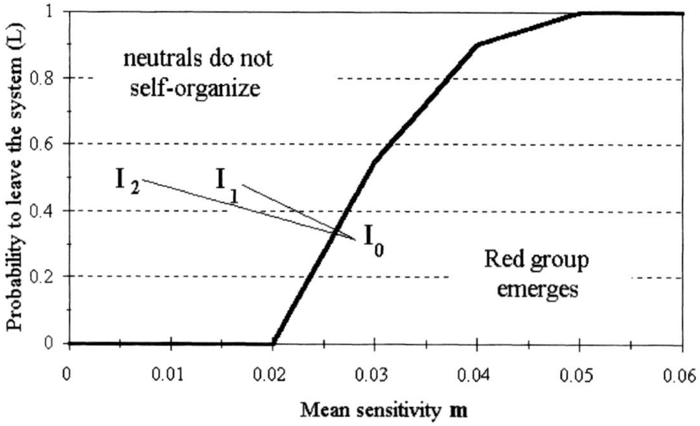

Fig. 7.11. The domain where neutrals self-organize and transform into Reds, and the domain where they do not, in the phase-space of the probability of leaving the system and the mean sensitivity

transform into Reds, and an area where they do. This phase-space diagram can be used as an instrument to generate hypotheses regarding changes in the values of L and/or m. For example, what would happen if the city as evolved above in Fig. 7.7, became subject to a large wave of segregative Green and Blue individuals? The hypothesis generated by Fig. 7.11 is that the city will move, or rather reverse, from position I_0 to position I_1 or I_2, depending on the number and composition of the immigrants. That is to say, the result might be the elimination of the newly created Red cultural group. Given the complexity of the model and the processes it simulates the answer to this hypothesis is not obvious and will have to be found by experimenting with City-2. This is a task we will have to leave for a subsequent study.

The above conditions by which neutrals change their identity and become Red implies that while cognitive dissonance alone can transform the intention of individuals, it is only in conjunction with spatial dialectics that a new socio–spatial entity can emerge. This spatial dimension of the process can be clearly observed in Figs. 7.6, 7.7: it can be seen that while neutral individuals are created and spatially distributed all over the city, the process is more intense on boundaries between the homogeneous Green and Blue areas. The areas in between – the boundaries – are thus the most critical areas in the city for socio–spatial changes (and in society at large). Similar results were obtained by Batty (personal communication) in his studies on Fractal Cities, and also in in Chap. 5 by means of our model City. In the latter we have devised a graphic description (and an index) representing the city as a surface of stable and unstable cells (see Figs. 5.2, 5.7, Chap. 5). As can be seen in Figs. 5.2, 5.7, the boundary areas are the unstable areas and they are also the areas where the gap between intention and behavior is at its highest level. In fact, similar results were obtained in the present study, using City-2, with the addition that it is in these boundary areas – 'the areas in between' – that socio–spatial bifurcation has a high probability to occur and a new socio-spatial entity to emerge.

Now compare the above discussion with the time–space evolution of the third map from the left in Fig. 7.7 – the (imagined) cultural identity of the cells: It can be seen that substantial areas in the city become neutral (i.e. Yellow), some time before the inhabitants of these areas become neutrals (i.e. Yellows) and then change their identity and become Red. Our interpretation is that this is the collective cognitive map representing the spatio–cultural image of the city. The map is collective, in the sense that it is as if the city as a whole 'cognates' these areas as potentially neutral and Red. That is, in the city a house or a larger area might acquire a certain cultural identity, say Red, even if the inhabitants of the house or some of the inhabitants of the area have a different cultural identity (Green or Blue). This is a consequence of a complex interplay between the spatial organization of the Reds, in relation to the other cultural groups and the city as whole. This situation characterized what happened in Adgami – the Arab quarter of Tel-Aviv-Yaffo: for several years during the 1960s and 1970s this sub-quarter of the city was known as 'the Arab neighborhood', despite the fact that until 1976 the majority of the population in that area were Jewish and not Arab (Portugali 1993, Fig. 7.2). From the location point of view these maps can also be interpreted as representing 'expectations'. Indeed, in the above empirical case of Adgami this imagined, or expected, character of the neighborhood played an important role in the location decisions of individuals and in the subsequent evolution of the area.

We now turn in Fig. 7.7 to the first map from the left – the original cultural identity. The interesting phenomenon observed here is that despite the above noted changes in the socio–spatial and cultural composition and configuration of the city, a Blue–Green socio–cultural spatial segregation is maintained. It is as if the city as an entity "remembers" and preserves the old, original, cultural divisions. This is, if you wish, another collective cognitive map, now representing the long-term memory of the city. The importance of this property of the city – that it preserves past socio–spatial and cultural divisions – cannot be exaggerated. The recent re-emergence, in the ex-USSR or Yugoslavia, of old boundaries and ethno–nationalistic conflicts, in cities and regions, just illustrates what has been identified above by means of City-2: That cities and regions enfold in them, as long term memories, the old boundaries and past divisions, and that these past socio–spatial and cultural entities are potential forces which might re-emerge.

The 4th column of maps in Fig. 7.7 shows how the land-value maps are evolving parallel to, and in connection with, the evolving socio–spatial surface of the city as discussed above. This is the only representative we show in this study of the economic variables included in City-2. Observing these maps it can be seen that the socio–spatial processes of cultural identity are intimately associated with economic processes. For example, the various games or scenarios discussed above started with relatively rich Blue and relatively poor Green individuals. The Red individuals as a new socio–spatial cultural entity, emerged out of a process by which Blue and Green segregatives turned into neutrals (Yellows). As noted, this process was specifically intensive along the boundaries between the segregated Blue and Green areas. As can be seen in the land value maps, this emerging new socio–cultural spatial entity captures 'the area in between' also in the economic sense of the words. They become the 'middle class' between the rich Blues and the poor Greens. This implies that the emerging new group has acquired not only new socio–cultural spatial properties and identity, but also spatio–socio–economic identity and status. This interplay between the spatio–cultural and the spatio–economic is rather interesting and we hope to be able to further develop this topic in future studies.

7.4.2 Internal Versus External Spatial Information and Cultural Emergence

In this section we further elaborate on the question of socio–cultural emergence: our aim is to investigate the conditions by which a new cultural group can emerge in City-2. In particular, we focus here on the way individual mobility L is related to the emergence of a new cultural group – the Red people. Note that the higher is a individual's mobility the higher is also the individual's ability to respond to the incoming flow of external and internal information and thus to resolve cognitive dissonance.

In addition to description of segregation by means of Lieberson index LS_X, we describe the time evolution of the fraction of Neutrals in City-2 by the dependence of $\Delta N(t) = N(t+1) - N(t)$ on $N(t)$ for $m = 0.03$.

As shown in Fig. 7.12, for each value of L the fraction of the Neutrals in City-2 converges to equilibrium. The resulting fraction of the Neutrals remains high (above 35%) irrespective of change in the mobility from 0.1 to 0.9: at $L = 0.1$ we have close to 100% Neutrals, at $L = 0.5$ the fraction of neutrals stabilizes around 60–70%, whereas at $L = 0.9$ the stable fractions of neutrals are between 35% and 50%. We do not discuss here the fact that there exist two accumulation points for $L = 0.5$ and $L = 0.9$ in Fig. 7.12.

A high fraction of Neutrals is thus a necessary, but not a sufficient condition for the emergence of a new social group. For the value of $L = 0.9$ the fraction of neutrals is high, but they mostly occupy houses in culturally mixed neighborhoods. As a result, the value of Lieberson index LS_X remains below the threshold level of 0.6 (Fig. 7.13) and a new group does not emerge at the meso-level (see Fig. 7.14).

From Fig. 7.12 we can conclude that there exists a boundary value of L that defines the region of self-organization, and this value is above $L = 0.5$. The numerical estimation gives L approx. 0.6 (Fig. 7.15). For L below this value the Neutrals self-organize into a new, spatially continuous, group (Reds).

In Figs. 7.12–7.15 we can see the spatial implications of changes in individual mobility. At $L = 0.1$ the entire city becomes Red and the original Blue and Green groups almost disappear. This is an extreme case of a 'melting pot' situation by which extreme restrictions on free movement reduce the cultural variability in the city. With time the city population becomes one homogeneous intermediate cultural group.

At $L = 0.9$ the situation reverses. Now most, but not all, individuals can resolve cognitive dissonance by moving to a new location or by leaving the city. The city is now highly pluralistic in the sense that there are randomly appearing and disappearing local neighborhoods of specific cultural identity, while most of the individuals of the various groups live side by side in culturally mixed areas.

$L = 0.5$ is typical of a culturally segregated city. The city has produced a new cultural group (the Reds) and this newly born group segregates in its own area. The whole city is highly segregated and its territory is spatially divided between the three main cultural groups – the previously existing Blues and Greens and the newly self-organized Reds. The Neutrals are now spatially distributed along the boundaries between the large segregated areas. From the above follows, that spatio–cultural emergence and segregation are inherent properties of the city dynamics.

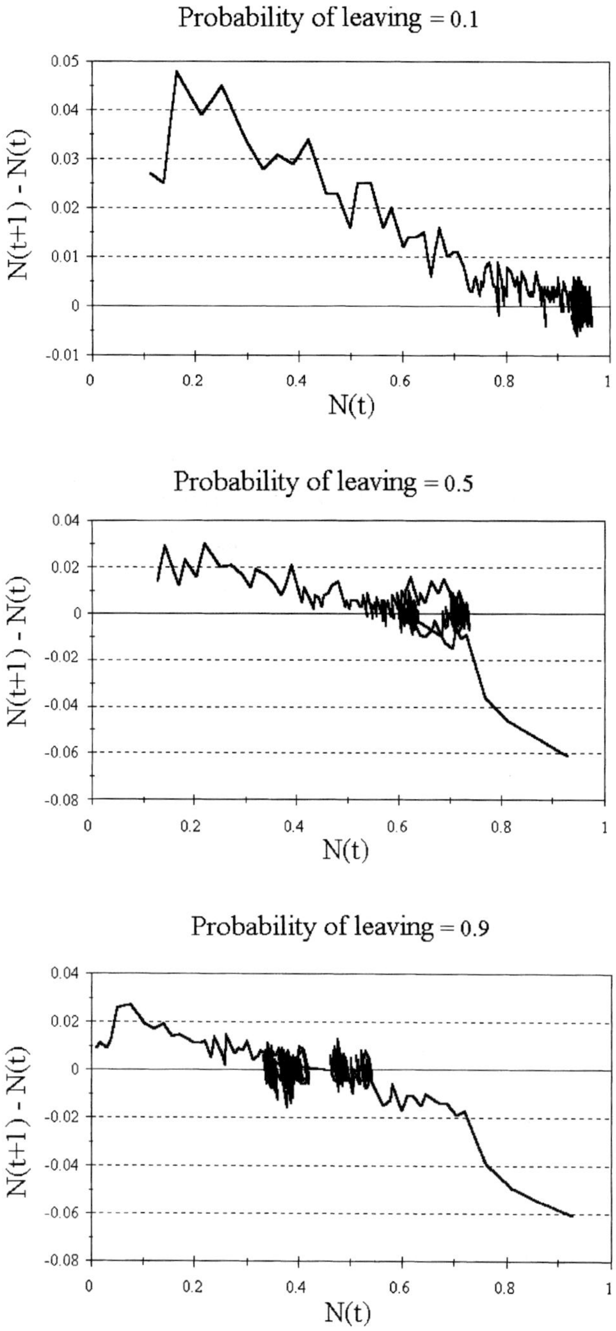

Fig. 7.12. Dependence of $N(t+1) - N(t)$ on $N(t)$

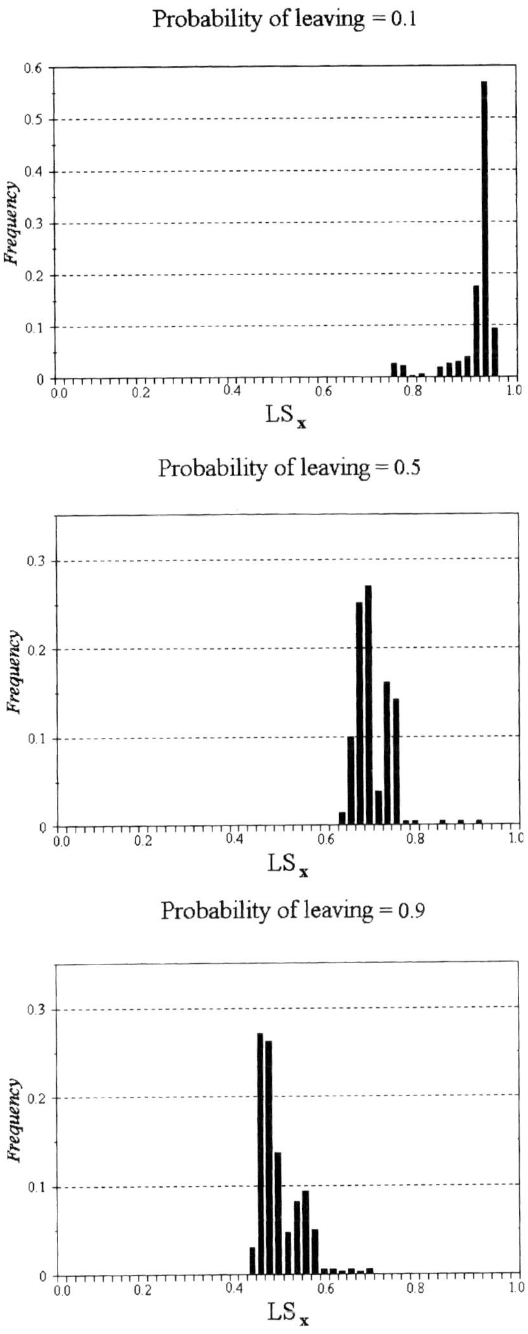

Fig. 7.13. Distribution of LS_X for the fraction of Neutrals near equilibrium

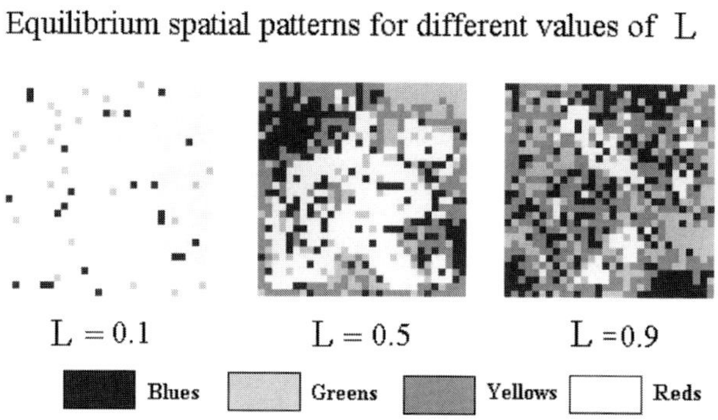

Fig. 7.14. The effect of individuals' mobility on spatial cultural segregation in the city

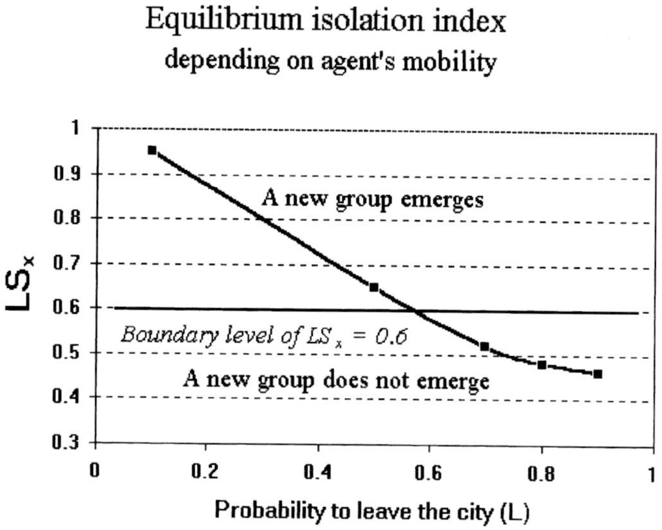

Fig. 7.15. The effect of individuals' mobility on spatial cultural emergence

7.4.3 City-3: Human Agents Between the Local and the Global

In Chap. 4, Sects. 4.1.3–4.1.5, we have discussed at some length the nature of internal and external information, in conjunction with local–global relations which are central to self-organization in general and as such to self-organizing cities. In the previous sections of the present chapter we have theoretically justified the notion of spatial cognitive dissonance, operationally established the possibility to model this process, and investigated some of the conditions by which spatial cognitive dissonance emerges and entails the birth of a new cultural group. In the model we present below, we explicitly define global external information and examine its role in the process of cognitive dissonance and the emergence of a new socio–spatial entity.

In City-3, in addition to the economic status and cultural identity, the internal information of individual A includes A's sensitivity L_A to the local cognitive dissonance and sensitivity G_A to the global city structure. G_A is the main feature that differentiates the agent of City-3 from the agent of City-2; it enables us to explicitly investigate the role of global external information in the evolution of cognitive dissonance and the subsequent emergence (or not) of a new cultural group.

To describe the action of the global city population structure on the cultural identity of the individual, we suggest that a spatially segregated group, whose members' cultural identity equals the original identity of an individual A, influences the tendency of A to accept the identity of the local neighbors. We use the Lieberson (1981) isolation index $LS_{X,Y}$ to characterize the level of segregation of a certain group X relative another group Y:

$$LS_{X,Y} = 1 - \sum_i (x_i/N_x)(y_i/t_i) \tag{7.14}$$

where N_x is the total number of agents of group X in the model city, x_i and y_i indicate the number of agents of groups X and Y within neighborhood $U_i(H)$ and t_i is the total population of the $U_i(H)$. The sum in (7.14) is taken over the coverage of the City by the lattice of 5×5 neighborhoods $U_i(H)$.

In most situations we consider the level of segregation of a given group relative all the other individuals taken together, provided that $t_i = x_i + y_i$. In this case we denote the index as LS_X.

In terms of the Lieberson index the above two-fold condition means that relative to agents in the other states LS_X for neutrals is sufficiently high. In the discussion below we fix LS_X on 0.6 with the implication that the Red cultural entity will emerge first, when LS_X for neutrals will pass this value. Second, when at least, 60% or more of the neighbors of the neutrals are themselves neutrals.

Using LSX as an index of segregation, we extend (7.10) to:

$$\begin{aligned} C_A(t+1) = C_A(t) - L_A \Delta C_A(t) + G_A(O_A - C_A(t)) \\ \max\{(LS_X(t) - LS_{\text{threshold}}), 0\} \end{aligned} \tag{7.15}$$

where x denotes a group of individuals whose original cultural identity equals that of A.

The value of $LS_{\text{threshold}}$ is introduced to reflect the visual impression of segregation. From experience we can say that a pattern looks segregated when LS_X equals 0.5 and above, thus, we take 0.5 as the $LS_{\text{threshold}}$.

7.4.3.1 Results and Interpretation

In order to study the influence of local and global forces in the city we consider the following basic scenario:

Original cultural identity of individuals is either Blue or Green. For Blues $O_A = 0$, for Greens $O_A = 1$. Current cultural identity of the individual that enters the city for the first time equals his/her original identity. Initial distributions of S_A and T_A are normal, truncated on $[0, 1]$, initial distributions of L_A and G_A are uniform on $[0, L_m]$, $[0, G_m]$, $L_m < 1$, $G_m < 1$, $L = 0.5$.

According to (7.15), current cultural identity of an individual can vary in time. As in City-2 we say that current cultural identity of an individual A is Blue when $C_A < 1/3$ and Green when $C_A > 2/3$. Individuals, whose current cultural identity lies within the interval $[1/3, 2/3]$ are defined as Neutrals and are marked Yellow.

We run each scenario on a 50×50 lattice. Initially, a small number of Blue and Green individuals are randomly located at the center of the model city. At each time step a constant number of external immigrants enter the queue and, consequently, try to settle in the city. All the parameters of the model for Blues and Greens are identical.

The typical outcome of a basic scenario, that accounts for local relations only ($G_m = 0$ for all individuals), is a fast spatial segregation of Blues and Greens after about twenty iterations (Fig. 7.16, first row). Afterwards, the cultural identity of the individuals, occupying houses within the mixed neighborhoods (mainly along the self-organizing boundary between Blues and Greens), is changing, and Neutrals start to appear (Figs. 7.16, 7.17). The increase in the fraction of Neutrals is accompanied by their segregation at the boundary between Blues and Greens, thus minimizing the overall cultural gaps. After about 200 iterations, the Neutrals as a group become segregated – a situation that can be observed quantitatively, by the value of LS_X (Fig. 7.17). In the long run the system tends towards equilibrium with only Neutrals all over the city (Fig. 7.16, 7.17); and it can be seen from the simulations and proved on the basis of (7.14).

We say that Neutrals become a new socio–cultural group of Reds when three conditions are simultaneously satisfied at different levels of the system organization:

1. Individual level – a certain number of individuals are transformed into Neutrals.
2. Local level – most of Neutrals are located within Neutral neighborhoods.
3. Global level – Neutrals are spatially segregated.

Below in addition to Figs. 7.16, 7.17 we present the dynamics of the city structure in the phase space of two Lieberson indices (Fig. 7.18). The first is LS_X for the group of Blue individuals, and the second for the group of Neutrals. For each L_m there exist two critical values of G_m, that define three intervals of qualitatively different stable outcomes of the model runs. For low G_m, all the individuals in the city become Neutrals (Fig. 7.18a). With the increase of G_m, three segregated groups – Blues, Greens and newly emerging Reds – coexist (Fig. 7.18b). For higher G_m the fraction of Yellow individuals

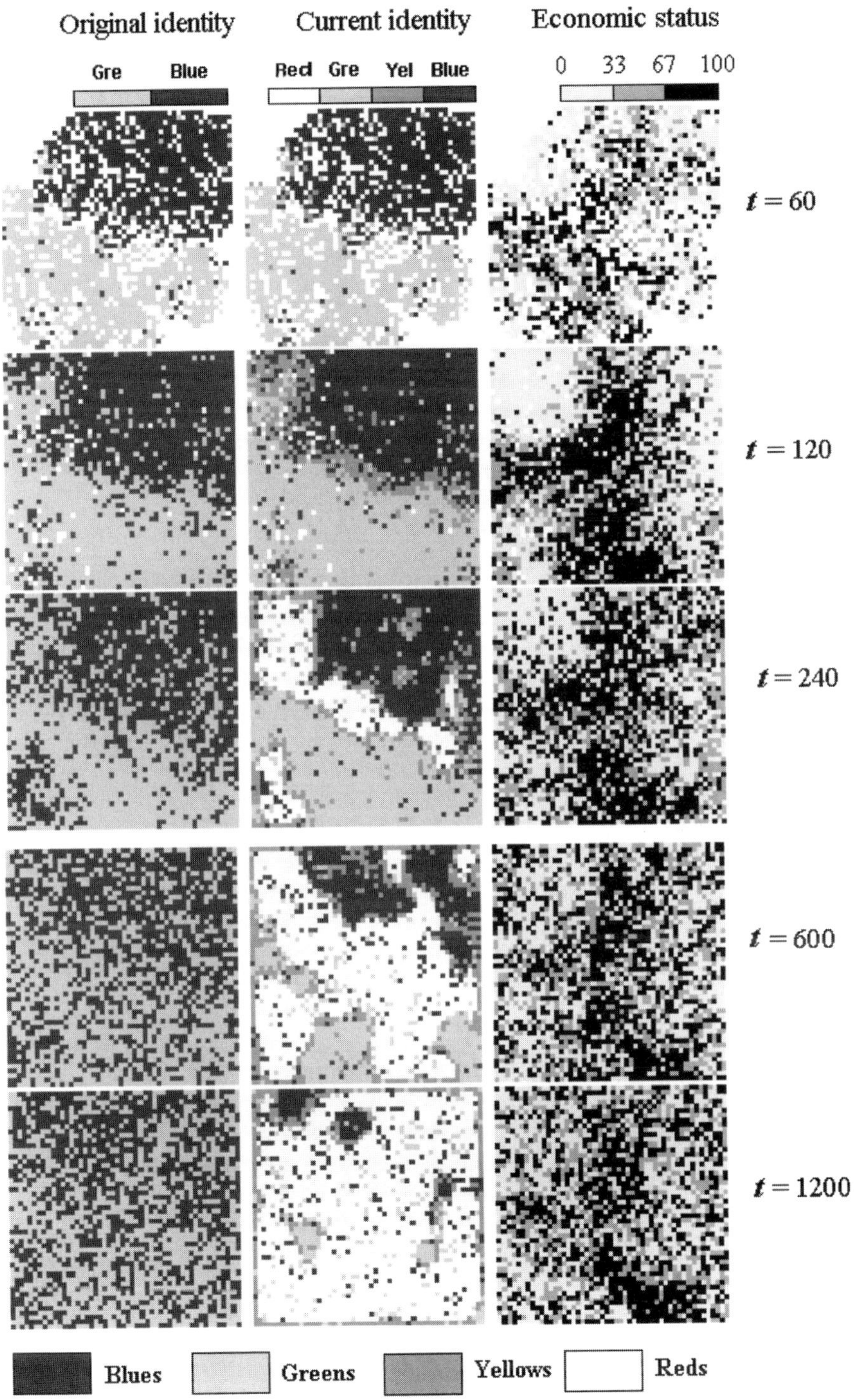

Fig. 7.16. Five snapshots at a typical evolution of City-3

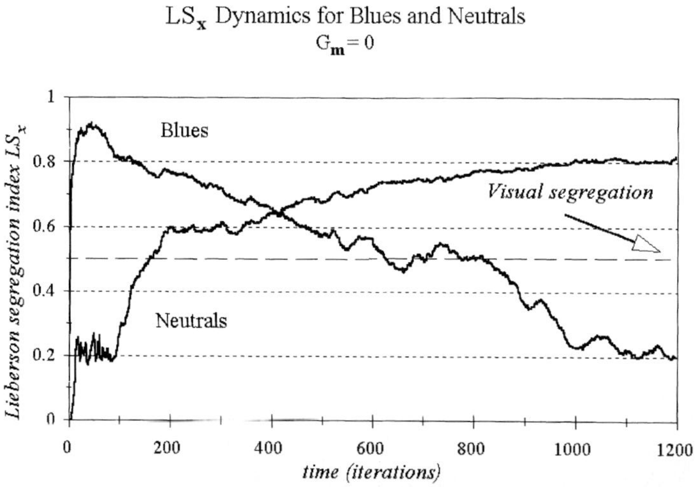

Fig. 7.17. Time evolution of spatial cultural segregation of Blues and Neutrals in the evolving City-3

remains low, and the Red group does not emerge (Figs. 7.18c, d). These results coincide with the analytical investigation of (7.15) that shows that Blue and Green cultural identities are preserved for sufficiently high G_A.

The above results seem to be in line with the empirical findings noted above in Sect. 7.2.3 and Fig. 7.2. According to the latter, after some 17 years of living together, the distribution of attitudes of the Jewish inhabitants of Adgami (the Arab neighborhood of Tel-Aviv-Yaffo) towards their Arab neighbors becomes bimodal: most people were found to be either positive or else negative. In the context of Sect. 7.2.3 and Fig. 7.2, we have interpreted this as a possible indication to the operation of cognitive dissonance. Here we further suggest that the fact, that as the time passes some Jewish inhabitants have become more negative toward their Arab neighborhood, might indicate the effect of the global structure of the city and society.

7.5 Conclusions

Considering the city as a self-organizing system we have shown how the dynamics of the city leads to a gap between individuals' intentions and their behavior and then to the emergence of a spatial cognitive dissonance, which in turn leads to the emergence of a new socio–cultural spatial entity. As emphasized throughout the discussion, we have concentrated in this chapter mainly on the cultural and social dimensions of the process and their properties. In particular the economic forces which were active in the various scenarios above have not as yet been examined and discussed, nor their conjunction with socio–cultural processes.

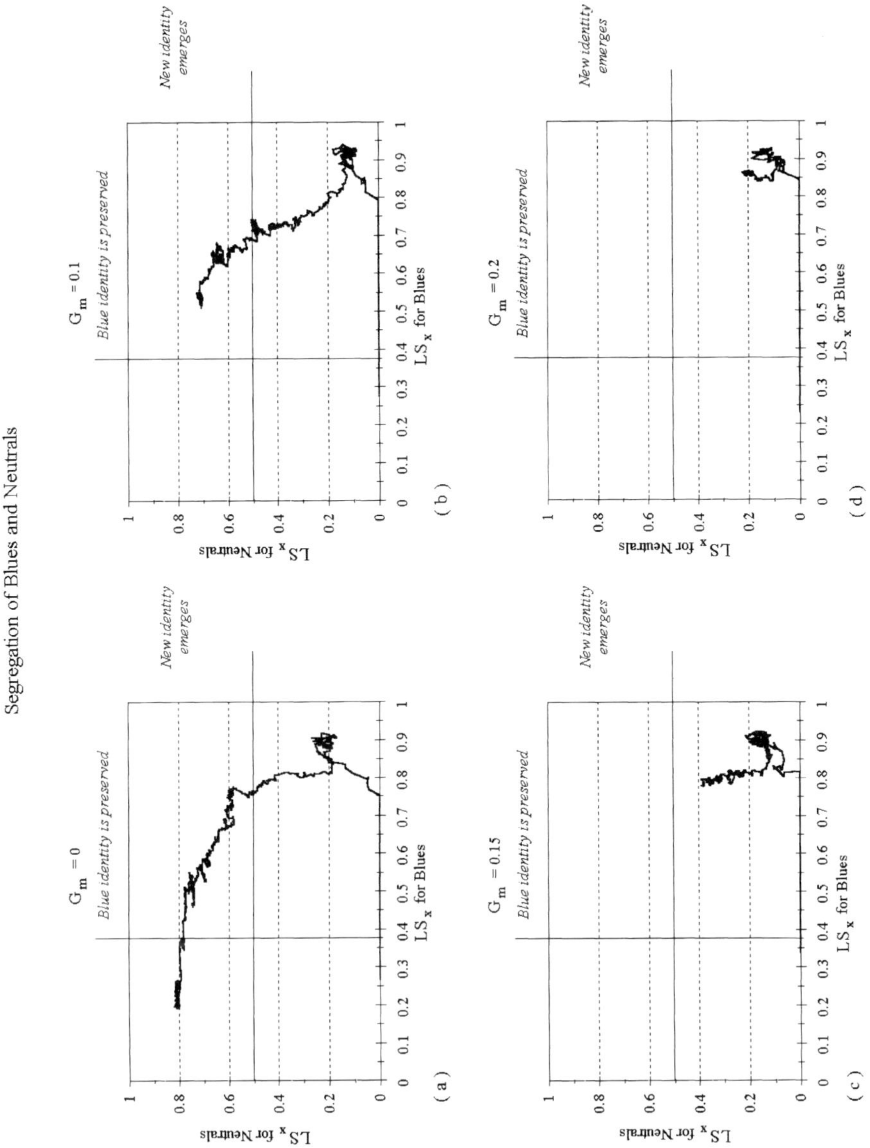

Fig. 7.18. City dynamics in the phase-space of LS_X for Blue and Neutral individuals, depending on sensitivity to the global cultural structure of the city ($L_m = 0.05$)

8. Individuals' Cultural Code and Residential Self-Organization in the City*

8.1 Introduction

In previous chapters we presented the individual agents in a somewhat oversimplified manner. Their 'personal' properties (economic status and cultural identity) were defined as one-dimensional quantitative variables. Each individual agent was defined, first, along a one-dimensional economic scale as rich, poor, or some intermediate degree in between the two poles. Second, and in a similar way, each individual was defined also by a cultural measure as segregative Blue, on the one pole, segregative Green on the other, or Neutral in some degree i.e. Yellow in between. While such a uni-dimensional quantitative description might be appropriate with respect to the economic state of agents, it is by no means acceptable in the case of their cultural properties. Cultural groups, and even more so the cultural properties, tendencies and affiliations of individual agents, cannot be stretched along a single, uni-dimensional, continuum. Rather, in reality, the cultural properties of individuals and cultural groups form complex, multi-dimensional family resemblance structures, not very different from the categories and cities discussed in Chap. 1 (see further discussion below).

In this chapter we make a preliminary attempt to take account of the complexity of the cultural variable and to integrate it to the family of City models developed above. We call this new variant model City-4. In the latter, each individual agent is defined by, and thus comes to the city with, a personal cultural code, reminiscent in its nature of a genetic code. The genetic code is employed here as a metaphor. From the theory of genetics we learn that each human individual enters the world with an inherited genetic code, which affects, among other things, his or her potential to behave and interact with other individuals. Taking the genetic code as a metaphor we suggest that the above noted cultural code defines the potential of that individual to interact with other individuals and locations in the city, and in the process to create cultural groups.

* With Itzhak Benenson

8.2 The Memetic Code

To be sure, the metaphoric connection between genes and cultural traits that we propose here is not new and was advocated by several writers in the past. A rather amplified voice in this direction, is Dawkins (1976) in his *The Selfish Gene*. The central thesis of this 'gene-eye-view' on nature and evolution is that at the core of Darwinian evolution stand not whole animals, which are usually selfish but sometimes also altruistic, but the genes. According to Dawkins, the latter are the only biological entities that always tend to replicate themselves and as such are genuinely 'selfish'. 'Selfish genes' do not interact directly with each other, however, but indirectly through their phenotypes. In a second book – *The Extended Phenotype* – Dawkins (1982) further suggests that, regarding animals, the notion of phenotype should include in addition to the immediate bodily properties, also some of their products such as the bird's nest, the beaver's dam and the spider's web.

Towards the end of his *Selfish Gene*, Dawkins speculates with the idea that Darwin's theory of biological evolution enfolds the principle for a general theory of evolution, biological or otherwise. The fundamental evolutionary principle as he writes, is

> that all life evolves by the differential survival of replicating entities. The gene, the DNA molecule, happens to be the replicating entity that prevails on our own planet. There may be others. If there are, provided certain other conditions are met, they will almost inevitably tend to become the basis for an evolutionary process (Dawkins 1986, p. 192).

Are there such new replicators? Yes, answers Dawkins, "a new kind of replicator has recently emerged ... still in its infancy" (ibid), it is the *meme* – the replicator at the core of human cultural evolution. The memes are new replicators that only recently have appeared on the stage of the world, as a by-product of biological evolution, and are already moving fast with their own independent evolutionary process.

Examining the various examples of Dawkins' memes, it can be seen that they are, in fact, concepts, categories, cultural traits, schemata, generally held ideas, and the like. Common to all those entities and notions is that they are all public – all were selected and accepted by culture or society. The latter implies that in order to qualify as a meme, a personal idea, trait etc. must 'go public' – it must be publicly and culturally accepted. A case in point is the 'meme' which was originally Dawkins' personal idea, and has proved to be culturally very successful and became itself a meme: it has invaded many brains and has resided there as an internal representation, and has invaded the *New Oxford English Dictionary*, and has resided there as an external representation.

Dawkins' meme and extended phenotype were recently subject to an enthusiastic re-interpretation by Dennett (1991) in his book *Consciousness ex-*

plained (see also Dennett 1996). Dennett's main concern in this book is consciousness, but in the way to explain it, he elaborates also on the meaning of the *Self*. According to Dennett the *self* is the Homo Sapiens's center of gravity of its extended phenotype. Like the individual spiders who protect themselves by making a web, we humans create a Self:

> Out of the brain it (the Homo Sapiens) spins a web ... this web protects it [the person] ... stripped of it, an individual human being is as incomplete as a bird without its feathers, a turtle without its shell (Dennett 1991, p. 416).

Each person thus builds a web, which is the person's cultural and social extended phenotype. This web, then, becomes the medium through which the memes that have invaded and occupied that person's brain externalize themselves in public. Writing from the perspective of our FACS–IRN cities (above Chaps. 2, 3 and below Chap. 14) Portugali has suggested to add two points to Dennett's self. First, that internal and external representations (mimetic, lexical and artificial) are part of humans' extended phenotype, and as such part of the web that defines the human self. Second, that the innermost intimate element of humans – the one that makes the Self of each of us, is defined by means of our extended phenotype. That is to say, by means of our interaction with the environment around us, which includes our clothes, cars, houses, friends, neighbors, the neighborhoods we live in, our cities, countries and so on. And so, a typical answer to the question 'who are you?' or 'what is your identity?', might be, 'I'm an Israeli', 'Parisian', 'New-Yorker', 'citizen of the world', etc. My/Your individuality is defined by means of your connection to the world.

The preliminary theory and model that we propose here provides a conjunction between, on the one hand, the notion of self-organization with the family of City-models we've developed above, and on the other, Dawkins' and Dennett's memes, extended phenotype, and Self. The process we model below starts with a set of concepts, categories, cultural traits, schemata, generally held ideas, in short, memes, internally represented in an agent's memory and defining its cultural identity (i.e. Self). This set is termed *memetic-* or *m-code*. In our model, these personal memes do not interact directly, but only indirectly through the interactions between the agents who carry them. In the model these are the interactions between the free agents in their attempt to find an appropriate house in the city. These interactions give rise to cultural groups, which are groups of agents with identical or similar m-codes (see discussion below). Each group can thus be also defined by its collective m-code. This grouping process is the mechanism through which memes are environmentally selected, or in other words, 'replicate themselves'. This is so because when an agent becomes a member of a cultural group, the memes that compose its m-code, enter the group's m-code (or pool of memes). If, on the other hand, the agent lives in a 'foreign' neighborhood, it might either preserve its previous m-code (Self), or else acquire a new one – by means of spatial cognitive dissonance, for example. In the latter case some of the memes that compose this agent's m-code will disappear. This is thus a typical process of circular causality by which the the personal m-codes give rise to the group's one, which then determines the Selves, and so on.

In terms of synergetics the above can be described as follows: the interaction between individual agents with their m-codes gives rise to a few order parameters (cultural groups) which then, by various means, including spatial cognitive dissonance, re-define, i.e. enslave, the individual m-codes. In this process some memes survive and replicate, other disappear, and so on in a process of circular causality.

8.3 The Model

City-4, the model we construct below elaborates on our previous City models. It differs from past formulations in two important respects. First, in its definition of the cultural identity of the agents – this is the novel feature we study in this chapter. Second, and as a consequence of the above, in that the cultural groups are not pre-determined (as Blue/Green), but from the start of the game they emerge out of the dynamics that takes place in the city. Similar to the model of Chap. 7 it consists of two interacting layers – *an infrastructure* sub-model, which is an extension of cellular automata and represents the dynamics of the city's physical structure, and a sub-model of *free human agents*, which describes the migratory movements of individuals.

8.3.1 The Definition of the m-Code

In genetics, as well in studies of artificial life, it is common to represent the individual's genotype by means of a high-dimensional binary vector (Banzhaf 1994). Below, we define the m-code of an individual agent in the same manner. As emphasized in the previous chapters, we suggest that human agents are characterized by their ability to vary, and, consequently, self-organize, in line with the dynamics and evolution of the system they belong to. Here too, the m-code of an agent and its residential behavior can change through its interaction with its neighbors, neighborhood, and the city as a whole.

In City-4, the m-code of an individual A is described by the K-dimensional Boolean vector $C_A = (c_{A,1}, c_{A,2}, c_{A,3}, \ldots, c_{A,K})$, where $c_{A,k} \in \{0, 1\}$, $k = 1, 2, 3, \ldots, K$. As a consequence, individuals A and B have different identities when vectors C_A and C_B differ in at least one component. Quantitatively we measure this by the difference ρ between A's and B's identities

$$\rho(C_A, C_B) = \sum_k (c_{A,k} \text{ XOR } c_{B,k})/K. \tag{8.1}$$

8.3.2 Cultural Groups

Given the definition of individuals' m-codes as above, there are three basic ways to derive cultural groups. One is to say that a cultural group is a collection of individuals with *identical* cultural codes. Second, to say that it is a collection of individuals with *similar* cultural codes, and third, with family resemblance as their grouping process (see definition in Chap. 1). These three ways refer to three grouping principles, which might be termed the *identity*, *similarity* and *family-resemblance* principles. It can be demonstrated that in the spatial context of our model where the location decisions of individuals are taken at the local scale (5×5 cells) of neighborhood, the principle of family resemblance becomes a special case of the principle of similarity.

According to the first grouping principle there can be a maximum of 2^K different cultural identities in the city. In the case of the second and third principles, the number and nature of the groups depend on the similarity measure according to which individuals can join the group.

In City-4 (as in City-3), we explicitly consider the interplay between local and global information. The representation of *local cultural information* is related to the notion of *local spatial cognitive dissonance* of a free agent. Applying the general definition to the multidimensional m-codes of agents, we define local spatial cognitive dissonance CD_A of agent A, occupying house H_{ij}, as an average of the differences between A's identity and the identities of his neighbors:

$$CD_A^t = \sum_B \{\rho(C_A^t, C_B^t) \mid B \text{ occupies } H_{kl} \in U(H_{ij}), H_{kl} \neq H_{ij}\} / \\ (N_{\text{oc}}^t(U(H_{ij})) - 1), \qquad (8.2)$$

where $N_{\text{oc}}^t(U(H_{ij}))$ is the number of occupied houses in $U(H_{ij})$.

The role of global information is treated in the same way: If individuals similar to A in their m-codes are segregated in the city then beyond a certain threshold their spatial distribution might affect the behavior of A. For this purpose we define the *global cultural information* GD_A, available to agent A, about residential segregation of individual agents of identity C_A. As in previous chapters we use the Lieberson (1981) segregation index $LS_{X,Y}$ to characterize the level of segregation of a certain group X relative to the other group Y. $LS_{X,Y}$ is a probability for individual A that belongs to group X and located at house H, to meet a member of group Y within $U(H)$. The complete information on the residential segregation in the city at iteration t is given by the $2^K \times 2^K$ matrix of Lieberson segregation indices $LS_{X,Y}^t$ for each pair of cultural identities (X, Y). To reduce the enormous dimensions of this description we suggest below that agent A's behavior depends on the global level of segregation of its cultural group relative to all the other individuals *taken together*, and denote the corresponding value of Lieberson index as LS_A^t. The dimension of the latter description equals the number of identities, i.e. 2^K. The values of LS_A below 0.2 corresponds to visually random distribution of agents of identity C_A, while the values above 0.8 correspond to one or several domains occupied by these individuals almost exclusively. Quantitatively, we describe the global cultural information an agent A accounts for as:

$$GD^t_A = \frac{\max\{0, (LS^t_A - LS^*)\}}{1 - LS^*}. \tag{8.3}$$

Here LS^* is the value of Lieberson index that corresponds to a visually segregated pattern, and below we set LS^* equal to 0.4.

We suppose that local and global information influence the agent's cultural identity in opposing ways. High local cognitive dissonance CD^t_A pushes agent A to change its cultural identity. A's reaction to local cognitive dissonance is described in the model by sensitivity $L_A \in [0, 1]$. A high level of segregation of individual agents of identity C_A, per contra, pushes A to preserve its current identity and A's reaction to global segregation is defined by sensitivity $G_A \in [0, 1]$. In the following we assume that L_A and G_A are inherent properties of A and do not depend on t.

The change in an agent's cultural identity thus depends on two opposing tendencies. The cultural identity of an agent A can be changed when the local tendency to change an identity exceeds the global tendency to preserve it, i.e. when $L_A \cdot CD^t_A > G_A \cdot GD^t_A$. If the latter holds, then the probability that the i-th component of C_A will be changed is proportional to the absolute value of the difference between the fraction of this component among A's neighbors and its value for A. Additionally, we introduce the possibility for a 'cultural mutation' with probability r_m per a component of identity. As a result, the probability for an agent A of identity $C_A = (c_{A,1}, c_{A,2}, \ldots, c_{A,i}, \ldots, c_{A,K})$, occupying house H_{ij}, to change its i-th component $c_{A,i}$ of C_A from unity to zero or vice versa, is

$$p_{A,i} = \max\left\{\frac{0, (L_A \cdot CD^t_A - G_A \cdot GD^t_A) \cdot (|f^t_i - c^t_{A,i}| + r_m)}{\sum_k |f^t_k - c^t_{A,k}| + r_m \cdot K}\right\}, \tag{8.4}$$

where f^t_k is a frequency of NOT $c^t_{A,k}$ in cultural identities of A's neighbors at iteration t:

$$f^t_k = \sum_B \frac{\{c^t_{B,k} \text{ AND } (\text{NOT } c^t_{A,k}) | B \text{ occupies } H_{kl} \in U(H_{ij}), H_{kl} \neq H_{ij}\}}{N^t_{\text{oc}}(U(H_{ij})) - 1} \tag{8.5}$$

We suppose that only one component of cultural identity can be changed at a time-step.

8.3.3 Model Dynamics: Trade Off Between Migration and Individual's Change

According to the flow-chart (Fig. 8.1), at every iteration, each agent A in the city decides whether to move from, or to stay at, its present location. As shown in Fig. 8.2, with an increase in the cultural dissonance CD_A (8.2) the probability to leave a house increases monotonically, and the probability to occupy a new house decreases monotonically.

The conjunction between local and global factors can lead agent A to decide to continue to occupy house H in spite of high economic tension and cultural dissonance. The reason, for example, might be a lack of alternative attractive vacant houses in the city. The basic suggestion in City-4 is that in

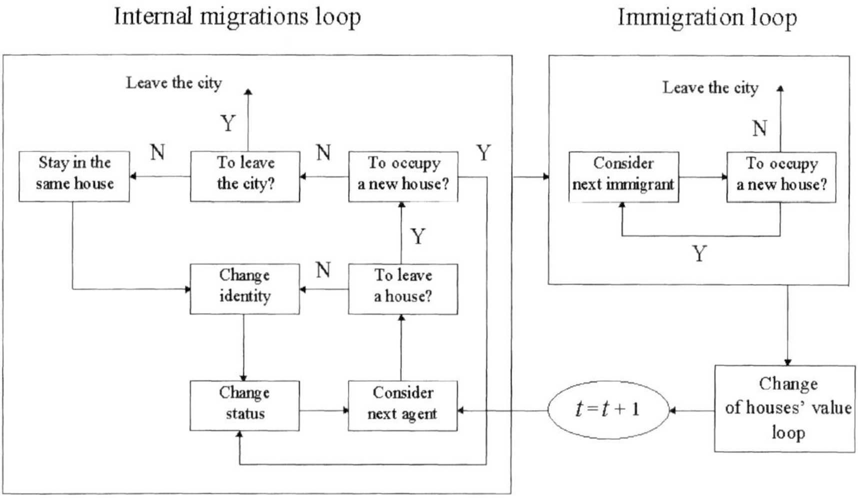

Fig. 8.1. The flow-chart of City-4

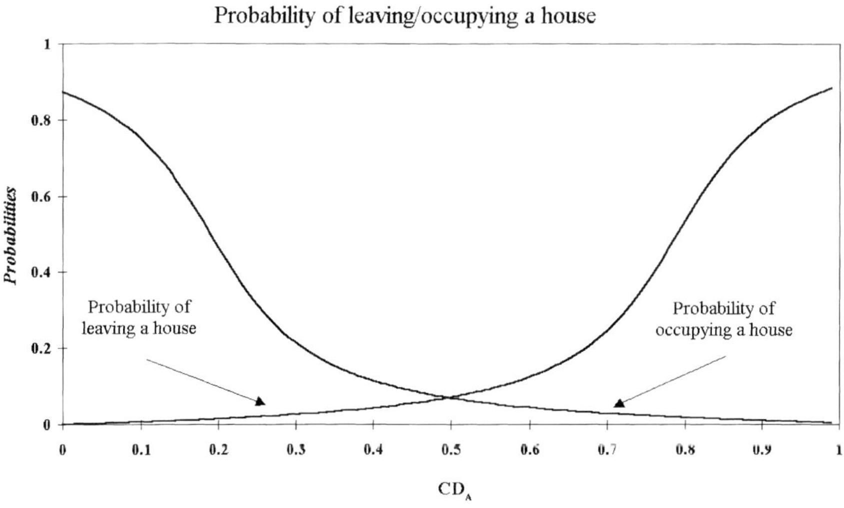

Fig. 8.2. Probability to leave or occupy a house in City-4

such a situation the dissonance is resolved either by leaving the city, or by *change in the properties of the agent itself.*

8.3.3.1 Free Agent's Behavior Under Increasing Cultural Cognitive Dissonance

An inherent source of the City-4 cultural dynamics is a mutation process that prevents it from becoming culturally homogeneous. An individual agent located in an heterogeneous neighborhood of non-zero dissonance, either succeeds in changing residence, or fails and, thus, either changes an identity towards the 'modal' identity of the neighbors (Fig. 8.1), or preserves its current identity due to a high level of segregation of agents of similar identity in the city (see (8.4)). Unlike the changes in the one-dimensional economic status, the changes of agents' cultural identity do not decrease the cultural diversity of the city when $K > 1$. As an example consider the agents located at a boundary between two segregated groups of individuals (0, 0, 0, ..., 0) and (1, 1, 1, ..., 1). According to (8.4), there is a high probability that the identity of, say, the (0, 0, 0, ..., 0)-agent will change to a new one with unity at one of the components and, thus, will differ from the identities of the agents of *both* groups. This salient consequence of a multidimensional representation of C_A determines most of the results below.

8.3.3.2 Emigration

We have stated above that an individual, whose economic status reaches zero, leaves the City. A free agent that failed to reside might (1) leave the city with probability p_U; (2) change cultural identity with the probability given by (8.4); and (3) stay at his current location and not change at all.

8.3.3.3 Immigration

At every time-step, a constant number I of individuals try to enter the city from outside and to occupy a house in it. The cultural identity of the immigrants is assigned at random, in proportion to the current fractions of agents of each of the 2^K possible identities.

8.4 Results

The aim of our model is to examine the process of socio–cultural emergence in the city, the inhabitants of which can vary in their cultural identity according to a potentially infinite number of traits. To qualify as a newly emerging socio–cultural entity, a group of individuals must fulfill simultaneously three conditions (Chap. 7). At the individual level the members of the group must have the same cultural identity, at the local level most of the group members should be located within neighborhoods of their own, and at the global level the number of group members and their spatial segregation have to be sufficiently high.

In Chaps. 5–7 above we have seen that different sets of parameters might generate three kinds of residential dynamics in City. One is a 'random' city, another is a 'homogeneous' city, in which most of the agents belong to the same group, and the third is characterized by a complex structure. All these regimes are observed in the present study too, and, below, we deal with the set of parameters that entails the most interesting 'structured' dynamics. In this chapter we are specifically interested in the question of whether the residential distribution of the individual agents in the city evolves towards a state that can be called 'persistent' in some respect and, if so, what are the characteristics of this state? In particular, what is the number and the level of segregation of the emerging socio–cultural entities? Are they fixed or do they vanish in time? What is their 'life-history'?

8.4.1 Parameter Value and Initial Conditions

The scenarios we run share the following conditions:

1. City-4 is a 40×40 lattice.
2. Initially, at $t = 0$, each cell within a circle of 3-cell diameter, located at the center of the city lattice, is randomly occupied by individuals of all-zero cultural identity $(0, 0, 0, \ldots, 0)$.
3. Immigration rate I equals 4, or 0.25% of the maximum number of the city residents – 1600.
4. Probability p_U of leaving the city, when failing to occupy a new house, equals 0.075.
5. Distributions of sensitivities L and G are uniform on $[0, 1]$. They are assigned to the agents independently of each other.
6. Mutation rate r_m is 0.02.
7. Threshold group size sufficient to recognize a group as an 'entity' is 40 individuals (enabling up to 40 different identities to exist simultaneously in the city).

At present, our computer allows us to study the system behavior when the dimension of the cultural identity vector C_A is less or equal to 5. The question of whether the case of $K = 5$ is representative of a higher-dimensional C_A, will be studied further in the future.

The results below are common for five repetitions of each scenario.

8.4.2 Presentation of the City Patterns

Figure 8.3 shows three different cultural landscapes of the evolving City-4 with three different m-codes, after some 500 iterations. In order to represent the changing properties of the city we use below three kinds of maps.

First, as in previous City models, we have here a map showing the distribution of agents' cultural identity, with each identity marked by its own color. This presentation is the most detailed one, but because of the high number and non-linear ordering of identities, it is difficult to visually inspect where $K > 2$. The second type of maps shows the difference $\rho(C_A, C_0)$ between identity C_A of agent A, occupying house H, and a pre-determined identity that equals, say, $C_0 = (0, 0, 0, \ldots, 0)$. This map shows effects which are not dependent on K. Its disadvantage is that several different identities C_A can equally differ from the one selected for comparison. The third map shows the spatial distribution of cultural cognitive dissonance of residents. This map is a surrogate to the Stability–Instability Surface (SIS) of Chap. 5, in the sense that the higher the dissonance is, the higher is the chance that the state of a given house will change.

Before proceeding to the analysis, let us point out that the dynamics of the distribution of cultural identity depends on the number K of its components. In general, an increase in K increases the 'resolution' of identity, but keeps the range of its variability. We mean here that according to (8.1) the maximal possible value of $\rho(C_A, C_B)$, i.e. difference between individual A of identity C_A and individual B of the opposing identity C_B ($C_A = (0, 0, 0, \ldots, 0)$ and $C_B = (1, 1, 1, \ldots, 1)$, for instance) remains equal to unity, no matter what K is.

8.4.3 Model Dynamics for Low-Dimensional Cultural Identity: $K = 1$ and $K = 2$

The case of $K = 1$ corresponds to our previous analysis of residential segregation between two cultural groups (Chap. 5). The city dynamics in that case entailed a fast self-organization of (0)- and (1)-identities within two or several segregated patches. The boundaries between the homogeneous patches remain the areas of instability, with intensive exchange of individuals (Fig. 8.3a, compare with Chap. 5).

When K equals two, the dynamics of the city still resemble some of our previous results (Chap. 7). At the beginning of the runs, in line with the restriction of a mutation process by one component per iteration, only (0, 1)- and (1, 0)-agents emerge. The numbers and the level of segregation of the initial (0, 0)- and of new (0, 1)- and (1, 0)-identities reach the levels satisfying the conditions of socio–cultural emergence, to $t \sim 100$, when the fraction of unoccupied locations in the city is at a level of 25%. The agents of (0, 1)- or (1, 0)-identities that change them to a (1,1) because of mutation or dissonance with the neighbors, still have the vacant houses to reside in. As a result, the (1,1) socio–cultural entity emerges in the city (Fig. 8.3b) in all of the model runs to $t \sim 400$. In parallel, the number of vacant houses tends to approach zero, and strong competition for houses turns out to be the factor that defines the survival of the entities. In general, the position and the size of the domains it occupies define the survival of a certain entity. The high

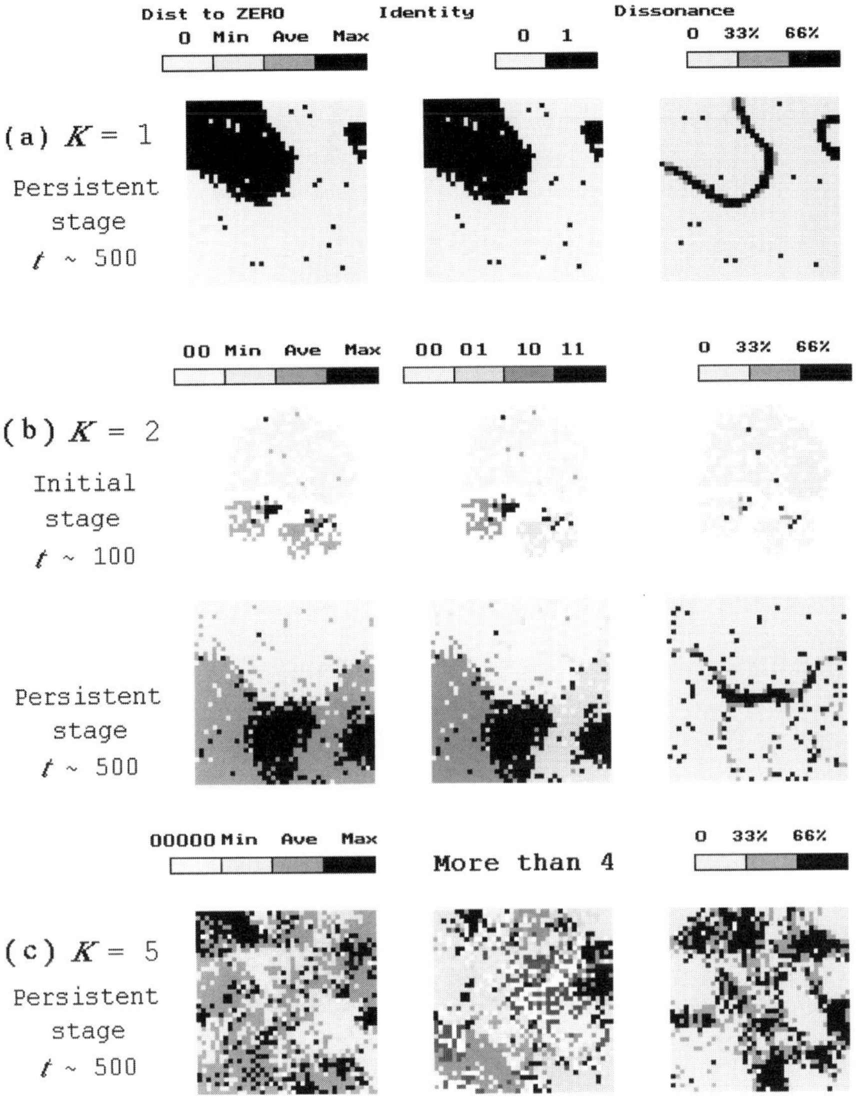

Fig. 8.3. Three different cultural landscapes of the evolving City-4, after some 500 iterations with three different definitions of the m-code: (**a**) In the first scenario $K = 1$, (**b**) in the second $K = 2$, (**c**) and in the third $K = 5$

value of the perimeter/area relation, as well as the common boundary with an opposite entity (e.g. (0, 1) for (1, 0) agents) decreases the chance that the entity will persist. As a result, in a long run (we stopped the simulations at $t = 2500$) the number of socio–cultural entities, existing simultaneously in the city for $K = 2$, fluctuates between three and four, and the life span of the entities is of the order of 500 iterations.

Let us now skip the intermediate cases of K equals 3 and 4, and proceed with $K = 5$.

8.4.4 Model Dynamics for High-Dimensional Cultural Identity: $K = 5$

8.4.4.1 Initial Stage of the Model Dynamics

The number of possible identities for this case is $2^5 = 32$. The first mutant agents belong to one of five 'close-to-zero' identities, which are characterized by unity at one of the components and zeros at the rest of them. Compared to $K = 2$, it is *not necessary* that all of them will emerge at the first stage of the city development. In the five runs we did, their number varies between two and four.

8.4.4.2 Persistent Dynamics of the City

The entities that emerge first determine the further dynamics of the city. In a way similar to the case of $K = 2$, the boundaries between two homogeneous domains (occupied by the entities that emerged at the first stage) and the heterogeneous domains, occupied by the agents of varying identities, are areas of instability. The agents located there, either leave their houses or change their identity. None of the properties of the certain socio–cultural entity currently existing in City-4 can be predicted in a long run. As a result, we cannot follow the qualitative fate of a certain identity, but still are able to understand and predict the properties of the model city as a complex self-organizing system:

1. The persistent city structure is characterized by a mixture of homogeneous domains, the population of which forms socio–cultural entities, and domains that are heterogeneous at different levels. The former cover about half of the city for $K = 5$ (Fig. 8.3c).
2. A limited number of cultural entities can exist in the city simultaneously (Fig. 8.3c, Fig. 8.4).
3. The life span of a socio–cultural entity is finite and the entities replace each other in the city space. About 20% of the entities persist in the city 11 iterations or longer and 10% persist 25 iterations or longer (Fig. 8.5).
4. The distribution of cultural differences $\rho(C_A, C_0)$ between the cultural identity C_A of agent A, and a certain 'basic' cultural identity C_0 ($C_0 = (0, 0, 0, 0, 0)$ for the maps we present here) is self-organizing as well (Fig. 8.3). This distribution has two opposing characteristics. First, in general, the difference $\rho(C_A, C_0)$ increases with the increase in the distance from the location of the agents of the C_0 identity. Second, *non-linear*

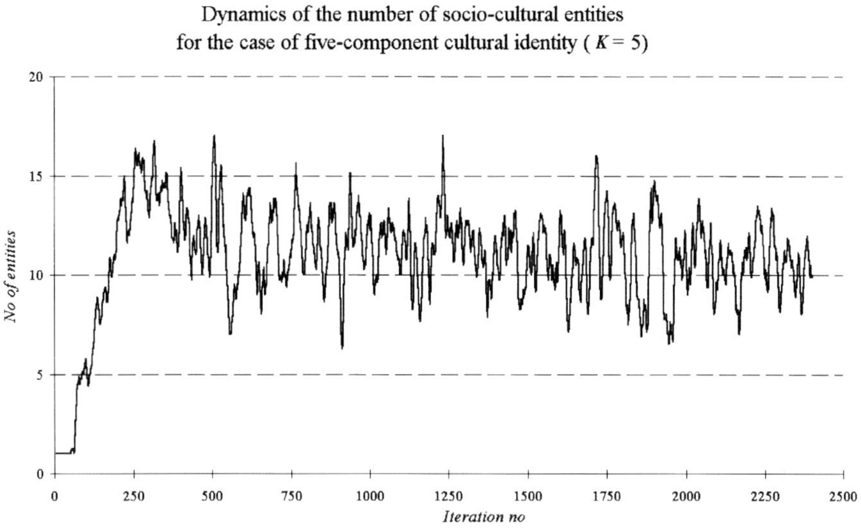

Fig. 8.4. The dynamics of the number of cultural identities in City-4 for a $K = 5$ m-code

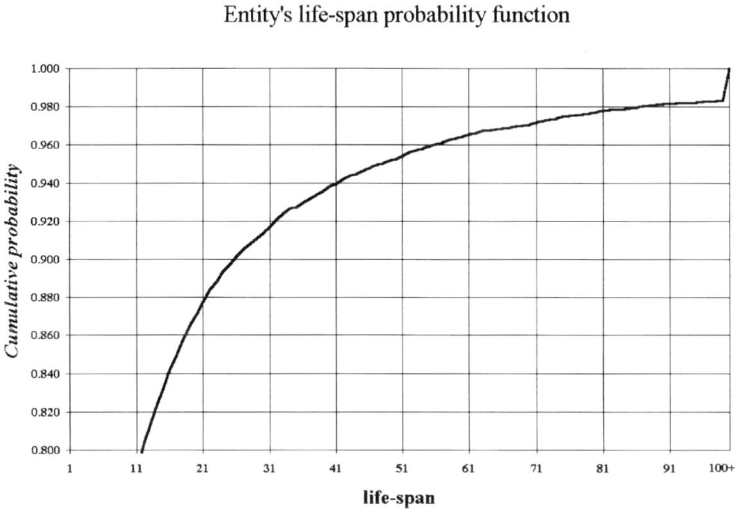

Fig. 8.5. The life-span probability of cultural entities (groups) in the evolving City-4

ordering of the identities implies the emergence of the adjacent areas of entities C_A and C_B, that equally differ from C_0 (i.e. $\rho(C_A, C_0) \sim \rho(C_B, C_0)$, but differ also among themselves (i.e. $\rho(C_A, C_B)$ is high). See, for instance, the bottom part of Fig. 8.3c, where the boundary between two similar groups domains is an area of high dissonance. This property, determined

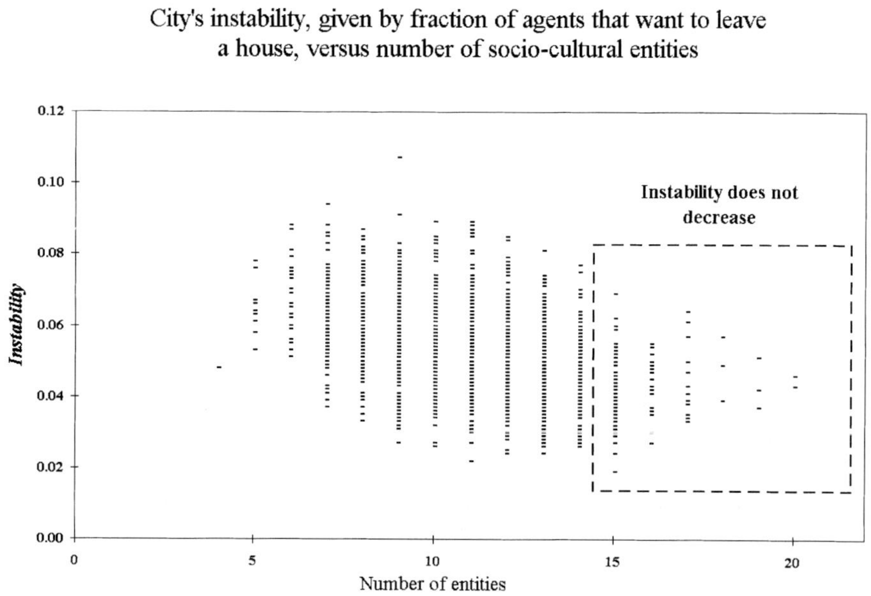

Fig. 8.6. Instability of City-4 defined by the fraction of agents that intend to leave their house, versus the number of spatio-cultural groups (entities)

by the multidimensional and quantitative nature of cultural identity of the model agent, limits the city's instability *from below*. With an increase in the segregation in the city, its instability does not converge to zero (Fig. 8.6) and several unstable zones are preserved. We can say, thus, that the city is self-organizing and evolving toward a *critical* internal structure, that preserves the ability to change.

8.5 Concluding Notes

In this chapter, we introduce the notion of m-code that describes the individual as a multidimensional and qualitative unit. From this perspective there follow three new qualitative phenomena. First is a phenomenon of recurrent self-organization, emergence and extinction of the socio–cultural groups in the city. Second, only a limited number of cultural entities (from a large number of possible ones) can exist simultaneously in the city space. Third, the city as a whole tends towards a self-organized critical state that preserves cultural instability. As a consequence, the city cultural landscape appears as a mixture of a few homogeneous domains, each one occupied by individuals of a certain socio–cultural identity, and areas of heterogeneous population. The identities of the existing socio–cultural entities and their further evolution depend on the emerging situation and cannot be predicted in advance.

One of the more interesting outcomes of this new definition of individuals by means of m-codes is to sharpen and emphasise the role of spatial dialectics in connection with the formation of cultural groups and possible changes in individuals' cultural properties (or m-codes). First, cultural groups are not pre-determined, as in previous models, but from the start they emerge out of the dynamics, that is, out of the interaction between individuals with different cultural codes as it takes place in the space of the city. This situation was typical of the cities in 'new worlds' (America, Australia, ... Israel) and is now characteristic of the cities of end 20th century. In both the 'old' and 'new' cities immigrants came, and are coming to cities as individuals or families, each with his/her own cultural background. It is here in the new city (and country), and out of its specific dynamics, that new cultural groups, and new personal identities are formed.

This new feature of City-4 enables us to open a discussion on complex interrelationships of circular causality that exist between individuals' and cultural groups' identities. In particular, first, the way cultural groups emerge out of the city dynamics and the way their internal structure and group properties are determined in the process. Second, the way, in a process of circular causality, the emerging cultural groups shape the personal identities (and m-codes) of individuals. A central issue in this context concerns the question of what constitutes a cultural group. In the above we have mentioned three criteria: *identity*, *similarity*, and *family resemblance*. In this chapter we have consider the first of the three criteria. Yet City-4, the model we have constructed here enables us to study the two other criteria. It also enables us to study related issues such as the order of the grouping process, whether it is simultaneous or sequential and the way socio-spatial cultural grouping is related to the general process of categorization. Given the preliminary nature of this chapter all these tempting and interesting issues will have to await future study and research.

9. From CA- to GIS-City*

9.1 Introduction

The above FACS models and their results were based, as we have seen, on a simple representation of city geometry: a square grid of elementary units – 'houses', each containing one agent only. The neighborhood of a house in FACS is a 5×5 square with the house at its center. The city's infrastructure is considered in the model as a three-level hierarchy of (a) houses, (b) geometrically identical overlapping neighborhoods, and (c) the city as a whole. Such a representation of urban geometry is obviously problematic and can qualitatively influence the outcomes of the simulations, especially if we want to weight them against real-world situations. In this chapter we report on our first attempt to introduce into our FACS models a real-world heterogeneity of the city's infrastructure, and to investigate the possible consequences of this new type of modeling.

9.2 The GIS (Geographical Information System)-City

There are two basic ways to introduce the infrastructure's heterogeneity into the model. One way is to continue with the FACS grid frame, and introduce different capacities to the cells, and varying spatial restraints to the influence of each of the $5^2 - 1 = 24$ neighboring cells on the central one. Another way is to use a high-resolution digital *vector* map of a real city as the infrastructure layer, and apply the theoretical scheme of the FACS to this environment. The latter way seems to us more attractive. First, a realistic city map explicitly accounts for the heterogeneity of the houses and their neighborhoods. Second, there are plenty of digital city-maps readily available, many of them as standard municipal Geographic Information Systems (GISs). In addition to maps one can retrieve from the GIS additional information about the inhabitants' characteristics, houses' visibility, street geometry, land usage, etc., all being important for the definition of the house's neighborhood. From the model point of view, the non-geographic part of the GIS gives us the opportunity to ascribe the naturally varying attributes of the houses and the inhabitants to

* By Itzhak Benenson and the author

the units of the model infrastructure and to the migrating free agents. Third, we can evaluate the simulation results *spatially*. A GIS-based model might thus be an additional step towards the application of agent-based modeling to urban, regional and environmental planning (see Chaps. 11, 12, 14 below).

As a first step toward realization of the above approach, we perform the agent-based simulations within the environment of MapInfo-GIS and use the MapBasic programming language for the model implementation. We name this GIS-City model City-5. The version we present in this chapter is realistic with respect to its representation of the infrastructure layer. In fact, we carry out the simulations on a section of 1:500 digital map of a section from the Tel-Aviv metropolitan area. We do not make use, however, of the data on the inhabitants' characteristics that are stored in the Data Base part of the GIS. Our model agents are thus, as before, artificial creatures, with features assigned by us. In this first version of the model we limit ourselves to a minimal number of characteristics of agents and infrastructure: two variables describe an agent's economic status, and one variable represents the value of the houses.

In GIS-City one-to-one correspondence between agents and cells is replaced by many-to-one correspondence between agents and houses. Each house now contains a certain number of apartments, and each apartment can be occupied by a free agent or remain empty. As a result, the infrastructure of GIS-City can be considered as a four-level hierarchy: (a) apartments, (b) houses, (c) neighborhoods, and (d) the city as a whole. In what follows we first reformulate our previous FACS models in terms of our new GIS-City. Second, based on this new model we examine the influence of two different definitions of neighborhoods in relation to houses, on the dynamics of the city infrastructure and its residential distribution.

9.3 Model Description

As in Chaps. 6–8 two economic variables describe the state of a free agent A in City-5. They are the agent's *economic status* S_A, and *rate of the status' growth* R_A. We do not account for the cultural identity of the individual in this initial version of the GIS-based model. To reflect properties of real houses we postulate that there can be several apartments in house H. We describe the value of an apartment by means of a value V^H related to house H only. For each house H we define a set of neighboring houses $U(H)$, that vary depending on the geometry of the city in the vicinity of H. The relation between the properties of house H and the characteristics of its inhabitant A, is given by the simplest equation of the dynamics of A's status, which differs from the equation we used in Chaps. 6–8:

$$S_A^{t+1} = \max\{1, ((1 + R_A)S_A^t - mV^H(t))\} \tag{9.1}$$

Here $mV^H(t)$ is a 'mortgage payment' for staying in a house H at a time-step t.

At each time-step every individual takes a number of economic decisions concerning residential choice (Fig. 9.1). An agent A, occupying an apartment in house H, decides to search for a new residence by computing the difference between its own status and the house's value:

$$D_A^t = |S_A^t - V^H(t)|. \tag{9.2}$$

We suppose that the probability to decide to leave a house and seek another residence increases with the increase of D_A (Fig. 9.2).

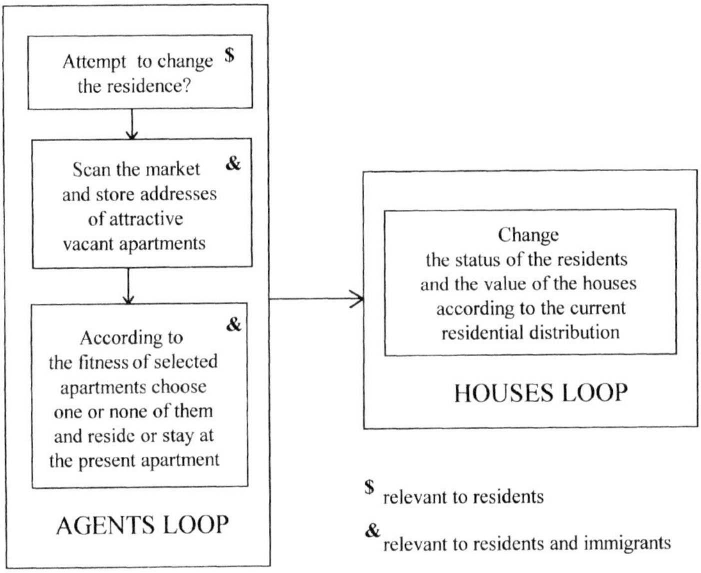

Fig. 9.1. The flow chart of City-5

If, at a current time-step, an agent decides to search for a new apartment s/he scans over *all the houses* in the city and constructs a list of 'attractive' houses with vacant apartments. According to our general approach (Chap. 13) at this stage we introduce an *attention parameter*, that is, a probability q that a certain house that is attractive in its characteristics, and has a vacant apartment, is included in the list of opportunities. The attractivity $E_{A,H}$ of a vacant apartment in house H for agent A depends on the difference between the agent's status and the weighted average of the values of the neighboring houses:

$$E_{A,H}^t = |S_A^t - F_H^t|. \tag{9.3}$$

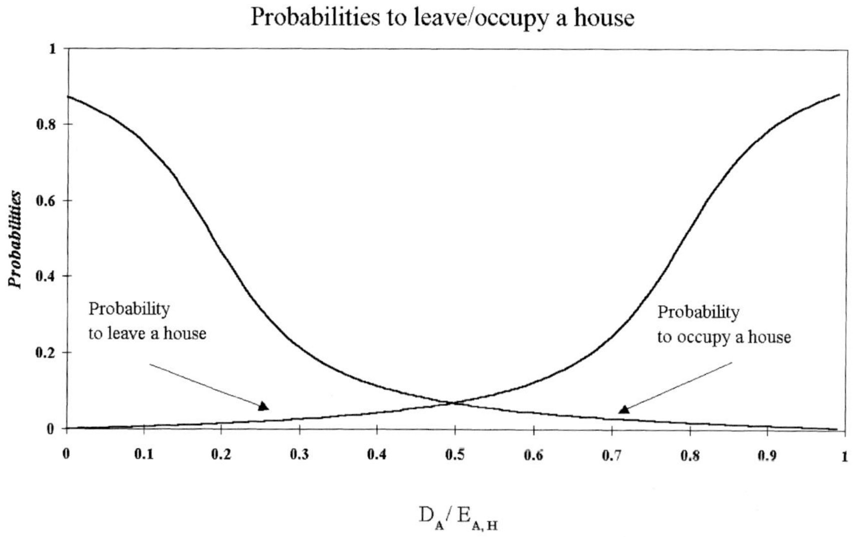

Fig. 9.2. The probability to leave and occupy a house in City-5

Here

$$F_H^t = \sum_G (w_H(G) \cdot V_G^t \mid G \in U(H)) \tag{9.4}$$

and the weights $w^H(G)$ of neighboring houses G relative to H reflect the relative influence of the houses within the neighborhood. With the increase of $E_{A,H}$ the probability $qL_{A,H}$, that A will pay attention on apartment in a house H and reside in it, when it is the only possible choice, monotonically decreases (Fig. 9.2). According to (9.3)–(9.4) the attractivity of the apartments in a house are equal. In case of several vacant apartments in a house we suggest that an agent accounts for one of them only. An individual that decided to change location in the city tries to occupy an apartment in one of K most fitting, according to (9.3)–(9.4), houses H_i, $i = 1, 2, \ldots, K$. Consequently, there are $K + 1$ mutually exclusive outcomes of the residential choice of the agent. The probability that none of K best apartments will be occupied equals

$$p_{A,0} = \prod_{H_j} (1 - qL_{A,H_j}), \tag{9.5}$$

where H_j belongs to the set of the B most attractive apartments.

The probability that an individual A will choose an apartment in house H_i from the best B houses equals

$$p_{A,H_i} = (1 - qL_{A,H_i}) \prod_{H_j \neq H_i} (qL_{A,H_j}) \tag{9.6}$$

Based on (9.5)–(9.6), we calculate the probability that an individual A fails to occupy a new residence as

$$Q_{A,0} = p_{A,0}/(p_{A,0} + p_{A,H_i}) \tag{9.7}$$

and the probability that an agent A will occupy apartment in house H_i as

$$Q_{A,H_i} = p_{A,H_i}/(p_{A,0} + p_{A,H_i}). \tag{9.8}$$

The value of house H is defined partially by its own value at the previous time-step, and partially by the mean status of its inhabitants:

$$V^H(t+1) = (\langle S^t \rangle_H \cdot P^H(t) + V^H(t) \cdot C_H)/(P_H^t + C_H) \tag{9.9}$$

where

$$\langle S^t \rangle_H = \sum_A (S_A^t \mid A \text{ occupies apartment in house } H)/P_H^t \tag{9.10}$$

is an average status of the H's inhabitants, P_H^t is the population, and C_H is the capacity (number of apartments) of house H; the latter does not depend on t.

GIS-City is an open system and at each time-step a constant number of agents enter it and try to occupy an apartment in a house according to the houses' attractivities and corresponding probabilities as given by (9.5)–(9.6). If, according to (9.1), an agent's A status decreases below zero, then A leaves the city.

9.4 Preliminary Results

The simulation runs we present below are based on real data concerning the settlement geometry (Fig. 9.3) and the houses' capacity. We study the steady distribution of the houses' value for two kinds of the neighborhood definitions. The first one is reminiscent of the uniform FACS City in the sense that it takes into account only the distances between the houses (scenario A). The second one accounts not only for the distances between the houses, but also for the heterogeneity introduced by the city's road network (scenario B). For scenario A, the neighborhood of a house H consists of the houses G, the centroids of which are at a distance of 50 m. or less from the centroid of H. The weights $w_H(G)$ of all neighboring houses G in (9.4) are equal to unit. The neighborhoods of scenario B includes only houses located within the 'quarter' of house H and at a distance 50 m. or less from it, when the 'quarter' is defined as the portion of the city that is limited by the closest streets. The weights $w_H(G)$ for scenario B are in inverse proportion to the distance of the neighboring house G from house H.

At the beginning of the runs the houses are assumed to be empty and immigrants thus populate the city. The immigration rate is constant: 20 individuals per time-step. We suppose that the initial economic status of the

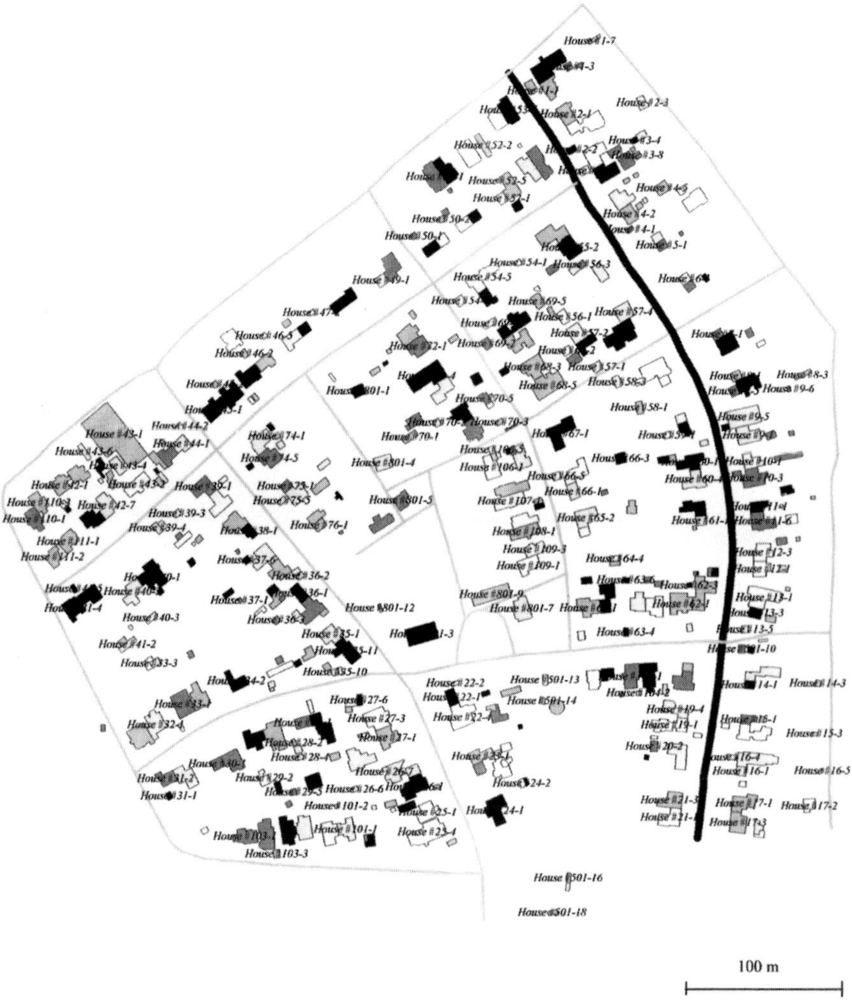

Fig. 9.3. The section of the city map (of Tel-Aviv) used in the construction of City-5, with its hypothetical, initial distribution of houses' value (random, uniform on $[0, 1]$)

agents varies less than their ability to improve it. At the beginning of the runs we assign the agent's status, and status' growth rate according to the truncated on $[0, 1]$ normal distributions, with the following means and coefficients of variation: $m_S = 0.5$, $CV_S = 0.10$ and $m_R = 0.10$, $CV_R = 0.30$. Additionally, we suggest that the individual's status and rate of status growth, correlate with $r = 0.8$. The status of an agent that immigrates into the non-empty city is assigned according to the normal distribution truncated on $[\min_{\text{city}}(S_A^t)/3,\ 3 * \max_{\text{city}}(S_A^t)]$, with mean equal to $\langle S_A^t \rangle_{\text{city}}$ and coefficient of variation CV equal to 0.10. The distribution of the immigrants' rate of status growth, and the correlation between S_A^t and R_A for them, do not depend on time. The value of m in (9.1) equals 0.05, the attention parameter v is set currently as unity, the number B of the houses checked by an agent during the time-step equals 2. Initial distribution of the houses' value is random uniform on $[0, 1]$ and the attention parameter q is set equal to one.

The GIS-City is an open and stochastic system. As a result, the distribution of house values does not stabilize – it is constantly varying due to the process of immigration. However, after some twenty iterations, the qualitative features of the distribution stabilize; below we are interested in these features only.

9.4.1 Scenario A: A City with No Internal Boundaries

The distribution of the house values does not differ qualitatively from the distribution obtained in the FACS models. Namely, both in FACS-City and in the current GIS-City simulations, after about twenty iterations the gradient of the house values in the model-city self-organizes itself and from that stage onwards it doesn't change much. Depending on the city's territory, the house value distribution has one or several peaks. The location of the peaks of the distribution, i.e. the areas where the value of the houses is at its highest, is defined stochastically and varies from run to run.

9.4.2 Scenario B: City of Quarters, with the Roads as Internal Boundaries

The road network that here creates the city quarters and constrains the neighborhoods of the houses, constrains also the dynamics of the houses' value distribution. This is illustrated in Figs. 9.4a, 9.4b for scenarios A and B respectively. Figures 9.5a, 9.5b are enlarged zooms of the right lower part of the city (i.e. of Figs. 9.4a, 9.4b), for typical runs of scenarios A and B. As can be seen in Figs. 9.4, 9.5, for scenario B the houses within the quarter are relatively homogeneous in their values. As a result, the self-organized distribution of the house values in the city is discontinuous, with abrupt differences between different sides of the street. For scenario B the quarter boundaries (streets) restrict the 'diffusion' of the areas of high (or low) valued houses and prevent the penetration of high- or low-status agents into the neighboring quarters.

The comparison between the two scenarios above demonstrates (a) the possibility, in principle, to implement agent-based simulations in a GIS environment and (b) the significant effects of the internal heterogeneity of the city on its dynamics and evolution. This is illustrated by investigating the influence of one of the infrastructure characteristics – the pattern of the city quarters – on the economic structure of the city. In the future we plan to study

Fig. 9.4. House values simulated by City-5. *Top*: scenario A; *Bottom*: scenario B

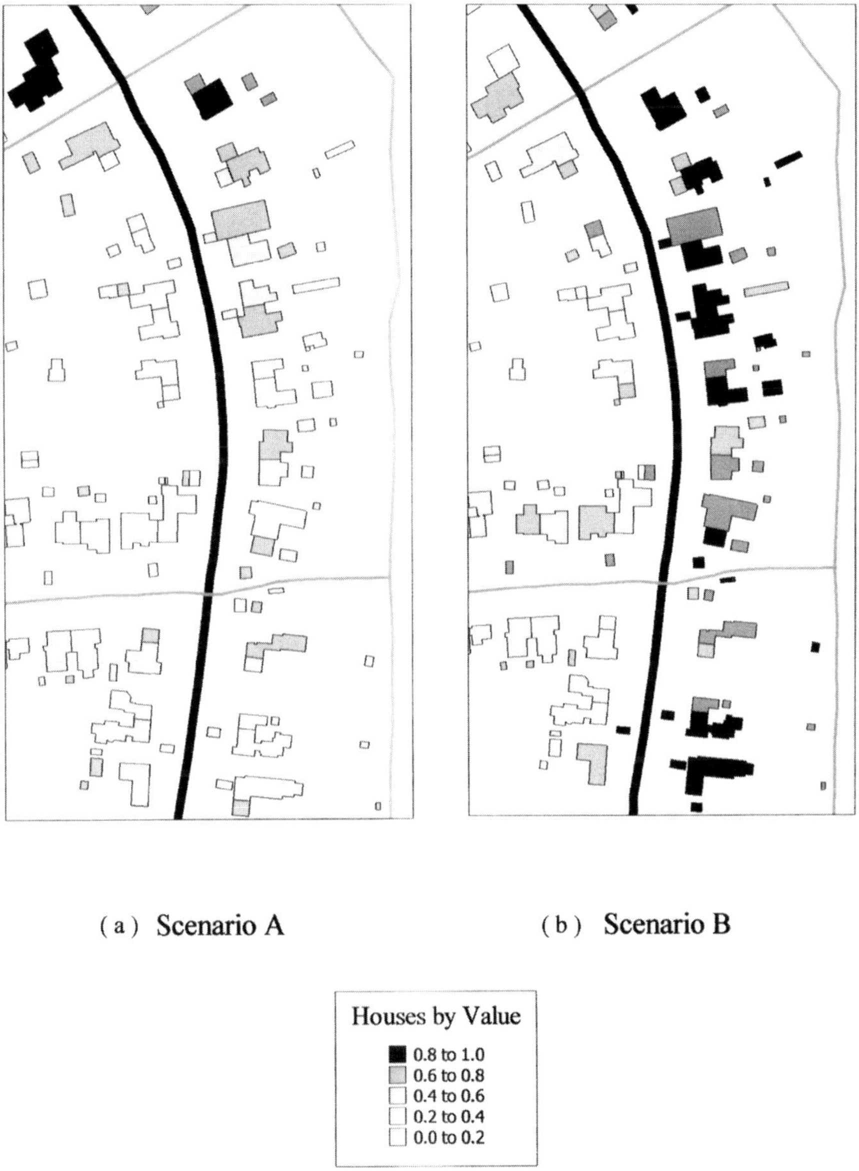

Fig. 9.5. A zoom on a section of City-5 showing typical distribution of house values along a street. *Top*: scenario A; *Bottom*: scenario B

the influence of some other characteristics of the infrastructure heterogeneity on the city dynamics. Two aspects seem to be most important in this respect. First, the influence of the varying characteristics of the houses (number of apartments, floors, etc.) on the emerging gradient of the house values. Second, the influence of different definitions of the neighborhood on the steady economic structure of the city. We have to consider here the visibility and accessibility of the neighboring houses relative to the given one.

10. Internal Complexity and Socio-spatial Segregation of Groups in a Self-Organizing City*

10.1 Introduction

10.1.1 Hierarchical and Non-hierarchical Cultural Groups

As noted above, in Chap. 7, modern, or rather postmodern, cities are characterized by spatial, social and cultural pluralism and can thus be described as mosaics of coexisting cultural and social groups. In previous chapters we have studied this spatial–cultural mosaic in connection with two types of cultural groups: 'old' groups who emigrated to the city as already established cultural groups, and 'new' groups which only recently emerged as the dialectical products of the dynamics of the city itself. In this connection, we have studied, on the one hand, processes of spatio–cultural segregation and on the other, the very process of socio–spatial emergence.

Implicit in the above chapters was the assumption that our cultural groups are non-hierarchical in their structure. This shows up, for example, in that the virtual cultural groups in our models were composed of individual decision-making units (households or families), with no intermediate levels between the local decision making individuals and the global cultural group. In fact, the notions of spatial cognitive dissonance, cultural code, and their connection (Chaps. 7, 8), illustrate, among other things, just how such hierarchies can emerge. However, whereas in previous chapters our aim was to show how the hierarchy emerges, here we want to study what happens when it already exists. This aspect is significant because while many of the cultural groups which make today's cities are indeed non-hierarchical in their structure (like the Blues and Greens in the above games), some groups, especially the 'old' ones, are characterized by a complex internal hierarchy and a vertical division in a social system.

It is common to assume that complex internal hierarchy is a property of 'traditional', non-Western, cultural groups, and that simple, non-hierarchical, internal structure, a property of 'modern' Western cultural groups (Fig. 10.1). Kin relations provide a typical example for the first, and atomistic–political relations for the second. According to the first, the group is perceived as an organic whole where each individual is a member in a complex family-like tree

* By Itzhak Omer and the author

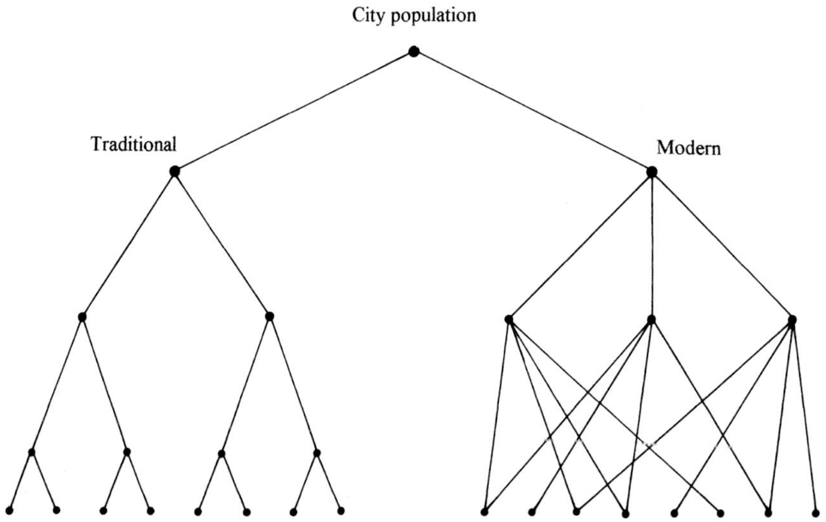

Fig. 10.1. Traditional (hierarchical) and modern (non- hierarchical) cultural groups

of nuclear family, extended family (e.g. clan), tribe, nation, etc. According to the second, the groups are perceived as a mechanistic structure where each individual, or at least each family or a household, is an autonomous and independent interacting entity.

The above difference between the traditional and the modern is often used to distinguish between traditional and modern cities. Such a distinction, we suggest, is rather problematic as now-a-day cities, in western and non-western countries, are composed of both types of cultural groups (and this is probably yet another aspect of the postmodernity of cities). In western cities this is to a large extent a consequence of labor migration and the fact that many of the immigrating groups did not culturally assimilate in their new cities but have kept in them their old traditional complex structure. A somewhat different case comes from Israel, the urban cultural landscape of which was from the start composed of hierarchical (mainly Arab) and non-hierarchical (mainly Jewish) cultural groups. A case in point is the city of Tel-Aviv-Yaffo. Its small Arab community, living in the neighborhood Adgami, is largely characterized by hierarchical kin relations, whereas its various Jewish cultural groups, by non-hierarchical atomistic relations. The above observation comes from a series of studies conducted by us in Adgami (Portugali 1991a, 1993, Omer 1996). These studies will provide the major empirical background and context to the more theoretical discussion we develop below.

10.1.2 Cultural Generative Orders

The internal structure of cultural groups can be regarded as a kind of ordering principle according to which group members self-define their personal and collective identity, perceive the urban or rural space around them, and take decisions concerning where in the city to live, when to migrate, and so on. From the perspective of self-organization, this ordering principle can be seen as a potential or a *cultural generative order* for the residential location decisions and actions of group members. In line with the discussion in the previous section one can thus distinguish between simple-non-hierarchical and complex-hierarchical, cultural generative orders. In the first, each individual decision making unit (i.e. a family) has a single group identity and as a consequence the main criterion in its location decisions is whether the neighbors belong or do not belong to its cultural group. This is, as noted, the case with the Green and Blue individuals in our previous models: their basic location criterion is whether or not, or the extent to which, the neighbors of a vacant cell are Green or Blue. As we shall see below, each of the virtual individuals in the game developed below has a multiple group identity and consequently its residential location decisions are determined by a hierarchical and complex cultural generative order.

10.1.3 Aims

Our main aim in this chapter is to study the way a complex hierarchical structure of cultural groups participates in, and shapes, the dynamics of the city's spatio–cultural evolution. An additional aim is to use the above study as means to further extend our discussions in previous chapters concerning the association between the conceptual framework of self-organization and that of social theory, and the possibility to integrate the two. In particular we will show that the conjunction between self-organization and social theory implies that cities, and spatio–cultural and economic entities in general, should be regarded not as the spatial expressions of social or cultural structures (however defined), but rather as one of the many facets and properties of the very culture and society. Furthermore, we will show, first, how hierarchical cultural structure entails unintended consequences in the residential segregation process. Second, how residential distribution, as an explicit form of hierarchical cultural structure, self-organizes itself and shapes the relation between cultural groups at differing levels and dominance.

The discussion below thus develops as follows: we first present the model, introduce the notion of Q-analysis and show how we integrate it to the model. Then we describe the cultural hierarchical structure dynamic and its significance to the unintended consequences and to the relations between cultural levels in the context of residential segregation.

10.2 The City-6 Model

We develop and elaborate the discussion about the above aims by means of a heuristic model developed by Omer (1996). City-6 is principally similar to the models in previous chapters, especially to City and City-1, but differs from them in the identity structure of the agents. While in the previous models, each individual had a single group identity – one could be Blue or Green or Red, etc. – here each individual has a complex identity. In addition of being Green or Blue etc., that individual also has a family identity, an extended family identity, a religious identity, and so on. Being subject to such a complex cultural generative order implies that residential location decisions are affected by this complexity in a way that will be elaborated below. Its general form, therefore, represents migration changes:

$$P^A(t+1) = F_t(A^H(t), P^A(t), A^{U(H)}(t), E(t)), \tag{10.1}$$

where, $P^A(t)$ is the current location of an agent A, $A^{U(H)}(\mathrm{t})$ denotes states of agents within a neighborhood of A's current location and $E(t)$ represents external parameters (immigration).

In order to be able to follow and study the role of multiple identity and complex cultural generative orders, the Q-analysis method is applied to the model. The result of the above is a two-layer model plus the new feature: a cellular space as the infrastructure layer, above it a layer of free agent individuals, and in addition to these two the instrument of Q-analysis. A description of the model now follows.

10.2.1 The Cellular Space

Consider Fig. 10.2, which represents a typical computer screen of the model. Its basic infrastructure is a 30×30 lattice of *cells*. Each cell represents a building that can house a single nuclear family. The cell might be either occupied by a family, or empty. Superimposed on the lattice of cells is a lattice of *places*. Each place is a square of 3×3 cells. As we shall see below, it is in light of the properties of the various places that the location decisions of individual families are determined.

Fig. 10.2. The heuristic simulation main screen. The cellular space depicted here shows the emergence of a minority residential spatial pattern, based on three attributes: community, sub-community and extended family. There are 2 communities, 4 sub-communities (2 sub-communities per community) and 16 extended families (four extended families per sub-community); every group of cells with the same markings (color and shape) comprises one extended family. (**a**) *Top*: communities, (**b**) *Middle*: sub-communities, (**c**) *Bottom*: extended families

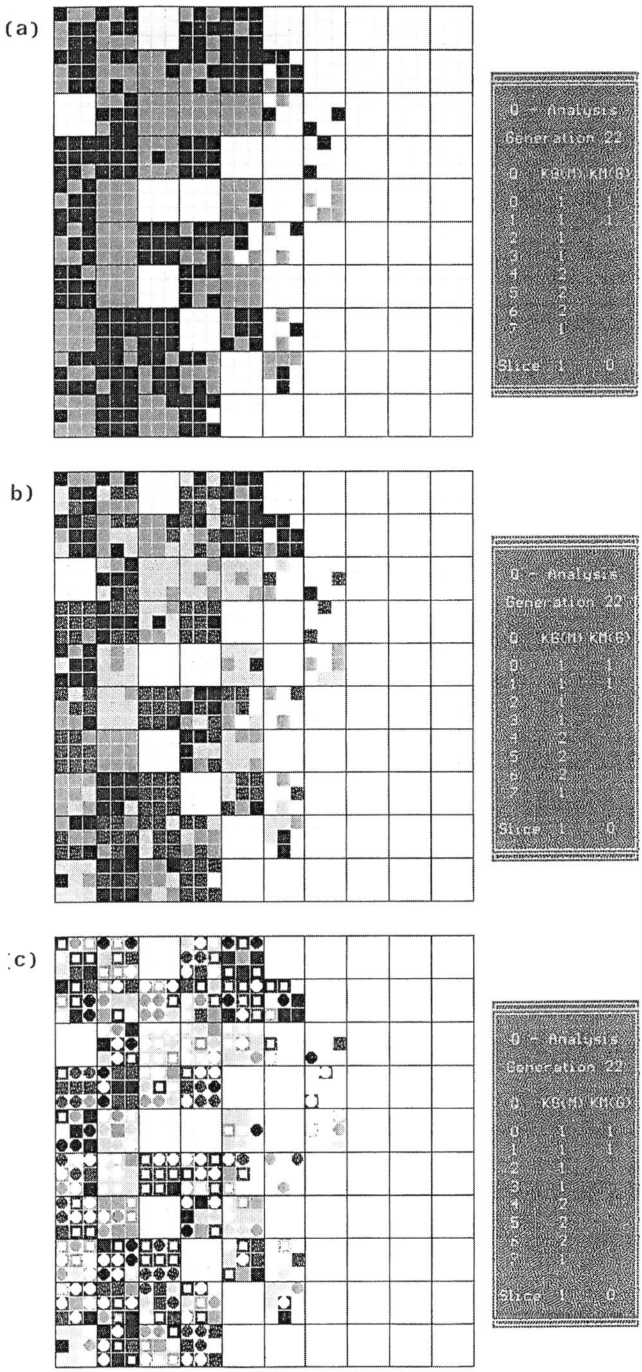
(a)
b)
(c)
Q - Analysis
Generation 22
Q KG(M) KM(G)
0 1 1
1 1 1
2 1
3 1
4 2
5 2
6 2
7 1
Slice 1 0

10.2.2 The Agents

The agents in the model are subdivided into two groups: a majority group with a non-hierarchical internal structure, and a minority group with a hierarchical internal structure. Cells occupied by members of the majority group are white, whereas cells occupied by minorities have non-white colors. Each simulation game starts in a situation where the majority group occupies the entire space. The role of the latter in the games is passive, however: they are just there, and it is into their area that the members of the minority group enter. The active agents in the game are thus the nuclear families of the minority group. They move in the cellular space in search of a proper house, or reside in one. The model thus simulates the diffusion of a minority group into the residential area of a majority population. The specific form of this process is determined by the linear probabilistic function of Fig. 10.3. The linear tolerance function components – the slope 'b' and the constant 'a' – are calculated as follows: $a = (y_2 - y_1)(x_2 - x_1)$; $b = y_1 - ax_1$, where: x_1 indicates the number of individuals in the minority group that constitutes the threshold, above which desertion of majority group members may be triggered. y_1 and y_2 respectively indicate the minimum and maximum values for the probability of desertion. These values are set by the user, while the value x_2 is determined by the size of the minority population at the place at which the function is implemented. In the latter, the probability that a majority family will leave its cell/house is directly proportional to the rate of minority families in its place (the 3×3 cells around it). The linear form of the curve in Fig. 10.3 is based on our empirical findings in Ajami. Other studies have reported (or suggested) different behavior, mainly logistic form (Woods 1980).

Each minority family can be regarded as a memory unit that memorizes its group's structure and is thus dominated by its group's cultural generative order. That is to say, each family 'remembers' its specific multiple identity and this is the source for its location decisions in space. Multiple identity implies that each family participates, in fact, in the evolutionary dynamics of a multiplicity of spatio–cultural areas in the city: the spatio–cultural pattern and spatial segregation of extended families, religious, ethnic, national groups and so on.

To describe formally a hierarchical cultural identity, let us denote an identity of an agent A by means of a vector $(a_i, b_{ij}, c_{ijk}, \ldots)$, where the first component, a_i, designates a group at the highest hierarchical level, the second component, b_{ij}, designates a group at the second level of hierarchy, and so on. If $a_i \in \{a_1, a_2, \ldots, a_I\}$ (that is, there are I different identities at the highest level of hierarchy), then for an agent with identity a1 at the highest level its identity b_{1j} at the second level belongs to the set $\{b_{11}, b_{12}, \ldots, b_{1J(1)}\}$, for an agent with $a = a_2$ its identity b_{2j} at the second level belongs to set $\{b_{21}, b_{22}, \ldots, b_{2J(2)}\}$, etc., where $J(1)$, $J(2)$, $\ldots$ is the number of sub-groups

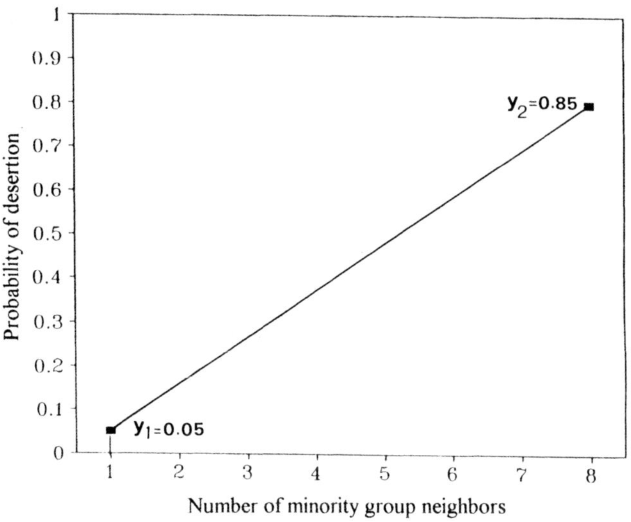

Fig. 10.3. Tolerance function – the probability that a majority group family evacuates its house is determined in every place by the number of minority group neighbors (in each place a total of 8 neighbors cells)

within each highest-level group, and the values b_{ij} are all different. Similar relationships hold for the second and third levels, etc.

10.2.3 Decision-Making

An a priori hierarchy of identities implies a nested process of agent's housing search and a hierarchy of resulting residential distributions. Each minority family enters the city and starts to search for an appropriate house/cell. The basic element in this house search process is the *decision tree* (Fig. 10.4). The family starts its search at the bottom of the tree. Accordingly, it first attempts to find a location in a place (i.e. 3×3 cells) where at least one other member of its extended family lives. If this attempt fails the family moves one level up the tree and tries to find a house in a place where at least one member of its religious group lives. If this attempt also fails, it moves further up and tries to find a house in a place where a member of its cultural group lives, and so on. If this hierarchical search process fails at all its levels, the family goes back to the bottom of the tree and starts the whole process again, but now in neighboring places (and not neighboring cells as before). Let D_i denote cells, which are occupied by families having identity a_i at the highest level of hierarchy, D_{ij} denote cells, which are occupied by families having identity b_{ij} at the second level of hierarchy, etc. Once families occupy the cellular space of the city, the domains, occupied by spatio–cultural entities are fully nested and co-exist in full 'implicate relations': $D_{ijk} \subset D_{ij} \subset D_i$. Moreover, the following condition is always true for first and second hierarchical levels:

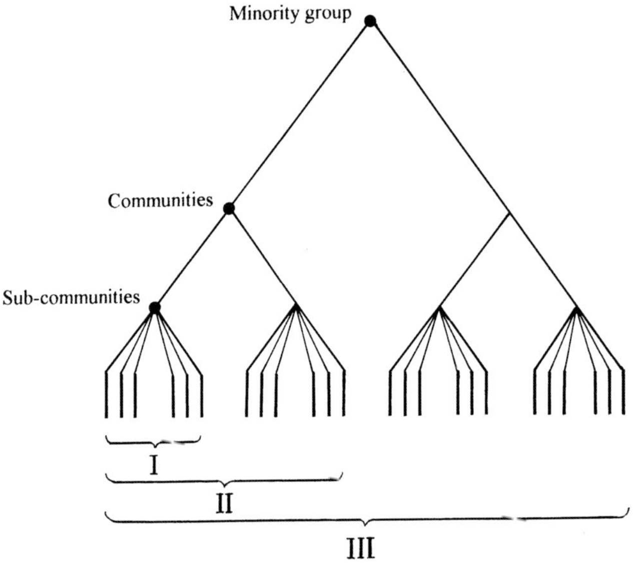

Fig. 10.4. Choosing residency according to the principle of proceeding along the identity axis – a minority group including two communities and four sub-communities

$D_i = D_{i1} \cup D_{i2} \cup \ldots \cup D_{iJ(i)}$, and a similar one is also true for second and third, third and fourth and further levels. That is, the distribution defined according to a general attribute contains the distribution defined according to the less general attribute.

Such a housing search process can be implemented by means of two alternative decision-making strategies. The first is known as the *satisfier behavior*, whereas the second, the *optimizer behavior*. In the first, the decision making family chooses the first place that satisfies its requirements (e.g. a place with at least one of its family members); in the second the decision maker chooses the 'best place in town' – the one that has the largest number of family members in it. See Lierop (1986) for a discussion.

Each of the model's components described above is defined when the model is initialized, like the structure and size of the cellular space and the places, the form of the spatial diffusion and the cultural structure, which defines the levels of identity. During the simulation, the user may intervene and change certain control parameters like an apartment inventory, a rate of minority group diffusion and a rate of residential mobility within its spatial pattern, and so on.

In this type of model, we can frame our hypothesis and assumptions at the individual level, directly in terms of behavior, intentions, abilities, and explore the implication of each for the patterns that emerge at the macro levels.

10.2.4 The Q-Analysis Device

Q-analysis is a topological method developed by Atkin (1974, 1977, 1978) to expose and represent structures in observations of events – specifically in an irreversible, dynamic, social reality. Its basic principle is to 'let the data speak for itself' (Gould 1981), without subjecting it to pre-determined statistical models. The method is, therefore, non-statistical, and interpretation of the various structures revealed, depends on the investigator's viewpoint. In Q-analysis the 'data speaks for itself' by means of *simplexes*, *q-chains*, *shared-face matrices* and a *structural vector*. Each simplex represents direct interaction or connection between entities, whereas the q-chain and the shared-face matrices, two levels of indirect connections or interactions between the elementary units of the data. The structural vector, as the name indicates, represents the general structural properties of the set of observations. We shall further clarify this method below by reference to actual analysis and interpretation of the simulation results.

The integration between Q-analysis and CA simulation models, as originally suggested by Omer (1996) and as illustrated here, is novel and it improves considerably the potential and power of both methods. It gives Q-analysis a new dimension, and most importantly, it enables a much deeper analytical insight into the results of computer simulations.

In order to study the role of multiple cultural identity and the relations between levels of a cultural structure, we examine the dynamic of residential distribution from two aspects. In the first, we study the role of multiple cultural identity in the creation of 'unintended consequences' in residential distribution, as an explicit form. In the second, we examine the implicit structures which emerge from the residential dynamics and which stand behind the agent's action.

10.3 Game 1: Forms of Unintended Consequences

Human beings are intentional entities in the sense that their decisions for action are determined by their intentions. However, as we have seen above, the outcome of their decisions, that is to say, their actual behavior and the structure of the system they produce quite often differ qualitatively from their intentions. Thus, two of the most significant properties in the self-organizing cities we have studied above were, first, the gap between intentions and actual behaviors. Second, the fact that in the process of self-organization the socio–spatial system (i.e. the city) acquires properties which are qualitatively independent of the intentions that have guided the decisions of the individual agents. As already noted, the latter properties, which are central to our interpretation of human systems in the context of the theory of self-organization, are central also to several social theories. In particular to Marxist perspectives which are very dominant in the social theory discourse on urbanism, and more recently in Giddens' (1984) theory of *structuration* and its notion of *unintended consequences.* Our concern here would be the association between the notions of 'unintended consequences' and 'structuration' as developed by Giddens in the context of social theory, and the notions of 'intentions versus behavior' and 'spatial cognitive dissonance' as elaborated above in the context of self-organization.

The difference between our approach and those of social theory is twofold: first, whereas in social theory (and other studies on self-organizing cities) the main focus of interest is the tension between the local agent and the global system (society or city), here we make a distinction, and focus on the dialectical tension, between *intentive decision*, and *actual behavior*, at the local level, and the overall behavior and structure of the city, at the global level. Second, in social theory the city is usually treated as a spatial representation of the more subtle social forces. 'Our' cities are not an expression of society, but one of its facets, properties and forces. With this perception we come close to Giddens' notion of structuration. However, Giddens has developed his theory at a very abstract level and with no reference to the dynamics of cities and urbanism. Here we follow Omer's (1996) recent demonstration that Giddens' structuration can be related to our self-organization theory of the city.

In previous chapters we have studied the phenomenon of unintended consequences in relation to relatively simple, non-hierarchical, single identity, cultural groups. Here, in this first game, we employ the above model for the study of this same phenomenon but in the context of hierarchical cultural groups with multiple identities.

10.3.1 Initial Conditions

The opening conditions for this game are the standard conditions of the model as described above: a minority group with a hierarchical multiple identity is entering an urban area occupied by a passive majority group. The elementary units of the minority group are nuclear families, with an hierarchical multiple identity structure of extended families, higher order religious identities, and still higher order identities (say, national or ethnic). The specific definitions of the game are as follows:

- Each family entering the city belongs to single minority groups, to one of two sub-communities, and to one of 16 extended families (Fig. 10.5a). This cultural identity structure is also the decision tree the families use in their housing search process.
- At each iteration ('generation') 3% of minority families enter the city and start their housing search process until they find a house and locate in a specific cell.
- The parameters of the tolerance function (curve) are $y_1 = 0.15$; $y_2 = 0.30$. These parameters define the probability that a majority-group family will

evacuate its house in face of a growing proportion of minority families in its place.

- The probability for a minority family to change location in the city is 0.1 (once in every 10 iterations/generations).

Given the above initial conditions, we run two simulations in which the families take decisions in line with the optimality principle. In the first, all the identities of the cultural order are active, in the second just one of them.

10.3.2 Results and Discussion

The results of the first simulation, based as they are on the optimality principle, are presented in Fig. 10.5b.

The first and immediate outcome that can be detected is a gap between intention and behavior, which indicates unintended consequences at work. There is very little correspondence between the spatial distribution of families and their multiple identity structure – no cultural spatial segregation reflecting the complex, hierarchical, internal structure of the minority group can be identified. As in previous chapters here too, the gap between the behavior and intentions of agents is a characteristic of the complexity of the system and its associated processes of self-organization. From the perspective of structuration theory, this is a consequence of the space–time *distanciation* of agents' actions.

It is interesting to compare the form of unintended consequences in the present game with those obtained in previous chapters. In the latter, the resultant global structure of the city was largely segregated and the city spatially

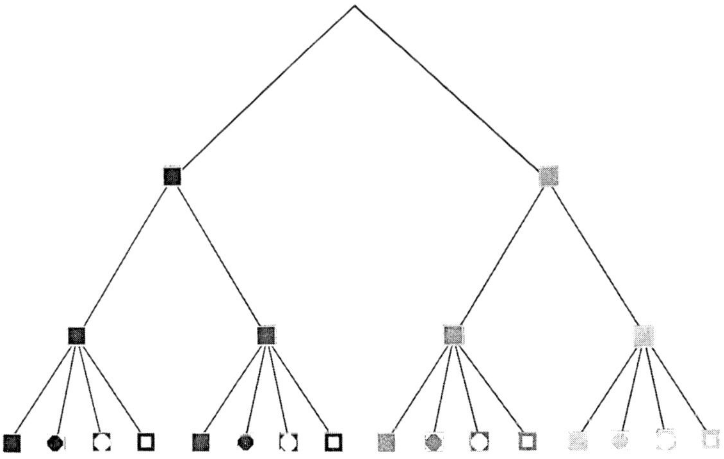

Fig. 10.5. a. Residential segregation with cultural hierarchical structure. The cultural hierarchical structure

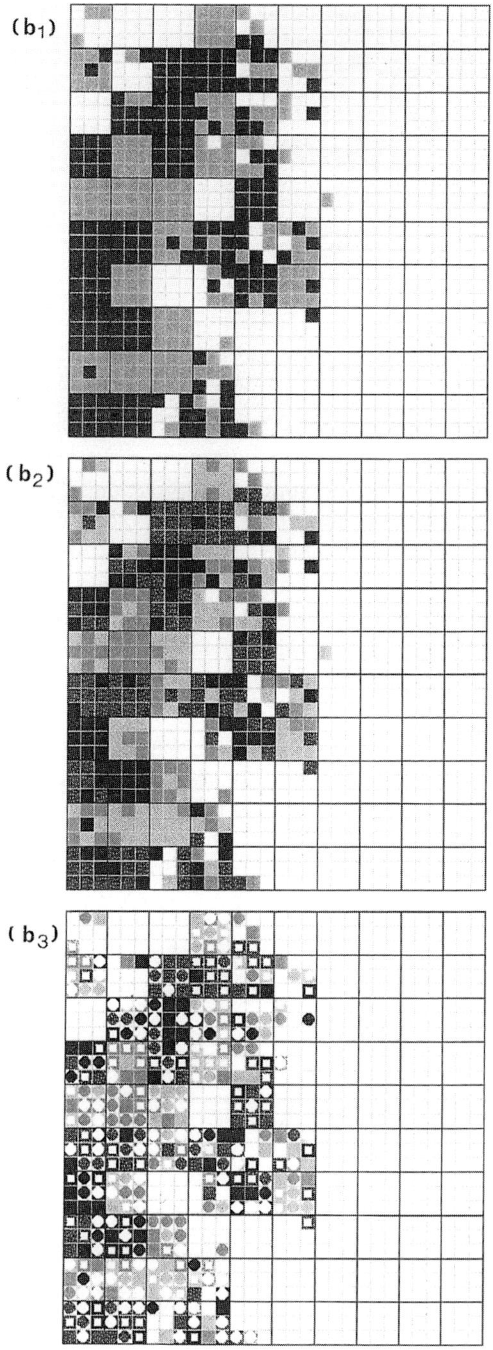

Fig. 10.5. b. Simulation of the residential segregation in generation 20. (b1) communities, (b2) sub-communities, (b3) extended families

represented the cultural composition of society. The gaps between intention, behavior and city structure, did exist, but mainly on the boundaries between otherwise relatively homogeneous cultural areas. In fact, the boundaries were the result of the above gap. In the game presented here, the gap is a property of the entire cultural area of the minority group. Our interpretation is that this difference stems from the hierarchical structure of the cultural groups and from the constraint that families have no choice but to find a location in the city. (In previous chapters' games, families had the option to improve their location in space or to leave the city (i.e. to internally or externally migrate).

In the simulation, like in a real-life situation, the above constraints act together and it is difficult to evaluate the relative contribution of each of them. In order to isolate their effects we have conducted a second simulation process, this time with a minority group with three cultural levels (extended family, sub-community, and community), and with individuals that take location decisions in line with their middle-level, sub-community, identity only.

The simulation outputs (Fig. 10.6) show that action according to middle-level identity leads to a certain degree of spatial segregation in all three identity levels: the upper-level of the whole community, the middle level of sub-communities, and the lower-level of extended families. This is significant since it implies that when the various cultural identities have a nested structure, as in our case, action at one identity level not only affects other levels, but affects them in a similar way.

What are the implications and findings of the above model on the interpretation of real self-organizing cities? First, a heuristic model as above can be used to answer a question such as 'what if in a culturally hierarchical society one component, i.e. religion, has become the cultural order parameter'? That is to say, 'what if families take location decisions in line with this sub-identity, and ignore the extended-family identity, on the one hand, and the cultural sub-identity, on the other'? Second, in the specific circumstances specified above, the city has self-organized a segregation between the various levels of the minority cultural group. Namely, it has created inter-group differences as an unintended consequence. This implication, can be appreciated in light of the notion of *spatial dialectics* developed by Portugali (1993, Chap. 6) in connection with the Israeli–Palestinian conflict, and in light of the notion of spatial cognitive dissonance discussed in Chap. 7 above. In both cases, it was shown, how, by means of the self-organized tensions between intentions and behaviors, old differences between individuals diminished and new ones were created. In the reality of the Israeli–Palestinian conflict this process has played an important role in shaping the Israeli and Palestinian national identities. In the virtual City-3 of Chap. 7 above, it has resulted in the emergence of a new cultural group – the Reds. This very same phenomenon also plays an important role in Adgami – the Arab neighborhood of Tel-Aviv that we use as an empirical background for our study. In the latter, self-organization

(a)

(b)

(c)

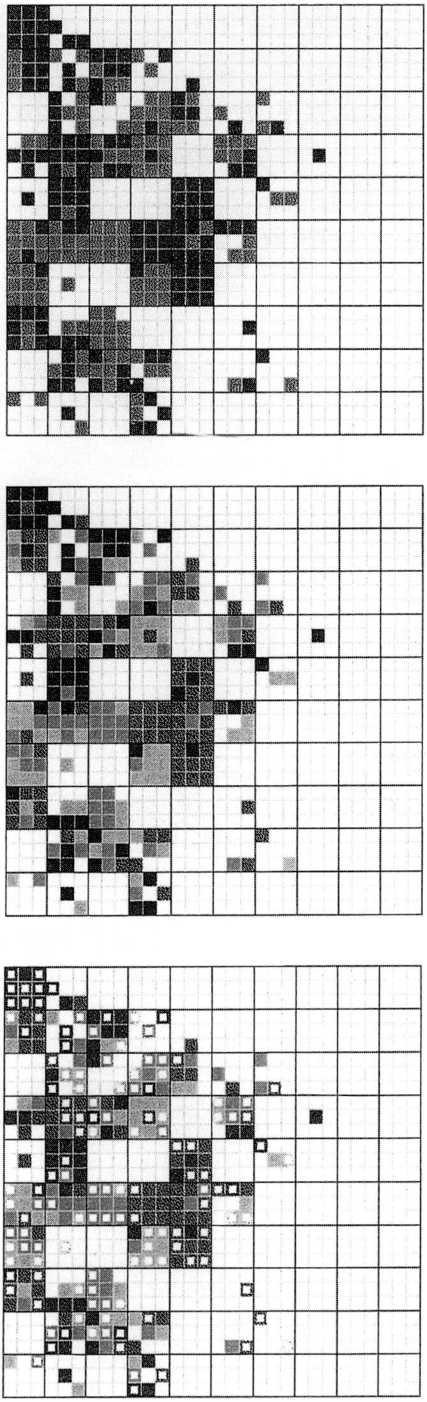

Fig. 10.6. Hierarchical cultural structure as a source of unintended consequences. (**a**) communities, (**b**) sub-communities, (**c**) extended families

of the city of Tel Aviv–Jaffa entailed a spatial mix of Christians and Muslims inside the area of Adgami, and a spatial segregation between the Arabs and the majority Jewish population of Tel-Aviv. This is one of the processes that help to reproduce the Arabs of Tel-Aviv as a distinct urban–national group. Similar processes can be observed in other cultural mixed cities around the world.

10.4 Internal Relations Between Cultural Orders

In previous chapters and at the beginning of this present chapter, a distinction has been made between *intentive decision*, and *actual behavior*, at the local level, and the overall behavior and structure of the city, at the global level. This distinction elaborated the view that people take intentive decisions, but that their behavior and the resultant city structure are qualitatively independent of their intentions. To this distinction corresponds the notion that in taking their decisions people are conscious or exposed to only part of the complex network that makes the structure of the city within which they intentively act.

The distinction between the exposed and the hidden has a long tradition in the various structuralist and post-structuralist schools of social theory. It plays a role in the Marxist distinction between base and superstructure, in Levi-Strauss' structuralism, in his distinction between the exposed and *deep structure*, and in Giddens' (1984) distinction between social versus systemic integration. *Systemic integration* according to him, is the total network of connections between individuals and their socio–cultural and political entities, whereas *social integration* refers to 'face to face' interaction and connections, and to that part of the network which people are conscious of. Intentive decisions in the city are made in line with the exposed network, whereas the city as a whole operates in line with the total network. The notions of systemic versus social integration was developed by Giddens in connection with his studies on social change and reproduction, and was further elaborated by sociologists and geographers (Gregory 1989), again in the same thematic connections. While these studies have contributed immensely to our understanding of the contextuality of the social and its relevance to the spatial domain, neither Giddens, nor these subsequent studies have explicitly studied the actual spatial appearance of these processes of social change and reproduction.

In a recent study, Omer (1996) considers the relation between system integration and social integration in the evolution of an urban residential spatial pattern. This study includes examination of the way the mechanism of system integration, including its connections in space–time, serve as a potential for the reproduction, first, of social form (institutional parameters etc., as in Giddens), second, of the spatial pattern itself and third, of the internal relations between cultural orders, as levels of cultural structure, and their spatial patterns in the urban domain. The main methodological tool applied in studying the concrete urban appearance of the relations between social and systemic integration, was Atkin's Q-analysis.

10.4.1 Q-analysis

As noted above, in Sect. 10.2.4, Q-analysis was specifically designed by Atkin to study and expose direct and indirect connections which create structures, in particular, in social or cultural systems. The basic concepts that underlie Q-analysis are sets and the relation between them. In terms of our context, let C be the set of m cultural groups, so that $C = \{c_1, c_2, \dots, c_m\}$, and P a set of n places, that is $P = \{p_1, p_2, \dots, p_n\}$. Let μ indicate that a pair of elements (c_i, p_j) is related. If several members of cultural group c_i live in place p_j, then c_i is said to be related to p_j by the relation μ: (c_i, p_j). These presence/absence relations between pairs of elements from sets C and P are described by an *incidence matrix*, an example of which is shown in Fig. 10.7a. Here, cultural group c_1 is present in places p_1, p_2 and p_3 and absent in p_4 and p_5; c_2 is present in places p_1 and p_2, and so on. Since Q-analysis is based on a binary language, one has to define the *slicing parameter.* The latter is denoted by θ, and it defines the number of group (c_i) members needed to be present in place (p_j) in order to maintain the relation (c_i, p_j). The relation is represented by unity. Such places form the residential choice area of that cultural group.

In geometrical terms, each cultural group is a simplex, with places P as its vertices (Fig. 10.7b). Therefore, the relations between the sets (places and cultural groups) and the connectivity in each set are represented by m simplexes which together create the simplicial complex $KC(P; \lambda)$ (Fig. 10.7c).

In this manner the various simplicial complexes represent the interaction between cultural groups, that is, between the groups that share the same place(s). In the framework of Q-analysis, we can observe the interaction in each dimension, denoted by q. When $q = 0$ the groups are said to be 0-near, in such a case they have at least one common place. Groups c_1 and c_3, for example, are in a 0-nearness relation (Fig. 10.7c) and c_1 and c_2 are in a 1-nearness relation, which means that they have 2 places in common and so on. It is important to note that simplexes can be connected even if they have no places in common (or faces in the simplicial complex). Two cultural groups (simplexes) are q-connected if they share $q + 1$ places (vertices) e.g. c_2 and c_3 are 0-connected.

10.4.2 Game 2: The Emergence of a Cultural Order Parameter in Self-Organization Process

In the discussion above (game-1), we considered a situation by which our minority group families, with their multiple-identity generative order, took

(a)

	p_1	p_2	p_3	p_4	p_5	p_6
c_1	1	1	1	0	0	0
c_2	1	1	0	0	0	0
c_3	0	0	1	1	1	1
c_4	0	0	1	0	0	0

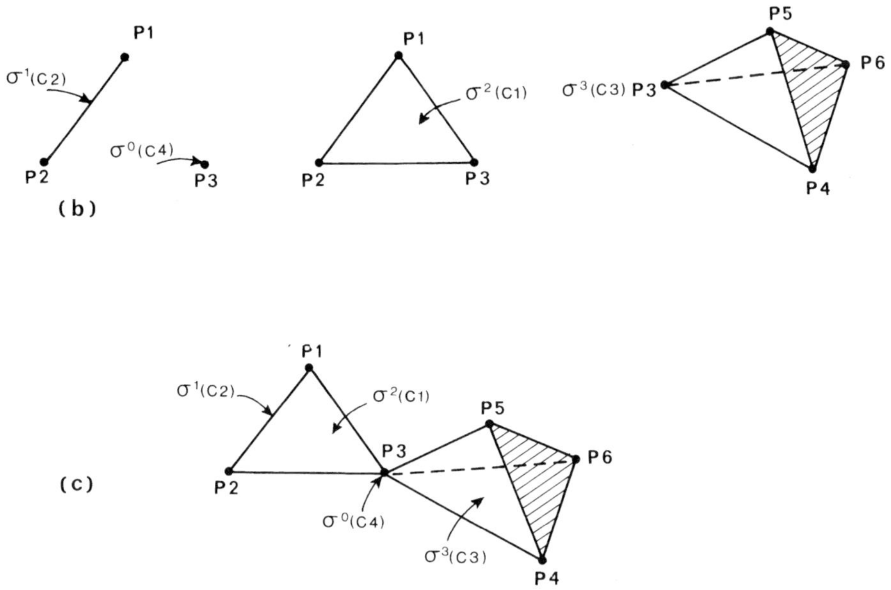

Fig. 10.7. A Q-analysis example. (**a**) *Top*: Incidence matrix for $\lambda \subseteq C \times P$. (**b**) *Middle*: Cultural groups as simplexes of different dimensionality. (**c**) *Bottom*: The simplicial complex $KC(P; \lambda)$

a location decision in the city in line with one identity component only. In terms of synergetics, such a situation can arise when one identity level in the hierarchical group structure has emerged as a *cultural order parameter* and enslaved other levels so that individual decision-makers decide according to these ordering principles. Our aim in game 2 is to clarify how this happens. For this purpose we have defined our simulation model so that:

- The probability that a family will desert the majority population is given by the tolerance function ($y_1 = 0.10$; $y_2 = 0.80$).
- An individual's search process is implemented on a cultural structure as presented in Fig. 10.8.

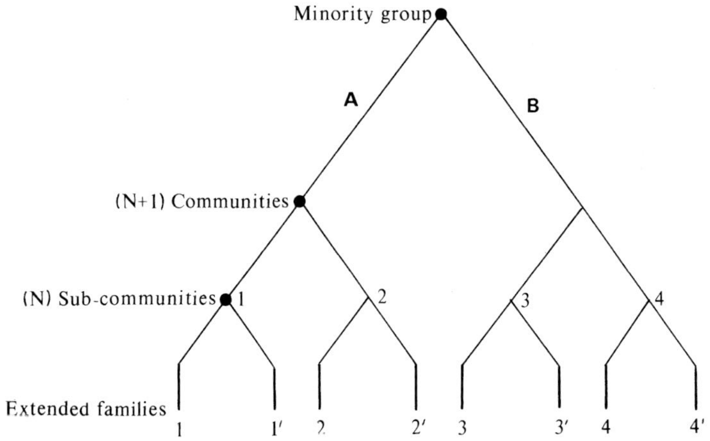

Fig. 10.8. The cultural hierarchical structure for the search process of game 2, based on three cultural orders: community, sub-community and extended family

- The search algorithm is formulated according to satisfier behavior, that is, a family chooses one place (9 cells in total) out of many possibilities that satisfies its requirements.
- The slicing parameters θ, which define for each identity level the number of agents required for a place to be part of the choice-space of a group, are as follows: four families with the same community identity ($\theta > 3$); three families with the same sub-community identity ($\theta > 2$); two families with the same extended family identity ($\theta > 1$).

10.4.2.1 Results and Discussion

The two Q-analysis sets in Fig. 10.9, created as they were in the frame of the two simulations defined above, will serve for studying how the socio–spatial system self-organizes the relative dominance of its cultural orders.

Both simulations start from the same initial structure, meaning they were constructed as a single game until generation 19. From this point, their development is contingent. Both thus use the same 'cultural structure', with the same search algorithm for a place of residence. Both are also subject to chance events, which determine who will change residence, when and where.

From Fig. 10.9, it is evident that the structures of the two simulations are quite similar until generation 22, probably as a result of their common history. From generation 22 onwards one can observe that the simulated structures evolve in different forms. This is because some of the agents' movement at the micro level can amplify or weaken some of the cultural orders at the macro level. In simulation 2, for example, the structure's transformation in generation 23 amplifies the community level – it creates a connection between sub-communities 3–4, which belong to community B. Moreover, in the

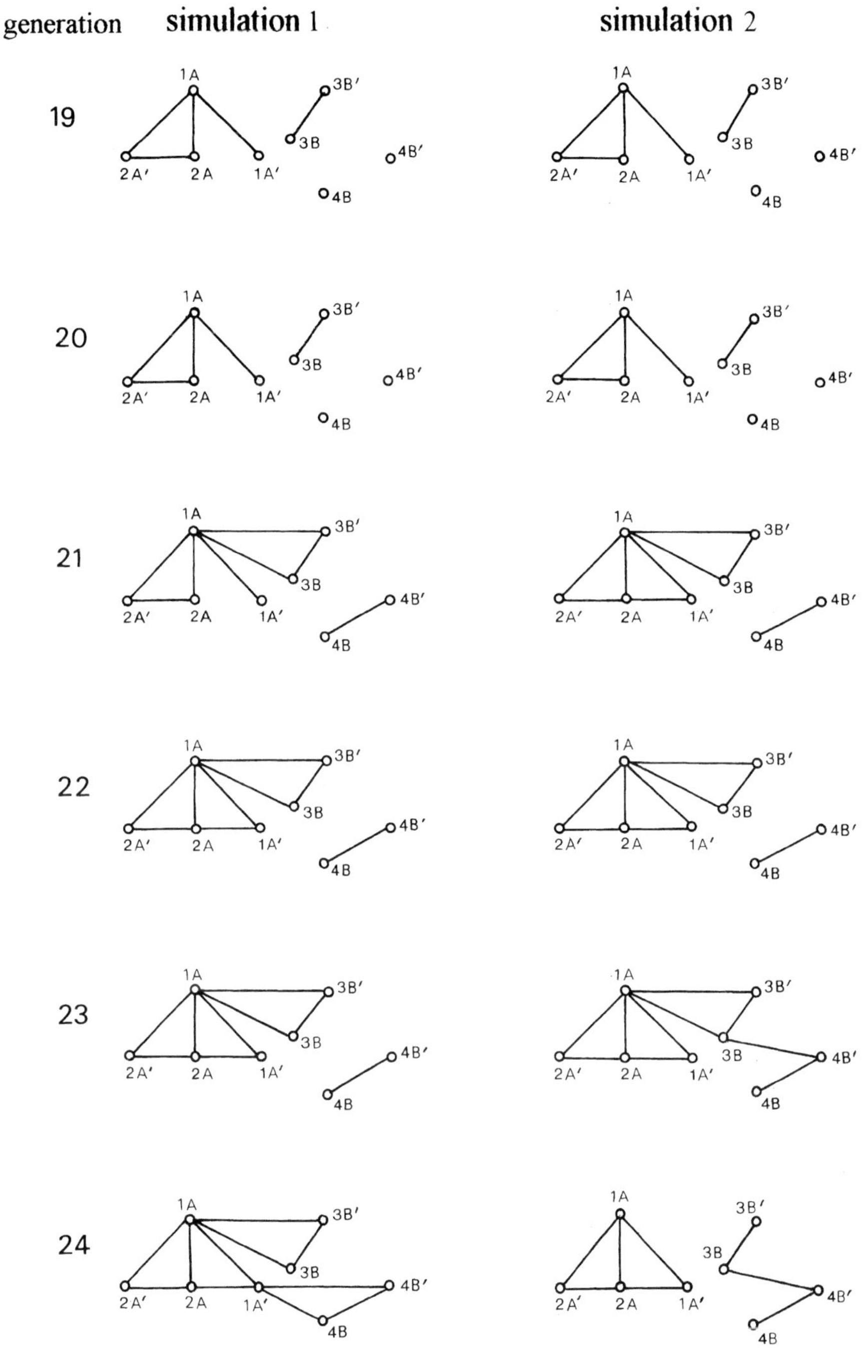

Fig. 10.9. The emergence of the cultural order parameter. $KC(P;\lambda)$, $q = 1$. extended family ($\theta > 1$); sub-community ($\theta > 2$); community ($\theta > 3$)

following generation, this community already has no significant interaction with the second community – they are in complete segregation.

On the other hand, the transformations of the simulation could lead to a weaker order. The structure's transformations in generations 22–24 of simulation 1, for example, weaken the community order. We can observe that sub-communities 3 and 4, which belong to community B, are without any direct interactions. That is, the community order has no explicit influence. Hence, in simulation 2 the two cultural levels are explicit at the same time, but in simulation 1, only one of them is dominant, at the sub-communities level.

From the viewpoint of the agents, the system integration structures that evolve in a socio–spatial system are essentially implicit – the agents do not see or experience such structures. And yet, these structures support the social integration, that is, the actual spatial segregation that the agents experienced in the various cultural contexts. Thus, as illustrated above, the actual dominance of a cultural order is in part a product of the specific contextual developments of the social system. The varying extensions of the cultural orders and their socio–spatial configurations create a potential for change in an agent's action and in the reproduction of the social structure.

In fact, we have already encountered such a situation, in Chap. 7, with the notion of *original identity*. In that discussion it was shown that despite the fact that many of our Green and Blue agents have changed their intentions, identity and location, and became Yellows and Reds, their spatial distribution in the city remained segregative and reflected their original Green/Blue identity. The city, so to say, has enfolded, and thus memorized, their original identity in its spatial implicate order. This phenomenon is specifically significant because many socio–spatial cultural groups are composed of sub-groups and sub-identities that have a potential to become a socio–cultural order parameter by themselves. For example, many national groups with their order parameters emerged by enslaving others, often competing cultural identities such as religious, ethnic etc. Their evolution was characterized by what we have defined above (Chap. 4) as *furcative change*. The recent history of Iran, ex-Yugoslavia and ex-USSR shows that they often re-emerge to become the dominant order parameter of society. In Chap. 7 we mentioned this phenomenon in reference to two distinct cultural groups; here we explicitly study it in the context of a cultural multiple identity.

10.4.3 Game 3: The Relations Between Cultural Levels – Synergetics and Dissonance

How can we explain the emergence of a cultural order, with specific relations between communities and sub-communities? To answer this question it is essential to study the basic relations or interaction between the various cultural orders. For this purpose, we run the simulation in game 3 with the same definitions as in game 2, except for the definition of the cultural structure.

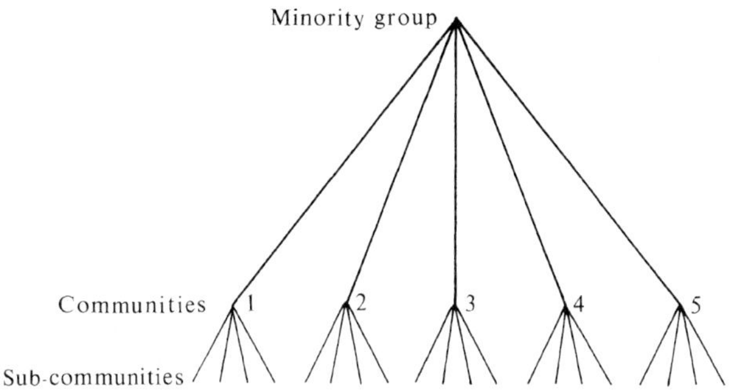

Fig. 10.10. The cultural hierarchical structure for the search process of game 3, based on two cultural orders: community and sub-community

The structure we study here is presented in Fig. 10.10, whereas the evolving Q-analysis structures of communities and sub-communities are in Fig. 10.4.3. Unlike Fig. 10.9, which was based on two simulation runs, the structures in Fig. 10.4.3 are based on a single simulation. These Q-analysis structures allow us to elaborate on the dissonance and cooperation (or synergy) between these relations.

The dissonance between the various cultural levels is illustrated in Fig. 10.4.3, by the dramatic change in the community level from generation 18 to 19. This change occurs with no parallel change in the extended family or sub-community's structures, with the implication that they have a certain degree of autonomy. These illustrations are interesting as they show that even if the residential patterns at the various cultural levels exist in physical nesting, their Q-analysis structures do not overlap and their development need not be identical.

The above dissonance, therefore, does not prevent a cooperation between the cultural order of the Q-structures. Cooperation exists not only in a steady state, but as long as the process continues. The similarity between the structures, from the start of the simulation up to generation 15, can be seen as an expression of a cooperation in a steady state, that is, the structures reproduce each other in a circular causation. Cooperation can also occur in an unstable state. As an example, consider the interaction between communities 1 and 4, and their sub-communities, in generations 16–19. The link between these communities is created as a result of the connection between their sub-communities, and vice versa, (namely, between communities 1 and 4). That is, one identity level creates the condition for the existence of the other identity level.

Generally speaking, a given identity level in a hierarchical cultural structure does not determine the development of the upper or lower levels – it only creates the conditions, or environment, for the others. This can be seen as an additional example of unintended consequence of action in the context of another cultural identity.

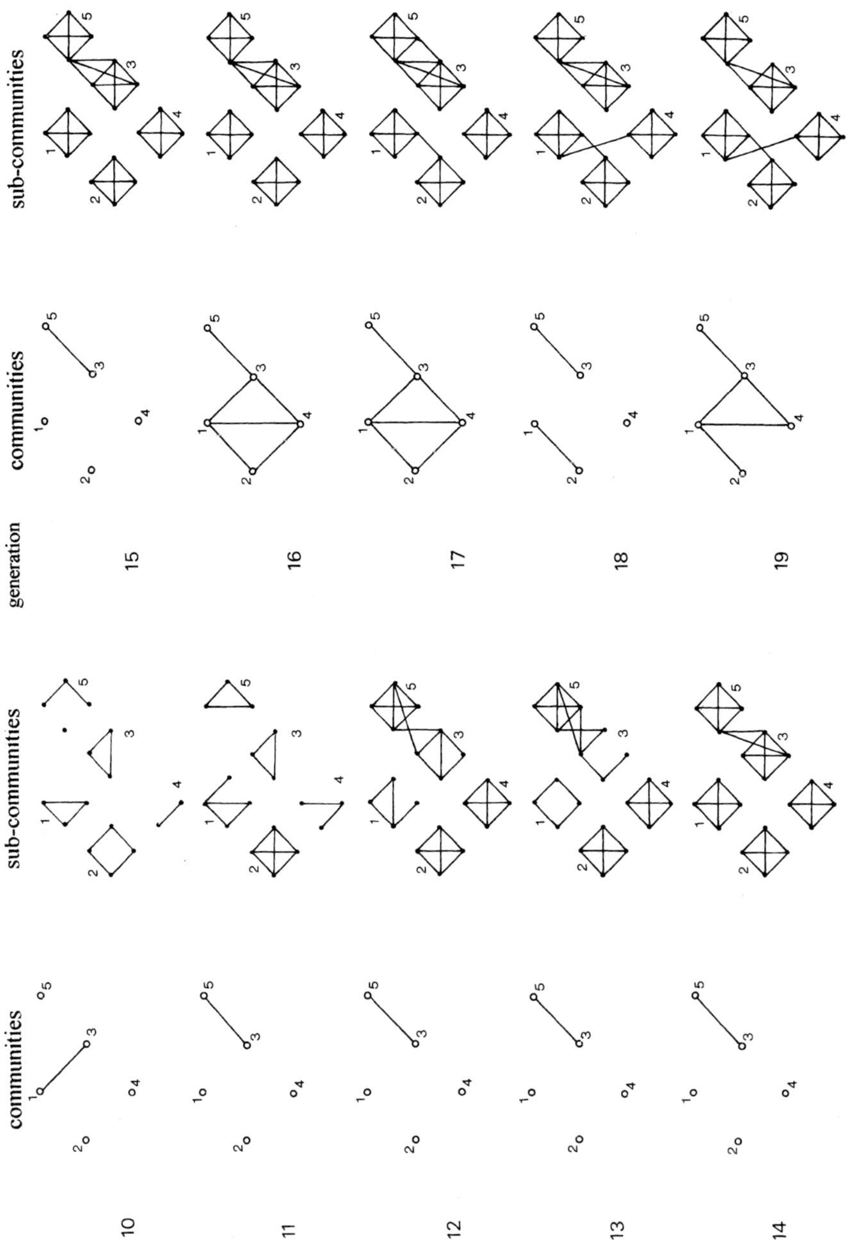

Fig. 10.11. Structural changes in different cultural levels of residential distribution throughout segregation process – synergetic and dissonance. $KC(P;\lambda)$, $q = 1$. extended family ($\theta > 1$); sub-community ($\theta > 2$); community ($\theta > 3$)

Part III Self-Organizing Planning

11. Planning the Unplannable: Self-Organization and City Planning*

11.1 Introduction

The issue of urban and regional planning has already appeared in the above discussion in several contexts. First, in connection with the positivist–mechanistic eco-city which has entailed what we've termed the *first planning dilemma*. Second, in connection with the (in)justice, humanistic and Marxist cities with their consequent *second planning dilemma*, and finally, in connection with the postmodern city which implies, as we shall see in a moment, a *third planning dilemma*. In the following we re-present the above three planning dilemmas as the theoretical context against which we develop *the fourth planning dilemma* which forms the contents of the present chapter – this is the planning dilemma of self-organizing cities.

11.1.1 The First Planning Dilemma

As described above, the 'eco-city' has developed hand in hand with 'rational comprehensive planning', when the division of labor between the two is in line with Faludi's (1973) distinction between *planning theory* and *theory in planning*. The eco-city was to supply the theory in planning, that is to say, an insight about the development and structure of the city and the way it should scientifically and rationally be, whereas 'the rational comprehensive' was the favorable planning theory, that is, the planning procedure which will enable us to plan and implement the good city in an efficient and rational way (Fig. 11.1). At the basis of both was the positivist mechanistic logical–deductive scientific method – first, during the 1950s and early 1960s, its verification approach, and later its Popperian falsification version (see Camhis 1979 for a discussion). As noted above, by the end 1960s and early 1970s came the disillusionment,

> it became evident that 'rational comprehensive planning' ... is an irrational assumption, that planning is a political, incremental ... and essentially 'non-scientific' and non-technical process; it became apparent that ... [the] spectacular scientific instruments we've developed fail to tame

* With Nurit Alfasi

> the city, the metropolis, the megalopolis, the environment... that beautiful scientific instruments such as the gravity, interaction, or entropy maximization models ... can hardly scratch the complexity of the urban scenario, and that so are the 'rent bid curves' of ... urban land use theory, and the 'factorial ecology' of Chicago's 'urban ecology' and the 'location triangle' of ... industrial location theory, and the hexagonal geometrical landscapes of ... 'central place theory'. All this scientifico – mathematical arsenal seemed 'incapable of saying anything of depth and profundity about [the real problem of society and]... when we do say something, it appears trite and rather indecorous' (above p. 32-3).

11.1.2 The Second Planning Dilemma

The first planning dilemma was one of the forces behind the 'qualitative revolution' in human geography and urban and regional planning and its consequent structuralist–Marxist–humanist (SMH) urbanism and planning. Two lines of thought emerged out of the SMH approaches with respect to a 'theory of the city' and a 'planning theory of the city'. One is the humanistic approach to cities and planning. Its central message was and still is *awareness*: humanistic studies of cities will expose the significance of cities to the subjectivity and individuality of people, will distinguish between Relph's (1976) *Place and Placelessness*, between humane and ... non-humane cities ... The cumulative effect of this discourse about the qualitative aspects of cities and landscapes will eventually enter the awareness of planners and architects when they are practically working in and on cities.

The Marxist–structuralist stand was to altogether reject the distinction between planning theory and theory of planning as ideological (false consciousness), with the implication that both the rational comprehensive planning theory and the above naive humanistic stand, are but an integral element (part of the superstructure) of the overall socio–spatial structure of the modern capitalistic city. Any genuine change in planning is thus conditioned by a total transformation – a revolution – in the structure of society. Despite their good will, planners are therefore structurally doomed to play to the hands of the bad guys – the politicians, the ruling classes and the multi-nationals that *control the system*. The combined result of both stands was the second planning dilemma: That despite the insight gained by the SMH approach it left you in a situation that as a planner you have nothing to do

> what are you to do with it when as a planner you have to make a decision about urban renewal, or road networks; what would you say? start talking about base and superstructure? The labor-process? how this beautiful theoretical insight becomes praxis? Gradually it became evident that SMH geography and planning are remote from reality and social relevance even more than positivism. Thus, since the mid 80s, we hear once again

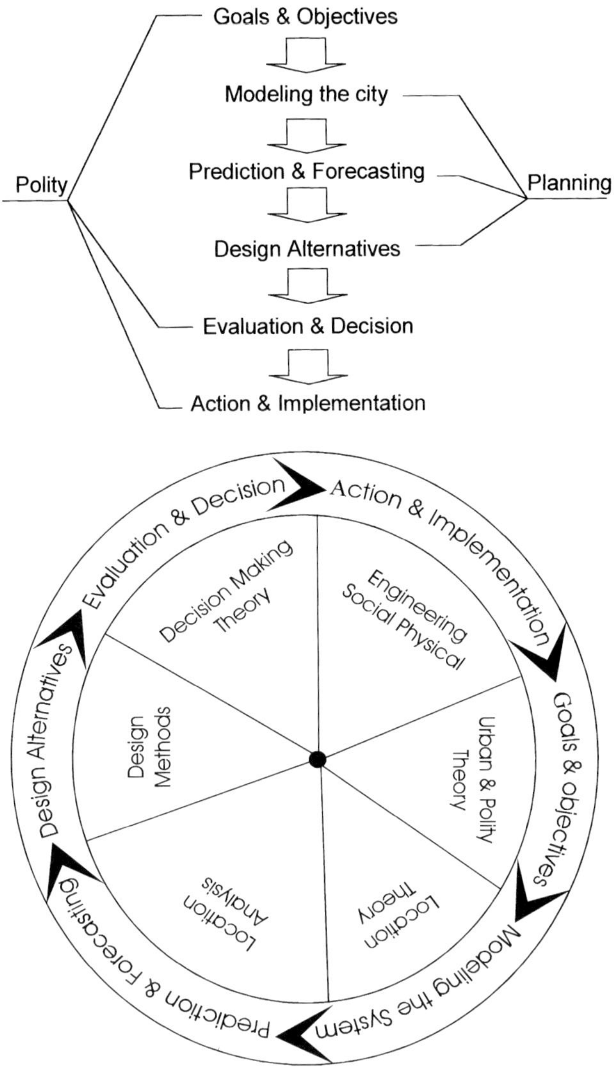

Fig. 11.1. *Top*: A common view in planning theory on the relation between the political and professional levels in the process of planning. The politicians determine the goals and objectives, the professional planners prepare alternative plans in a scientific, rational–comprehensive manner, and the politicians then choose the best alternative. The result is a circular, on-going process as described in the bottom diagram. *Bottom*: A possible division of labor between planning theory (outer circle) and theory in planning (inner circle)

the very same question: 'how can we account ...'; but this time not only for the coexistence of great scientific achievements, on the one hand, and the failure to apply them to society, on the other hand, but also 'how can we account for the failure of the alternatives' (above p. 42-4)

11.1.3 The Third Planning Dilemma

As elaborated above, this general feeling of discomfort from the SMH approach to planning is the ground on which the postmodern view of cities and planning originated. Postmodern urbanism has transformed the impotency of the rational–comprehensive and SMH approaches to control and plan the city, to an ideological platform:

> Not only that science cannot control society and its shrew environment, but that it should not attempt to do so. Let society, and its artificial products the city, the metropolis, the megalopolis, be what they have come to be: uncontrollable, unpredictable and unplannable. Let us urbanists and planners make them more so by deconstructing all that has been constructed; let us deconstruct disciplinary boundaries inside science and between it and art.

On the face of it this new vision sounds highly desirable and creative: an ever changing, ever moving city. However, as noted above, the reality of the emerging postmodern urban landscape and architecture shows a somewhat different scenario: Indeed the postmodern city started with free and creative quotations from the ancient past and from futurist visions, but very quickly it turned into a uniform style – into a kind of neo-conservatism – into the very opposite of what postmodernism was all about. This dissonance between the decided intentions and the daily praxis forms the deadlock of the postmodernist city of the 1990s and the first aspect of what we call here as the third planning dilemma:

> You can't tame, plan, engineer, the environment, since you are trapped in its chaos, and you cannot participate in its chaotic play since you are trapped in its structure, fashion and style.

This postmodernist cultural spirit with respect to planning coincides well with two other phenomena which characterize the last quarter of the 20th century: the crisis of the welfare-state and the associated trend toward privatization. While the complex interrelationships between these three phenomena have yet to be studied, their common denominator with respect to planning is apparent: all three point toward the elimination, or at least minimization, of public intervention and by implication of urban and regional planning. This sounds a straightforward conclusion yet it is not as simple as it may sound.

On the face of it, planning is the exact opposite of free-market and liberal pluralistic post-modern society. Yet this is not so. Urban and regional planning as a practice, discipline and theory, originated mainly in Western society as the logical conclusion of its liberal values and capitalist market economy. It originated out of the so called 'market failure' – the inability of the 'invisible hand of the market' to cope with one of the most dramatic consequences of the industrial revolution: the urban revolution of modernism. As later developed in economic theory, market failure, be it in the form of cities, public goods or externalities, provides the economic rationale for public intervention and planning. So privatization does not imply the elimination of planning, but rather a new balance between the planned and the unplanned. As for the nature and practice of planning, for the lessons from the first and second planning dilemmas, for the implications of the new postmodern city to the nature and practice of planning, there is no message here but sheer (neo)conservatism.

11.1.3.1 The Facade

The central element of urban and regional planning at the age of postmodernity, post-welfare-state and privatization is *the facade*. With respect to postmodern architecture and urbanism this is literally so: look behind the facades of the most extravagant postmodern buildings and you see nothing but the old modernism that postmodernism has promised to challenge and deconstruct. Look at the city scale and you'll find that urban design, that is, the planning of facades at the city scale, has become the dominant mode of urban planning. Look behind the facade of the postmodern city and you'll see the old (not so good) city of modernity with its alienated city centers, endless boring suburbs and slums. Look at the administrative–economic facet of city planning and you will find slogans of 'city selling' and cultural pluralism. But look behind the slogans and you will find that the modernist phenomena of urban blight, socio–economic and cultural deprivation are still there, behind the slogans. The difference is that in the high days of modernism there were great hopes and optimism about the ability of science, technology and humanism to improve the city; with postmodernism the hopes and optimism have gone – let us play with the city etc., etc., ... And this (if it is not already self-evident) is the other aspect of the third planning dilemma.

11.2 The Planning Dilemma of Self-Organizing Cities

The situation thus far is as follows: the structuralist–Marxists have nothing to say about the practice of planning since in their eyes planning is part of the ideological superstructure of capitalist society. Humanist geographers and city planners have nothing to say about the practice of planning since they put all their cards in the notion of planners' 'awareness'. Postmodernists and privatizationists have nothing to say about the practice of planning because for them 'planning' is a dirty word. The only explicit discourse about how planning should be studied and practiced comes from the various positivist approaches. The result is simple: planning around the world is practically conducted and implemented in line with the principle of the rational–comprehensive planning approach and its subsequent elaborations in light of the principles of general system theory. This is so irrespective of the ideological–theoretical orientation of the planners themselves.

To a large extent the above holds true also for many of the studies which have approached cities from the perspective of the paradigm of self-organization. The (often implicit) assumption in these studies is that the theory of self-organization provides, on the one hand, a deeper phenomenological insight into the dynamics and operation of cities, and on the other, some of the most sophisticated mathematical models, simulation techniques and statistical methods to analyze complex systems. This, as we've seen in the previous chapters, is indeed the case, but, there is a big *but* here: one of the central insights of our interpretation of cities as self-organizing systems, that is to say, as open, complex, far from equilibrium, systems, is that they are unpredictable and thus uncontrollable; and yet, predictability, control and 'social engineering' are the pillars of modern town and regional planning. Furthermore, in Chap. 2 and above in this chapter, we've discussed the failure of modern urban and regional planning to solve urban problems and control cities by means of the scientific method and rational planning. Our interpretation was that one reason for this failure is misappreciation, on the side of planning theory, of the nature of cities and metropolises as self-organizing systems. In light of this the planning dilemma of self-organizing cities is this: *in the absence of predictability and control can there be a meaning to planning?*

Perceiving cities as self-organizing systems raises doubts as to the whole rationale of urban and regional planning, because the elementary principles that make a system (in this case a city) a self-organizing system, diametrically oppose the theoretical principles and practice of the rational comprehensive planning. Modernist–rational planning started from two premises: First, that while a free society, free economy, and non-interventive policy, is the ideal, some degree of intervention and control, in selected domains, is necessary and unavoidable. Second, that science, by its capability to explain the past and the present and on the basis of these to predict the future, will provide the proper means for a planned intervention and control. Planning in general and urban and regional planning in particular have thus become the academic disciplines within which such premises should be elaborated (Portugali 1980, Breheny and Hooper 1985, Beauregard 1989, Goodchild 1990, Pasty 1992).

Now, in the case of cities as self-organizing systems, science herself tells us that she cannot predict their behavior, that this inability is not a technical shortcoming which science will overcome with time, but the very nature of cities. By their very nature as self-organizing systems, cities are unpredictable, uncontrollable and in this respect are unplannable. This situation might lead to two conclusions. One, to eliminate planning or at least to minimize it as much as permitted by the market – this, as we have seen, was and still is the postmodernist–privatizationist strategy which implies that the old, anachronistic, mechanistic planning approach still prevails. Two, to develop new forms of urban and regional planning which start from the perception of the city as a self-organizing system. The reminder of this chapter is a search toward this aim.

11.3 In Search of a New Planning Approach

We start our search by introducing a distinction, suggested by Alfasi (in preparation), between two forms of planning: *just-in-case planning* versus *just-in-time planning*. The first refers to the traditional mode of planning as currently practiced in most (if not all) urban and regional planning agencies, whereas the second, to what planning in a self-organizing system might be. On the basis of this discussion we suggest preliminary principles for a self-organizing planning system and elaborate them in light of studies which are already existing in the field. We then turn to studies of self-organizing cities. Most of these studies simply do not concern themselves with the above planning dilemmas. Some, however, do suggest how the self-organization interpretation can provide a guide to planners and policy makers. These studies will be presented at the second half of this section as some of the elements in our proposal for a new planning approach which can cope with the nature of cities as self-organizing systems.

11.3.1 Just-in-Time Versus Just-in-Case

The key notion behind this distinction is the *kan-ban*, the Japanese for *just-in-time* – a method of production and management originated in Japan in 1948, designed by Ono Taiichi, a labor union leader who became a Toyota manager. *Kan-ban* is the slogan of a managerial revolution which according to Castells (1996, p. 167) is one of three trajectories that mark a transformation from Fordism to post-Fordism, or *Toyotism*, as it is also called. The two other trajectories are a transformation from mass production to flexible production, and from an economy dominated by large corporations to an economy of small and medium-size firms as the agents for innovation and change. These transformations, according to Castells, provided the infrastructural engine for a sequence of really 'grand transitions' from industrial to *information society* and from the latter to the currently emerging *network society*. *Just-in-case*, is the exact opposite of the *kan-ban* – it is the method of production, management and planning which characterizes Fordism.

11.3.1.1 The Supermarket Metaphor

According to the legendary story it all started in a supermarket in the USA, when a visiting delegation of Japanese managers from Toyota, suddenly realized the simplicity, efficiency and sophistication of the supply system in front of their very eyes. In the supermarket there is no 'job-shop' which tells the workers when and where to put products on the shelves, nor a long term planning and management of what products to buy and in what quantities. The customers 'pull' products from the shelves and the supermarket workers simply fill the gaps; for this purpose they order the needed products from their suppliers, in the exact amount required by (the customers of) their specific supermarket, and so it goes on in the chain of production and supply. All the products thus come 'just-in-time', fresh, exactly when they are demanded, with no need for long-term planning of inventory and stock keeping. The shelves form a medium for a fast, complex and at the same time simple, highly synchronized and efficient, interaction system between sellers and buyers.

The supermarket was thus a metaphor for a new method of management – Toyotism. Toyotism was first applied to the Toyota corporation, then to other Japanese firms, and then to many other firms around the world, including Volvo, Volkswagen and others. The method is based on a strong cooperation and interaction between workers and management, on multi-functional workers, on workers' initiative and participation in the processes of production, on team work, on decentralization and a flat hierarchy (Fig. 11.2). In fact, the logic of the just-in-time system is the exact opposite of Fordism – the system that still prevails in most of Western economies. The latter is characterized, as is well known, by a strong and rigid hierarchical structure, by what is often described as a structural antagonism between workers and management, by workers' specialization and labor demarcation, and, by long-term planning of supplies in the form of stocks – just-in-case. *Just-in-case* for Fordism is what *just-in-time* is for Toyotism.

11.3.2 A City Is Not a Tree; Nor Is It an Aristotelian Category

A nice illustration of the difference between the just-in-case and just-in-time systems, is the conjunction between Alexander's (1965) paper 'A city is not a tree', and our discussion of the nature of cities and urbanization in Chap. 1. In his study, Alexander makes a distinction between two ways of thinking about cities: one is in terms of a hierarchy or a tree, and the other in terms of a semi-lattice (Fig. 11.3). The two cities differ from each other in their structure – a tree versus a semi-lattice, and in the processes that created them and that take place in them. In the tree city each sub-system in the city is fully independent from all other subsystems of its level, and it can thus interact with them only via a higher order subsystem. In the semi-lattice city there are overlaps between subsystems of the same order, so that interaction

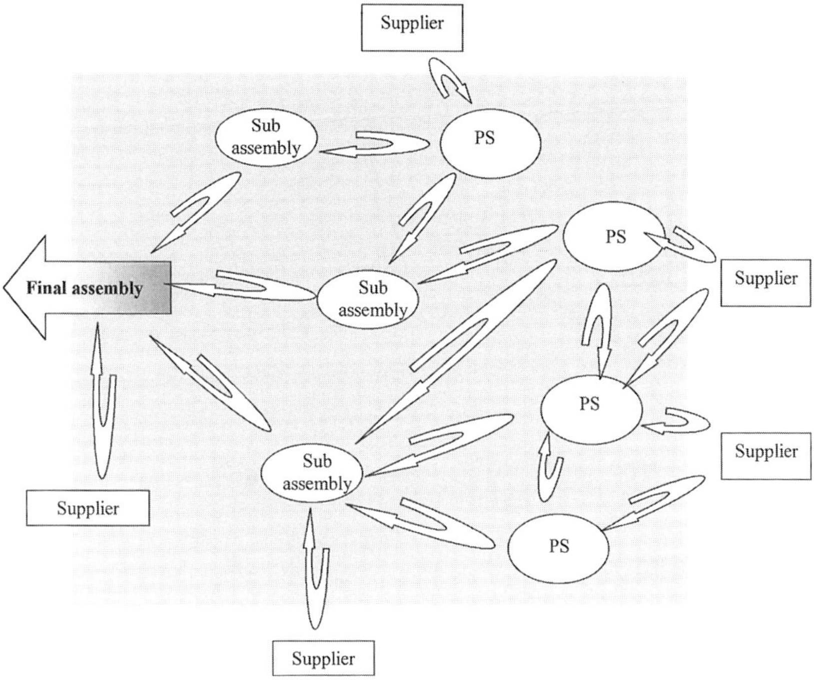

Fig. 11.2. According to the *kanban*, the various production departments of a given factory, for example, are relatively independent units. Each is both a producer and a supplier (PS) that pulls products from other PSs in the factory (the ovals in the diagram), as well as from external PSs (rectangulars in the diagram) when needed. The difference between internal (ovals) and external (rectangulars) PSs is rather fuzzy and quantitative: there are more interactions between PSs inside the factory then between them and external PSs. The production process is being initiated (i.e. pulled) by the demands at the final assembly cell(s), then proceeds to the sub assembly cells and to other PSs. The dark area symbolizes the production floor (after Hay 1988)

can occur vertically, horizontally and in oblique. As noted by Alexander, it is not only the overlap which makes the difference, but

> more important is the fact that the semi-lattice is potentially a much more complex and subtle structure that the tree ...: a tree based on 20 elements can contain at most 19 further subsets of the 20, while a semi-lattice based on the same 20 elements can contain more than 1 000 000 different subsets (Alexander 1965, p. 382).

The distinction between the just-in-case tree-cities and the just-in-time semi-lattice-cities, can be further elaborated in conjunction with the distinction made above, in Chaps. 1, 2, between the Aristotelian city and the Wittgensteinian, Prototypical, Ideal-type and IRN cities. In the Aristotelian

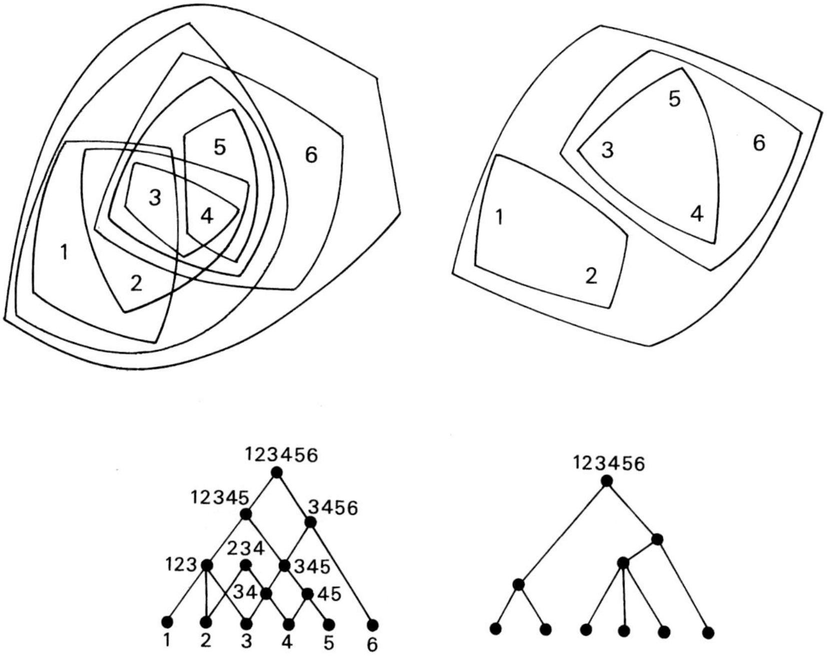

Fig. 11.3. The distinction between a tree structure (*right*) and a semi-lattice structure (*left*) according to Alexander

city, the city as a whole and every category in it, such as a house, a neighborhood or a freeway, are defined by some necessary and sufficient conditions. This implies that categories are closed systems which exist outside each other, and that they form relatively simple tree structures. In contrast, in the Wittgensteinian, Prototypical, Ideal-type and IRN cities, the city and the many categories that make it, form complex family resemblance networks with overlappings, articulated connections, prototypical and peripheral elements and so on. This implies that the city and the many categories in it are open systems that overlap, or rather enfold, each other, and that they form complex dynamic systems. In short, the just-in-case Fordist cities are simple, mechanistic tree structures, whereas the just-in-time cities are self-organizing systems.

In his work, Alexander criticizes the modernist mechanistic architecture and town planning for wrongly perceiving cities as trees, for planning and acting in existing cities as if they were trees, and for planning and building new cities with tree structures. As an extension to his criticism we add that not only planners have acted on cities as if they had simple tree structures, but they have also built their planning agencies and constructed their planning procedures as simple systems with tree structures and all the other ingredients of the Fordist just-in-case model: vertical–hierarchical integration, highly demarcated division of labor, specialization, and so on. We further suggest that the failures of, and the subsequent disillusionments from, urban and regional planning as described above under the titles of the three planning dilemmas, result, first, from planners' misappreciation of cities as open, complex, self-organizing systems, and second, from the organization of planning agencies, procedures and action, as just-in-case Fordist systems.

11.3.3 Push Planning

One of the main differences between just-in-case and just-in-time planning concerns stock and inventory. In a sense the whole of the planning machinery, as it is practiced and theorized today, is geared toward the preparation of planning stocks in various forms and scales. The process as a whole has a hierarchical tree structure, implemented step by step down the planning tree (or up, if you visualize it as a real tree). At the top (or trunk) of the planning tree are the overall, long-term, national/regional plans, which cover large areas and are supposed to prepare in advance the population, economic activities, transportation, housing ...stocks for the coming 2–3 decades. Then come smaller scale plans, say city land-use plans, which 'translate' the regional plans to a city scale and thus prepare the city for the coming 2–3 decades of development. Then development plans for quarters and below them construction plans for still smaller space–time sections of the city and so on. Once the stock of plans are there, that is to say, the needed areas for housing in the next decades are assigned, the land along the planned freeways is controlled (and if necessary confiscated), and so on, all that is left to do is *implementation*: families can build their home in the planned residential areas, firms can build or buy offices in the planned CBD (central business district), and industrialists or craftsmen can open their businesses in industrial areas.

In between the first part of the process which is often called *development planning*, and the implementation side of the process, which is also termed *development control*, there is an interval (called 'public hearing') during which the plans are made public and ordinary people and also organizations, have some limited time during which they can *submit an objection* to the plans or certain aspects in it.

The same logic holds true for the flow of ideas, innovations and initiatives along the planning tree. The first planning initiative starts at the top of the hierarchy, by politicians who know what people want or need by virtue of the fact that they democratically represent the people, and/or by national–regional planners who know what goods are needed on the basis of scientific long-term predictions. They set the national or regional goals, which are then translated by urban planners to smaller scale goals and objectives and so on. As for ordinary people, they are not supposed to initiate anything, nor to innovate or suggest ideas, but to obey – to act in accordance with the plans, and if they feel harmed by the plans, they can, as just noted, submit an objection.

Planning can thus be seen as an on-going process of pushing down the hierarchy plans, goals, initiatives and ideas, that altogether form the planning stock. In this respect urban and regional planning is not different at all from the process of production in most industrial and business firms in Western society. In the latter the dominant form of management is termed 'push management' (Hay 1988). It starts with a central plan, based on some forecasts, which determine overall production, which is then translated into job-shop instructions, which are delivered to work cells in the factory...

> Now begins the pushing. Each operation in the chain does its own thing and pushes it on to the next operation by a particular date. That operation knows it is coming, does its thing, and pushes it on by a particular date. The expectation is that all these things being pushed along will arrive at the same time on a particular date, so that assembly or shipping can happen on schedule (Hay 1988, pp. 104–105)

For this centralized pushing management to work there must exist a sufficient amount of stock of both raw materials and products. The stock safeguards the process from unexpected obstacles in this pre-determined process.

Another important property of this push/just-in-case planning is that the system has an in-built antagonism in it: in the firm it is between workers and management, as well as between different levels and stations in the production process. In planning agencies it is first and foremost between the planning administrations and the people (who have the right to object and thus to interfere in the smooth operation of the plan), as well as between planning agencies at different levels of the hierarchy. For example, between the planning committee of a city which represents the interests of the city, and the planning board of a district which is supposed to represent the higher-level interest of the district.

11.3.4 The Food-Market of New York City

The supermarket which provided a metaphor for the development of Toyotism is just a miniature of the real big, large-scale case of just-in-time production and supply system – the food markets of large cities. A case pointed at by Ruthen (1993) is the food market of New York City: a large number of firms of all sizes supply food for over 7 millions people, 'without creating shortages or surpluses'. All this happens as if by itself, with no central planning nor public authority to regulate the process and safeguard it by producing stocks and keeping of food stores. Furthermore, none of the millions of the interacting elements in the system – individual households, firms of all kinds and restaurants, keep stock for more that a day or two. And yet it works and very efficiently so (especially if you compare it to centrally planned, publicly supplied, products). It works not by means of a Fordist just-in-case push-planning, but by means of a kind of a Toyotist just-in-time 'pull-planning' – as in the supermarket.

Very much like the flow of the just-in-time plant, shown in Fig. 11.2, the food market of NYC begins at the end, driven by the actual demand and needs of its customers. Short term stocks are replenished by suppliers, which in turn go to their suppliers, etc. Each element in the system operates as a customer for its supplier and as a supplier for its customer. The system, containing numerous elements, is forever complex, open, and well connected to its environment, due to a large scale of flow passing through it (Fig. 11.4).

What happens in the food market of New York happens also in the food markets of cities around the world. And, what happens in the food markets of cities is but one facet, indeed a beautiful illustration, of a more general property of the city as a self-organizing system.

11.4 Preliminary Principles for a Self-Organizing Planning

As we have just seen, self-organization in the supermarket provided a metaphor for the *kan-ban*, the just-in-time Toyotism which is today regarded as one of the most advanced management methods in the field. According to Castells (1996), it is not only an advanced management system in terms of economic efficiency and productivity, but also one which is better tuned with recent large-scale transformations of society, from an industrial-Fordist society to the newly emerging Toyotist network society which according to Castells marks the end of the 20th century and the beginning of the 21st.

The suggestion here is to take the whole of the self-organizing city as a metaphor for the development of a new planning theory and methodology. Such an approach has the potential to make urban, regional and environmental planning more innovative, and productive – like the Toyotist just-in-time management system, and better tuned with the nature of cities in general and the cities of end 20th century and the beginning of the 21st century in particular. That is to say, better tuned with the self-organizing property of cities in the past and in our time.

11.4.1 Parallel Distributed Planning ("PDP") in a Self-Organizing City

PDP or *Parallel Distributed Processing* is a title of a book (Rumelhart et al. 1986) and another name for the neo-connectionist approach to the study of brain functioning and cognition. The central thesis is that the brain is a huge network of billions of neurons and that the various cognitive tasks are implemented by the brain by means of parallel processing and complex interactions among the neurons. The central thesis of Haken's (1996) synergetics approach to brain functioning and cognition is that the brain is, and operates as, a self-organizing system. This view is shared by some other students in the

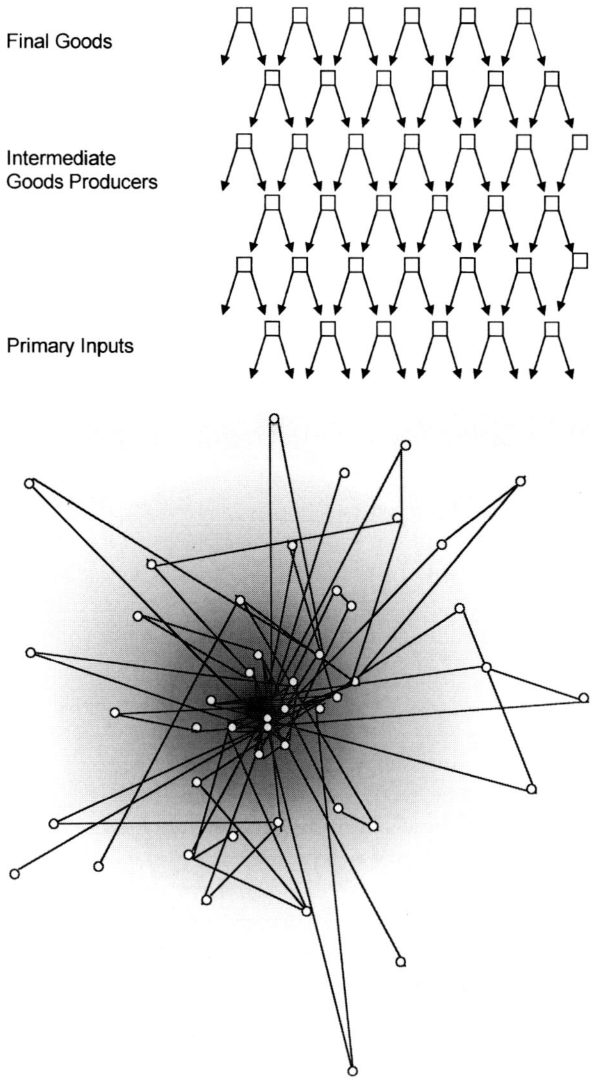

Fig. 11.4. *Top*: Bak, Chen and Creutz's (1993) diagrammatic representation of a self-organized market. Each producer receives orders from two customers. If he does not have the amount demanded in stock, he sends orders down the net, produces several items by himself, gets the ordered goods, and then ships the lot to the customer, and so it works in the entire network. As can be seen, the above presentation is rather schematic (each unit is connected to two units), and a-spatial (the spatial location of the units is ignored). *Bottom*: A non-schematic and spatial, self-organized market – a city, for example. Each producer/customer has a specific location in the city and each is connected to a not-fixed and changing number of suppliers/producers/customers. Note that the city as an entity is distinguished and characterized by the relative intensity of its internal interactions. This, plus population and built-up area densities, make the city a category and an entity distinguished from its environment. The dark area symbolizes the city and the degree of darkness, the intensity of interactions

field as well. Our usage of the notion 'PDP' comes, on the one hand, to point at the similarity between a central property of planning in a self-organizing city, and the operation of the brain as a self-organizing system, while on the other, the quotation marks we add to the PDP come to indicate that there are differences. The similar property we refer to, is the distributed parallel activities that take place in the brain and in cities.

From the perspective emphasized thus far in the above discussion, city planning and self-organizing cities stand in contradiction and opposition to each other. From another perspective that we'll elaborate in the remaining part of this chapter, however, cities as self-organizing systems exhibit a lot of planning – *parallel distributed planning* ('PDP'). The key feature in this new perception of cities is the fact that the various actors which form the parts of the city, are themselves a self-organizing system by virtue of the property that they are free agents. That is to say, they have memory, they act intentionally, and they are conscious about their memory, intention and action. In short, they have the capacity to take decisions, to plan and act according to their decisions and plans.

If you look from this new perspective at the supermarket, at the food-market of New York City, at the working of whole cities, and also at our virtual cities in Part II above, you'll see a huge number of planning and decision making agencies: individuals, families, firms ... you will see a huge number of relatively independent free agents and agencies *planning in parallel.*

11.4.2 Decision Making Agents as Self-Organizing Systems

What can self-organization theory tell us about the decision making and planning of the many planning agencies, which are distributed in the city and are working in parallel? First and foremost, that not only the city as a whole and its 'PDP' are self-organizing systems, but each decision-making agent is a self-organizing system too. This is so with respect to individual persons, families, firms, as well as the planning administrations of cities and nations. Each is a complex, open and thus self-organizing system. This follows, first, from what has just been said that the central thesis of the synergetics approach to brain functioning and cognition is that the brain operates as a self-organizing system. Second, it follows from our SIRN (Synergetic Inter-Representation Nets) approach to cognition and cognitive mapping. According to this view, not only individuals but also groups of individuals, such as households, firms or planning teams, are self-organizing systems by themselves. If this is so, if the individual, the household, the firm or the planning team, are free agents, and self-organizing systems, then decision making in the context of cities, must be seen as an ongoing interplay between self-organizing systems of several scales: individuals, families, firms, planning administrations, and so on.

The above perception of decision making in the city is implicit in the models we have presented in Part II above. In Chap. 14 below we explicitly consider this issue within the formalism of synergetics. More specifically, we suggest a decision-making theory and model which are based, first, on Haken's (1996) recent analogy between pattern recognition and decision making. Second, on the conceptualization of cognitive mapping in terms of synergetics (Portugali 1985, Portugali and Haken 1992). Third, on the reformulation of our SIRN model (Portugali 1996b, Haken and Portugali 1996) as a decision making model.

11.4.3 Self-Organization in Private and Collective Planning

Commenting on the food-markets of large cities, Holland (in Ruthen 1993) once noted that "from the point of view of physics it is a miracle that happens without any control mechanism other than sheer capitalism". The 'miracle' we want to address below is that out of this seemingly uncoordinated and chaotic parallel distributed planning emerge not just a steady-state balance between demand and supply, with little stock and so on, but overall collective and often innovative and creative, decision-making, planning and implementation, that is to say, self-organized and self-organizing planning and plans. Moreover, despite the fact that such plans emerge spontaneously, with no central and authoritative agency behind them, people act according to these plans that no one made. In order to describe the way such a collective planning emerges, we shall examine the 'PDP' described above from the perspective of some of the self-organizing cities of Chap. 3.

11.4.3.1 Planning in Sandpile Cities

From the perspective of the sandpile cities of self-organized criticality, the first step might be to treat the highly distributed and parallel planning decisions in the city as 'avalanches', and to look at their size distribution with the expectation that it will form a Pareto distribution. And indeed, preliminary experiments that we did indicate that in the case of Tel-Aviv, which is the metropolitan center of Israel, this seems to be the case, whereas in the case of Beer-Sheva, a centrally planned (by the government) development town, this is not the case (see Fig. 11.5 for details). Self-organized criticality thus tells you that the uncoordinated 'PDP' produces a highly coordinated *sandpile plan* – a large, collective plan which the millions of planning agencies operating in the city 'obey'.

Self-organized criticality shows that the highly coordinated operation of food-markets in cities, and of whole cities, is supported by a rich, chaotic and complex, array of overlapping planning activities and interactions between a huge number of planning actors operating in the city. (It also provides us with a good method to distinguish between a self-organized city such as Tel-Aviv, and a city such as Beer-Sheva.) It doesn't tell us, however, how this steady-state 'slope', this overall plan, emerges in the first place. This story is told by the 'grand' theories of self-organization.

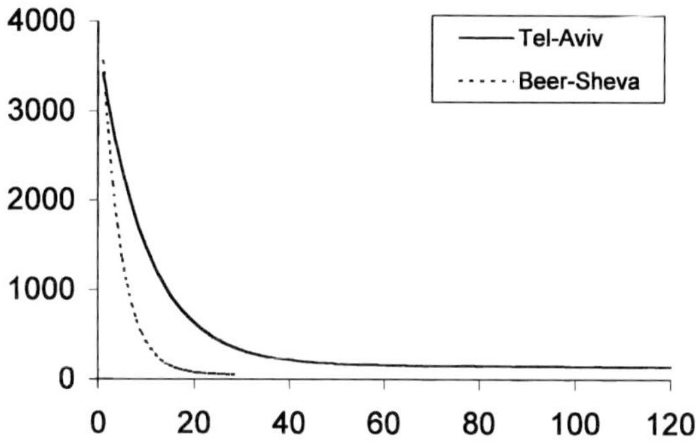

Fig. 11.5. A comparison between the size distribution of 'planning avalanches' in the cities of Tel-Aviv and Beer-Sheva. Tel-Aviv is characterized, as can be seen, by a normal, Pareto, rank–size distribution of plans, whereas Beer-Sheva by a dichotomized distribution. That is, in Tel-Aviv there are a few very large plans, more middle size, and many small size plans. In Beer-Sheva, there are a few very large plans (planned exclusively by the government), and more/many very small plans, made by citizens. The data for the Tel-Aviv graph are planning applications for housing submitted to the city's building committee, until December 1996 (133 plans for about 60 000 new housing units). The graph for Beer-Sheva is a reconstruction based on indirect information

11.4.3.2 Planning in Dissipative Cities

From the perspective of dissipative structures the parallel distributed planning which takes place in the city and its chaotic sandpile properties, provide yet another illustration to the role of dissipation in self-organizing systems. A flood of plans and in all sizes, with a lot of overlappings in content and scale, crosses the boundaries of the city, when only a few are eventually materialized and 'selected' as the city's overall plans. On the face of it, this is sheer waste and dissipation of human energy. However, according to the theory of dissipative cities it is this seemingly inefficient dissipation that keeps the city running in a far-from-equilibrium, creative, steady state. This seemingly inefficient dissipation is also the source of planning creativity of the city as a whole.

A case in point is the recent wave of Jewish immigrants from ex-USSR to Israel. The official forecasts, made by all governmental planning agencies in Israel, were that Israeli cities will not be able to cope with the coming needs for housing and jobs. Reality has shown a different story: the information about, and the wave of, the newcomers have triggered a short period of fluctuations during which many actors in the city have re-planned their routinized use of housing. The result was that quite suddenly a large number of vacant apartments appeared in the market and a new steady state dominated the housing market of Israeli cities. As we shall see below, this new steady state can be seen as a new housing plan for the city. The existence of 'waste' dissipation in the city was thus the source of its creativity – its ability to undergo structural change and phase transition and to adapt to the new planning situation in a creative way. Another interesting lesson from this story is that waste and dissipation at the level of the distributed planning actors in the city has proved itself as a source of creativity, whereas the housing stock prepared by the government according to the long and short term plans, became obsolete – much of this housing stock is still vacant and deteriorating ....

11.4.4 Planning Synergetic Cities

The more detailed story comes from the direction of the theory of synergetics. From this point of view, the highly coherent steady-state described by the sandpile plan was preceded by, and emerged out of, strong fluctuations and a chaotic stage. The latter is characterized by a competition between different configurations, which are eventually solved by means of the slaving principle and give rise to one or a few order parameters (and a specific rank–size distribution of avalanches). With respect to planning, this chaotic period, during which many configurations are being tested, is a situation by which the many actors operating in parallel in the city try to plan (= test different configurations) their individual activities. As we shall see immediately, the means to make such new plans, are memorized/experienced patterns of actions which seem to be most similar to the new situation at hand.

11.4.5 Plans as Patterns of Routinized Activities

The decision making and the plans themselves concern all aspects of life: how much stock and from what goods to buy and keep at home or at the firm, where to consume what, and how often, where in the city to live, to work, to bank, to go shopping, to movies, coffee shops, etc. All these decision makings and plans concern, on the one hand, ad-hoc decisions about specific single actions and, on the other, decisions about patterns of routinized activities, actions and behaviors. The latter are specifically important as most planning activities at all scales are aimed at finding an appropriate routinized pattern. This is so with respect to the relations between a building to its pavement and neighboring buildings, in which case the pattern is a certain architectural configuration, this is so with respect to commuting, in which case it is a pattern or schemata of movement, and this is so also with respect to how to apply for a new job, in which case it is a pattern of behavior. The chaotic stage which preceded the steady state is a stage of learning during which many configurations, that is to say, patterns of routinized activities are being examined. Out of this process of learning and decision making certain plans emerge – they are, in fact, *order parameter plans.* This process happens at the parallel distributed level of planning and also at the level of the city as a whole.

11.4.6 Enslavement to, and Emergence of, a City's Order-Parameter Plans

At the level of the city as a whole one can identify two basic situations. One, which is typical of a new actor who enters the city, and another, which is characteristic of a phase transition and the emergence of a new order parameter plan in the city as a whole. In the first case, for example when a new household or a firm immigrates to the city, the process of learning will involve learning the 'rules of the game' of that city, and taking decisions in line with intentions stored in memory. A typical result here will be *adaptation*, which in terms of synergetics can be regarded as slaving: the new actor is being enslaved by the city's overall order parameter plans. In some rare cases, however, some of the innovative plans adopted by this newcomer might attract attention and might entail a process of space–time innovation diffusion that eventually will result in a phase transition at the level of the city as a whole.

The latter scenario by which the plans and actions of the individual household or firm might trigger a global change, is just one possible route to the emergence of a new order parameter plan in the city. In fact, every event, internal or external, which acts on the city might in certain circumstances trigger a self-organized process of structural change. Examples might range from oil and financial crises, technological innovations, a new migration wave – as in Chap. 6 above, and even a car accident, as in the case of the Intefada. What is significant in these cases is that many actors simultaneously try to plan or re-plan their routinized activities in the city. That is to say, they simultaneously test many plans, many configurations, many space–time routinized patterns of activity. Out of these chaotic movements and strong fluctuations emerge new order parameters which are new spontaneous city plans, which by means of slaving entail a situation of circular causality: the plans and activities of the individual actors are determined by the order-parameter-plans of the city, and the actors by acting in line with these plans support and reproduce them. These ideas are further elaborated below in terms of the decision making model we now propose.

11.4.7 Between Fast Local Plans and Slow Global Plans

One of the main strategies of the synergetic approach to complex systems is to give a careful attention to the various time-scale rhythms that coexist in the system. According to synergetics, in many instances, the slow, spatially-global variables of the system act as order parameters for the fast, spatially-local variables. Accordingly, a large and complex system can be seen as a hierarchy of very slow and global space–time variables, which function as order parameters for lower scale variables, some of which are the order parameters of still lower scale variables and so on down the hierarchy. Looking at the large amount of parallel distributed planning that takes place in the city, it can be seen that quite often the slow, spatially-global plans function as order parameters for the fast, spatially-local plans. This is so with respect to a single individual, for example, when the fast-local routinized plan of 'buying milk and bread', is enslaved by the rhythm of the slower space–time plan 'commuting to work'. The latter often functions as an order parameter for the former. This is so also with respect to the relations between the relatively slow order-parameter plans of a single building (which might range between 50 to a few hundreds of years) and the relatively fast changes in the plan of the rooms inside the building. This is so also with respect to the relations between the very slow order-parameter plans of the city as a whole and lower scale order parameters.

As we shall see below, and as illustrated in more detail in Chaps. 7, 8, due to the complexity of the city, the parallel distributed planning in it, with its semi-lattice network and the plentitude of overlappings, the relations between the plans at the various levels and scales are far from being a simple hierarchy. On the one hand, the planning system is characterized by circular causality in that the fast-local plans are enslaved by the slow-larger ones, and by so doing they reproduce the structure of the city with its hierarchy of order-parameter plans. On the other, because of parallelism and overlappings, changes at small-scale fast and local plans, often very minor, might in certain circumstances trigger strong fluctuations, bifurcations and phase-transition of the city as a whole. In fact, every level and scale in the parallel planning network might be a source of global structural change and morphogenesis.

11.5 Planning the Ideology of the Unplannable

11.5.1 Planning the Ideology of Planning

'Planning the Ideology of Planning' is the name of a chapter, in David Harvey's (1985b) *The Urbanization of Capital*, in which he elaborates his Marxist criticism of urban and regional planning in Capitalist society. His thesis is founded on the Marxist dialectical tension between practice and ideology, that is to say, between planning as it is daily practiced and planning as it is perceived. As practiced, planning is part of the political superstructure which safeguards the reproduction of the capitalist mode of production: by means of its specific administrative structure and planning procedures, it ensures that capital will be redistributed is such a way that the rich will remain or become richer and the poor will remain poor, that the former will live in the more expensive quarters of the city whereas the latter in poverty areas and slums, and so on. As perceived, planning is commonly regarded as a mechanism to correct market failures with respect to public goods and externalities, as a method by which capitalist liberal society achieves moral goals that are not necessarily capitalist or liberal. That is to say, as an organization in charge of the construction of roads, on the allocation of open spaces for the good of society, on building housing for the poor, on rehabilitation of poverty areas in cities, and the production and allocation of social welfare services in general.

This gap or dissonance between the way things are and the way they are perceived is captured in Marxism by the Hegelian notion of *ideology*. Ideology, according to this view, is at the very same time an idea – a plan for the future, and a *false consciousness* that obscures people's vision from the reality in which they live. According to Marxism, ideology ensures not only that people accept, but also justify as 'natural', inevitable and necessary, all kinds of man-made injustice that exists in society. Urban and regional planning, says Harvey, is part of the ideological superstructure of capitalist society – it helps to reproduce capitalist cities and perpetuate and justify the many forms of injustice inherent in their structure. This is so not because the planners are 'bad guys', but because of structural determinism – whether they like it or not they are part of the capitalist mode of production and as such are determined to play according to its rules.

11.5.2 Planning Is an Ideological False Consciousness

We accept Harvey's observation about the gap between the practice of planning and the perception of planning, but we suggest to interpret it in terms of a cognitive dissonance between the state of cities and urbanism as self-organizing systems, and the insistence of the modernist rational comprehensive planning (both Marxist and capitalist–liberal) that cities are like machines that can be predicted, planned, controlled and engineered. Today's planning is indeed an ideological false consciousness, as Harvey claims, but not because its role in the reproduction of capitalism (which in capitalist societies it might indeed play), but because its role in the reproduction of a scientifico–mechanistic world view which is at the foundations of the various facets of modernism: first and foremost of nationalism, and then to its various appearances as capitalist–liberal society, socialist–communist society as well as the bizarre combination of 'communist-capitalism' that is emerging in China in front of our very eyes.

An ideology is at the peak of its strength when (i) its ideological claims are fully accepted by society as if they belong to the domain of natural phenomena, and (ii) as a consequence of the above, when they are embedded in society's political structure. Both aspects hold in the case of modern urban and regional planning: its (false-conscious) mechanistic view of predictable, controllable and thus plannable cities and urbanism is generally accepted as an unquestionable truism, and as a consequence urban and regional planning administrations, practice and law are built in line with this (false) truism. This is so in both capitalist societies with their free markets and authoritative–communist societies and their planned economy and society.

11.5.3 Planning Between Nationalism and Urbanism

The historical origin of modern urban and regional planning is associated with two global processes which can be regarded as two of the order parameters that dominated modern society: urbanism and nationalism. In his small and illuminating book, Benovolo (1971) suggests to see *the origin and growth of modern town planning* in the process of urbanization that followed the industrial revolution. This modern industrial–urban revolution, argues Benovolo, entailed urban problems previously unknown in old forms of urbanism. The free market could not solve these problems (because of market failure) and so urban and regional planning arose out of this vacuum. Urbanism has thus played a role in determining one part of the agenda of town planning: the planning and supply of roads, light, water, welfare services and the rest.

The other order parameter which has shaped the modern urban and regional planning, is nationalism. As elaborated elsewhere (Portugali 1993), parallel to, and in association with, the rise of industrial society, nationalism has become the order parameter of modern society with the nation state as its material content and the ideology of nationalism as its information content. The connection between nationalism, modernization (and modernism) and urbanism is well recorded (Gelner 1983, Smith 1971, Reisman 1970). What we would like to add here is that the origin and growth of planning in general, and of urban and regional planning in particular, is intimately linked with the origin and growth of nationalism and its associated projects of nation-building, modernization and the like. And if urbanization was the major process which has painted planning as a supplier of services, nationalism was the force that determined its hierarchical authoritative and often oppressive nature: its ambition to lead society (i.e. the city) to determine its goals, its future, to determine who should live and work where and how in the city.

This nationalist–oppressive connection of planning deserves, of course, a separate study which goes beyond the context of the present chapter. We bring it up here, however, in order to emphasize the ideological strength and power behind the hierarchical, mechanistic, tree-like, just-in-case, structure of current administrative, procedural and theoretical structure of urban and regional planning, and by implication, the difficulty in any suggestion for change. But we will suggest it anyway.

We will suggest it anyway not only in light of the example of Toyotism and its just-in-time *kan-ban*, and in light of the city as a self-organizing planning system, but also in light of a property or element that is inherent in modern urban and regional planning. The latter, as we have just seen, emerged out of the combined force of urbanism and nationalism. The latter two, while indeed have shaped modernist planning, have also installed in it an inherent structural contradiction: between the hierarchical, centralized and often oppressive nationalist tradition of planning, and the highly distributed, diffused and decentralized urban tradition. So far the planning pendulum leaned toward the nationalist tradition, will it now start its move toward the other side?

11.6 SOCity: A Virtual Self-Planned City

In this chapter we have portrayed the city as a huge self-organizing planning system, as a huge self-planned system. We have seen that the city is a self-planned system by virtue of its property as an open system which exchanges matter, energy, information and plans with its environment, and by virtue of its complexity which in turn is a consequence of its structure as a semi-lattice network characterized by parallel distributed planning. In this 'PDP' network plans can be initiated at any level of the network and by each of the large number of actors connected to it. And the fact that the net is a dissipative structure, that because of overlapping and parallel planning it produces dissipation, is not considered a waste, but rather a source of creativity: ideas about the very same planning task are initiated by many, in parallel and at various levels, and because of nonlinearities, each plan or idea has chances to participate in the synergetic process by which various configurations compete and cooperate until one or a few win, enslave the rest and become the order-parameter-plans of the city or some parts of it.

Will planners be prepared to tune their planning administrations and procedures in line with the self-planned city? Will they be prepared to give up some of the power given to them by the nationalist–oppressive tradition that has shaped their profession and reorganize themselves in line with the more diffused and decentralized urban tradition? In discussing 'the new urban revolution on the verge of the 21st century' (Chap. 15) we suggest that there are indications that urbanism is indeed moving toward this direction. Until this happens, however, let us imagine how it might look like, let us look at a virtual, self-planned city.

11.6.1 SOCity

Imagine a big city, not an isolated city on a uniform plain as in Thünen's Isolated State, but an ordinary city – one that you can find these days in all parts of the world. Let us call our virtual-imaginary city Self-Organization City, or in short SOCity. Imagine further that like all other ordinary cities our SOCity too exhibits all the properties of the self-organized/self-planned city that we have portrayed above. Imagine that it is characterized by parallel distributed planning, just-in-time planning and the like. Our SOCity is similar to now-a-days ordinary cities in all these respects, but one: its planning system is different. Unlike the ordinary urban-planning systems which are characterized by just-in-case, push, top-down planning, in line with the nationalistic–oppressive tradition of modern town planning, ours is designed in line with the decentralized urban tradition of planning. In harmony with the rest of the city it too has a non-hierarchical, just-in-time, pull, 'PDP' structure. Let us have a closer look at SOCity.

The principles of its design are rather simple and well known. They take the principles of the modern liberal democratic polity as its model and apply them to the domain of city planning. Planning in SOCity is thus a 3-layer system: the legislative, the judiciary and the executive. On the face of it, this is exactly what happens in now-a-days modernist cities. But note the differences.

The planning law. First look at the planning law. In our ordinary cities the core instrument of the planning law is a set of plans and policies, initiated, planned, and pushed down the planning hierarchy, by the various planning authorities. Conventionally, these are essentially land-use plans that determine residential and commercial areas, industrial zones, open spaces, and the like. Once approved, these plans become the law. In SOCity there are no such land-use plans. The planning law in SOCity is not based on plans, but rather on a set of regulatory planning principles or laws, that refer to qualitative local and global relations in the city: between different activities, buildings and objects, between people, firms and the other agents that operate in parallel in the city, including also the city's planner. These planning principles (laws) refer to the relations between the existing structure and nature of the city and the elements that the many agents/planners plan to add to it, and they apply to all parts and aspects of the city. An example to a planning law would be that one is not allowed to build a high-rise next to a low rise building (not even the City), because it offends the basic right of the people in the low rise for air and light. Or, one cannot locate a noisy factory in a residential area. But, one can locate a 'clean' high-tech factory near residential buildings, and so on.

The planning judicature. The people who act in this domain are professionals who have specialized in both law and planning. They are spatially organized in 'planning-courts' of various scales so that there is a court for neighborhoods, for the city, the metropolitan area, and so on. Their function is to evaluate, approve or reject, the plans prepared by the many agents that operate in parallel in SOCity. The latter, as noted, include individuals, families, firms of all sizes, and the planners from the legislature and executive planning authorities of the city. Each agent who wants to take action that might change the city (a new building, for example) must get approval at the 'planning court'. Once approved, the agent can implement its plan.

Separation of authorities. The above is completely different from planning as normally practiced today. In the latter the city's planners wear two, or even three hats. One day they are planners that prepare plans or initiate planning policies; for example to build high-rises at a certain area of the city. The next day they are the members of the city's planning committee that has to approve or reject the plan or policy they have made yesterday. The planners, in most urban planning administrations, prepare the master- or development- or land-use plans, they then approve their own plans that later become the planning law, and once this is done they are responsible for the so-called planning control, i.e. implementation. All these take place under the same administrative roof – the city's planning department, for example. There is no separation of power and authorities in the usual modernist town planning – a situation which makes planning structurally authoritative and oppressive. In SOCity there is a clear-cut separation between the planning judicature with its planners–judges, etc., and the planners at the legislature and executive authorities.

Planning hermeneutics. There is no need for a top-down, just-in-case, centralized planning in SOCity. Each agent plans at its own scale. Once approved, the plan can be implemented. The city is thus being self-planned in a 'PDP' manner which now includes all agents as decision makers and planners: individuals, firms of various sizes, as well as governmental planning agencies. This process is specifically significant because, as in ordinary judicial system, here too, the decisions taken at the planning courts have a feedback effect on the planning law and practice. Every decision taken by a 'planning judge' establishes a 'planning precedent' and as such plays a role in future planning decisions, taken by other 'judges' in different planning situations – a property which puts a high responsibility on the 'judge' (as in an ordinary court). The result is a commutative process of hermeneutics by which, first, plans and ideas get into the planning system and the city as a whole in a bottom-up manner. A planning initiative at a personal–individual–local level might have, in this way, an impact on the dynamic of the whole city, no less than a plan initiated by the government. Second, the planning law and the structure of the city are in an on-going feedback process of re-evaluation, updating, and reinterpretation.

The planning legislature and executive systems. Their role in SOCity is more modest than in ordinary modernist cities. In line with the urbanist tradition of modern town planning, they are first and foremost 'civil servants', and not 'visionaries' or 'leaders of the people' as implied by the nationalistic tradition which dominates today's city-planning. As civil servants they have two roles. The first role is to supply information to the many planning agencies that operate in parallel in the city. As an extension to their authoritative, nationalistic tradition, today's planning agencies tend to 'keep their cards (i.e., planning information) close to their chest'. In SOCity planning information is a public domain available to every agent who needs it. This is achieved by various means including a user-friendly GIS (geographic information system) installed in the Internet and in other public-domain communication–information systems. Such a public domain information city is essential for the operation of a parallel distributed planning.

The second role concerns the preparation of plans. Like the other planners that operate in SOCity, its city planners too initiate plans and policies and submit them to the planning courts for approval and implementation. However, compared to today's practice, in SOCity their domains of planning are restricted (by the planning law). They are expected to plan only those domains that are essential to the operation of the city, but that the other planning agents cannot or do not plan. The city's transportation system, its schools' network, or urban system of open spaces, are cases in point. An example for a possible principle for such a separation is the division between private and public goods, which in the context of cities take the form of private versus 'local' or spatial public goods, services and externalities. This division, which is partly technical and partly ideological or socio–cultural, provides, in fact, one rationale for planning in a liberal, free-market society (Portugali 1980).

11.6.2 A Concluding Note

The virtual city that we have constructed above is just a sketch, of course. Its aim is not to propose a complete program for an alternative, utopian, self-organized mode of planning, but rather to make the point that with a little bit of effort and imagination this is possible. It is possible and feasible because the principles of such a planning mode are already here, with us. They are enfolded in our cities – in the 'PDP' that already takes place in them and in the urban–civil tradition which has inspired the very origin and growth of modern town planning. Will that potential materialize? Will the implicate order which is enfolded in our cities become an explicate order?

12. Artificial Planning Experience*

12.1 Introduction

In Chaps. 2 and 11 we discussed the disillusionment of modern urban and regional planning following its attempts at the 'Taming of the Shrew Environment' (Portugali 1995). That is to say, the failure of modern urban and regional planning to solve urban problems and control cities by means of the scientific method and rational planning. Our interpretation was that one reason for this failure is misappreciation, on the side of planning theory, of the nature of cities and metropolises as self-organizing systems: A central property of such open, complex, far from equilibrium, systems is that they are unpredictable and thus uncontrollable; and yet, predictability, control and 'social engineering' are the pillars of modern town and regional planning. In light of the above the major question in Chap. 11 was: 'in the absence of predictability and control can there be a meaning to planning? ' In Chap. 11 we answered this question in the affirmative and started to outline a new, self-organization, approach to planning. In this chapter we further elaborate on this new approach by illustrating how our FACS models from Part II can be designed and used as heuristic planning tools. The specific model we use for this purpose is City-1 as developed in Chap. 6.

12.2 Planning by Means of Artificial Experience (AE)

As already noted, from the point of view of planning, perceiving cities as self-organizing systems implies a major and fundamental problem – in fact, it raises doubts as to the whole rationale of urban and regional planning. Modernist–rational planning started from two premises: First, that while a free society, free economy, and non-intervening policy, is the ideal, some degree of intervention and control, in selected domains, is necessary and unavoidable. Second, that science, by its capability to explain the past and the present and on the basis of these to predict the future, will provide the proper means for a planned intervention and control. Planning in general and urban and regional planning in particular have thus become the academic disciplines

* With Itzhak Benenson

within which such premises should be elaborated (Portugali 1980, Breheny and Hooper 1985, Beauregard 1989, Goodchild 1990, Pasty 1992).

Now, in the case of cities as self-organizing systems, science herself tells us that she cannot predict their behavior, that this inability is not a technical shortcoming which science will overcome with time, but the very nature of cities. By their very nature as self-organizing systems, cities are unpredictable, uncontrollable and in this respect are unplannable. Writing about the failure of modern urban and regional planning to plan such complex environments, it has been suggested (Chaps. 1, 2 and above), first, that this failure is due to planners' misappreciation of cities as self-organizing systems, second, that in such circumstances planners can still participate in the urban play (for a similar view see Jantsch 1975). In the following we shall attempt to be more specific and operative about this participation in the urban play.

Let us start by noting that while science tells us that self-organizing systems cannot be predicted, she further teaches us that such systems can still be modeled. Our family of models City, City-1, ..., City-6, are, in fact, artificial self-organizing systems exhibiting the properties of such systems: openness, complexity, non-linearity and of course, self-organization. While such models cannot predict the future, they still allow us to play futurist games: to explore and reveal the many potentialities, which are enfolded within a given starting urban situation, the sensitivity of the various parameters and so on.

All the heuristic city-games we have developed above in Part II are, in fact, simulation games. The association between simulation, games and planning is not new (e.g. Chadwick 1971, McLoughlin 1969, Wynn 1985) and in this respect our suggestion here to use them as planning games is not new. What is novel in what follows is the role and rationale we assign to planning games in the overall process of planning. Our aim is not to predict the future, as is usual in simulations, but to enable planners and decision-makers to artificially experience a certain planning phenomenon and in a relatively short period of time. We suggest, if you want, a device for *Artificial Planning Experience.*

Awareness to the importance and role of 'experiential knowledge' in the overall process of urban and regional planning arose out of a discontent with analytic rational planning. Writers such as Forester (1985), Goldberg (1985) and Yewlett (1985) have emphasized the limits of rational planning and advocated the need to develop new notions such as 'bounded rationality', 'subjective–experiential planning' and the need to integrate intuition with rationality.

The difference between rational and experiential planning is related to the difference between 'explanation' and 'understanding'. If explanation is a process which seeks to define the laws and common denominators of a multiplicity of phenomena, then understanding can be described as a kind of knowledge, or a process, which seeks to expose their variability and uniqueness. If explanation is gained by analysis, then understanding by experience. In line with this distinction we suggest a heuristic planning tool with which a planner can play, learn the various facets of a situation, acquire (artificial) experience and understanding and as a consequence be in a better position to take 'intuitive' decisions.

12.3 The Planning Game

The 'Planning Screens' provide the interface between the player/planner and the heuristic planning model (City-1 in our case). On them the player inputs and defines the value of the parameters, and observes the results in the various domains. There are two screens: the Starter Planning Screen (SPS), and the Playing Planning Screen (PPS). The first is to input information, while the second is to observe the output results.

To start a game the player has to decide about the state of the city at the outset of the game: its spatial infra-structure, the individuals that live in it (inhabitants), those who want to live in it (immigrants), and the rules of the game. As illustrated in Fig. 12.1, each decision is assigned a special window on the SPS into which the player types the needed information. There are three such windows:

Window 1: Spatial infrastructure

1. The width and length of the city's grid (and consequently the number of cells in it).
2. The initial distribution of houses/cells value. We can use any initially stored value distribution in space, or alternatively input parameters of some analytical formula that defines such a distribution.

Window 2: The individuals

1. The spatial distribution of individuals in the city (the inhabitants) at the outset of the game (i.e. at $t = 0$), including their status and tendency.
   The planner can begin with a previously stored spatial distribution of individuals or construct a new one. For the latter case the planner has to input the initial fractions of both groups (in relation to the overall number of cells in the lattice) and the means $\mu = \mu(O_A)$ and standard deviations (STD), $\sigma = \sigma(O_A)$ of status and tendency distributions of the Veterans and Olim. Each individual receives his/her status and tendency independently, according to the distributions given.
2. The immigrants to the city.
   The rates of Veterans and Olim immigrants to the city as well as means and STD of their status and tendency distributions.

Window 3: The rules of the game

1. Deterministic behavioral rules.
   According to (6.3), (6.4), (6.9) the planner has to input parameters δ_0, δ_1, τ_0, τ_1, τ_2, τ_3. The payment per time unit W^H is given in (6.13).

WINDOW 1. Spatial Infrastructure.

City size: Width = 20; Length = 80; (Number of cells = 1600).

Initial VALUE distribution (Press Enter to generate a new one): C:\CITY\VALUE.DAT

<ENTER: Linear growth from bottom to the top of the City

VALUE at the bottom = 15; VALUE at the top = 15;

\>

WINDOW 2. Individuals (0 < STATUS < 1, 0 < TENDENCY < 1).

Initial INDIVIDUAL distribution: C:\CITY\INDIVID.DAT

(Enter to generate a random distribution):

NEW IMMIGRANTS' parameters

VETERANS

Immigration rate = 1%

STATUS: Mean = 0.25000; STD = 0.2500;

TENDENCY: Mean = 0.00195; STD = 0.0078;

OLIM

Immigration rate = 1%

STATUS: Mean = 0.0625; STD = 0.06250;

TENDENCY: Mean = 0.0078; STD = 0.00195;

<ENTER

RESIDENTS' parameters

VETERANS

Initial fraction = 25%

STATUS: Mean = 0.25000; STD = 0.2500;

TENDENCY: Mean = 0.00195; STD = 0.0078;

OLIM

Initial fraction = 0

STATUS: Mean = 0.0625; STD = 0.06250;

TENDENCY: Mean = 0.0078; STD = 0.00195;

\>

WINDOW 3. The rules of the game.

BUYING LIMITATIONS.

TENDENCY > 0.3 + 0.0045 * VALUE.

STATUS > 0.5 * VALUE.

RENTING LIMITATIONS.

TENDENCY > 0.4 + 0.006 * VALUE.

Attractivity functions for VETERANS/OLIM.

Dependence on neighbors' STATUS | Dependence on neighbors' ORIGIN

$S_{neighbors}$	Q_A	$S_{neighbors}$	Q_A	$O_{neighbors}$	Q_A	$O_{neighbors}$	Q_A
0.0	0.0	0.0	0.0	0.0	0.0	0.0	0.0
0.5	0.005	0.5	0.005	0.5	0.005	0.5	0.005
1.0	0.950	1.0	0.950	1.0	0.950	1.0	0.950

Repel functions for VETERANS/OLIM.

Dependence on neighbors' STATUS | Dependence on neighbors' ORIGIN

$ABS(S_A - S_{neighbors})$	R_A	$ABS(S_A - S_{neighbors})$	R_A	$1-O_{neighbors}$	R_A	$1-O_{neighbors}$	R_A
0.0	0.01	0.0	0.01	0.0	0.01	0.0	0.01
0.5	0.10	0.5	0.10	0.5	0.10	0.5	0.10
1.0	2.80	1.0	2.80	1.0	2.80	1.0	2.80

STATUS threshold = 0.0 VALUE decrement = 0.98

Probability to leave the system = 0.05 Maximum length of homeless period = 30

Fig. 12.1. SPS: the Starter Planning Screen

2. Stochastic behavioral rules.
   These are the attractivity and repel functions of houses as defined by (6.1), (6.2), (6.5), (6.6). The value of R_A for

$$|S_A - S_{\text{neighbors}}| = 0, \qquad 1 - O_{\text{neighbors}} = 0, 0.5, 1, \text{ and for}$$
$$|S_A - S_{\text{neighbors}}| = 0, 0.5, 1, \qquad 1 - O_{\text{neighbors}} = 0.$$

   and the value of Q_A^H for

$$S_{\text{neighbors}} = 0, \qquad O_{\text{neighbors}} = 0, 0.5, 1, \text{ and for}$$
$$S_{\text{neighbors}} = 0, 0.5, 1, \qquad O_{\text{neighbors}} = 0.$$

   have to be determined for Veterans and Olim. Parameters α_i, β_i, γ_i ($i = 1 - 4$) are calculated for an internal representation of the functions presented in Fig. 12.1. To enable a qualitative tuning their graphs are drawn after the values are determined. So, one can choose appropriate repellency and attractivity functions in a 'visual' manner. Two more parameters are the probability to leave the system L (6.10) and the value decrement D (6.16).
3. Maximum number of time-steps an individual can remain in the queue.
   Once the above windows are decided, the player can start the game. That is to say, veteran and new immigrants enter the city, buy/rent houses, some of the city's inhabitants change their houses, some outmigrate from the city, and so on. The player can observe the urban dynamics as it evolves, on the PPS. As can be seen in Fig. 12.2, in the upper half of the screen there are four maps, each illustrating a different evolutionary process of the city. In this specific case, the first map, from the left, shows the evolving spatial distribution of veterans and new immigrants in the city; the second map shows the spatial distribution of individuals' status, the third, their tendency, while the fourth map shows the evolving and changing spatial distribution of land values in the city.

12.3.1 An Example

Once we have defined the initial conditions on the SPS as noted above, we could observe on the screen (as presented in Fig. 12.2) several scenarios which represented various configurations of the numbers of Olim and veteran immigrants to City-1, various configurations of status/tendency relations among the Olim immigrants, and various configurations of spatio–cultural behavioral relations between the veteran population and the new Olim immigrants (sympathy and/or antagonism with respect to residential location).

As noted above, and implied by our choice to use City-1 from Chap. 6 as the illustrative model, the aim in this chapter is to present a model which can simulate the city as a self-organizing system, and on which planners can gain 'artificial planning experience'. The actual case study concerns a planning issue which is at the focus of planners' interest in Israel since the early 1990s:

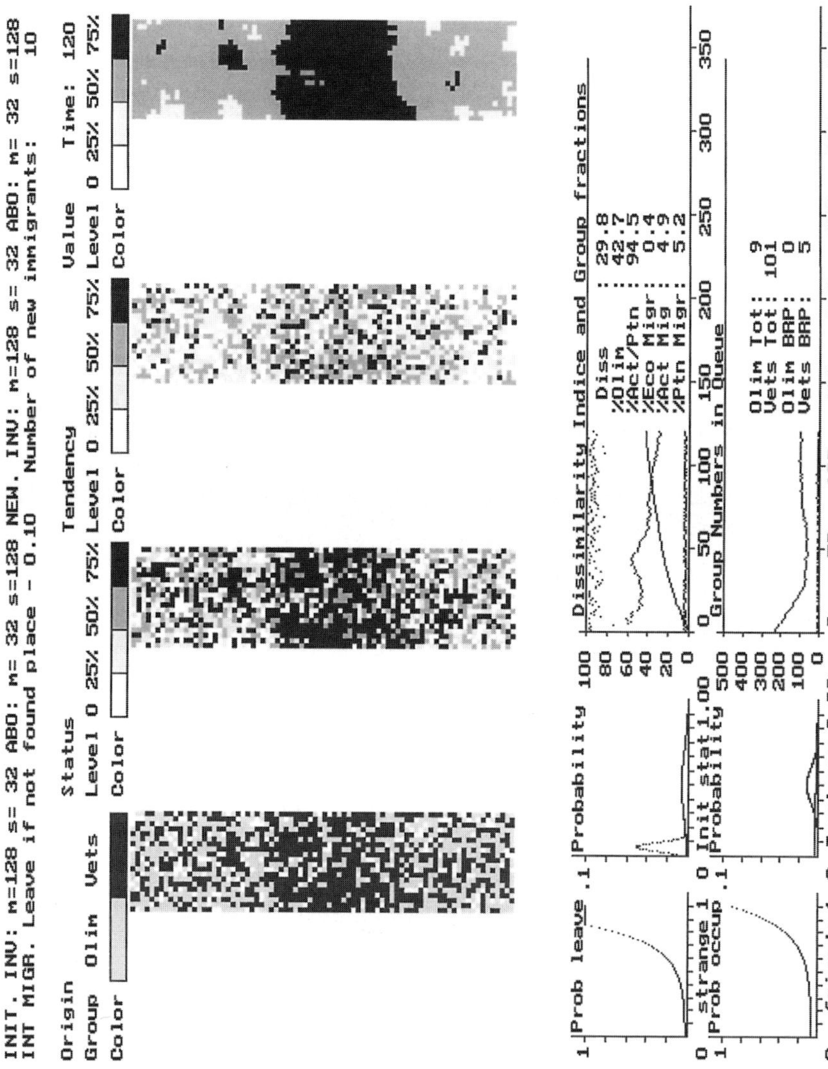

Fig. 12.2. PPS: the Playing Planning Screen

the relations between international migration to the country (mainly from ex-USSR) and the internal structure of its cities.

The example presented below illustrates a property, which is central to the perception of cities as self-organizing systems, and to the notion of artificial planning. That is, that a given set of initial conditions gives rise to a variety of urban scenarios. In Fig. 12.3 we show results from three urban games 'played' on City-1. The three started with identical starting conditions (1% veteran- and 0% Olim-immigrants and the rest as specified in Fig. 12.1) and yet, as

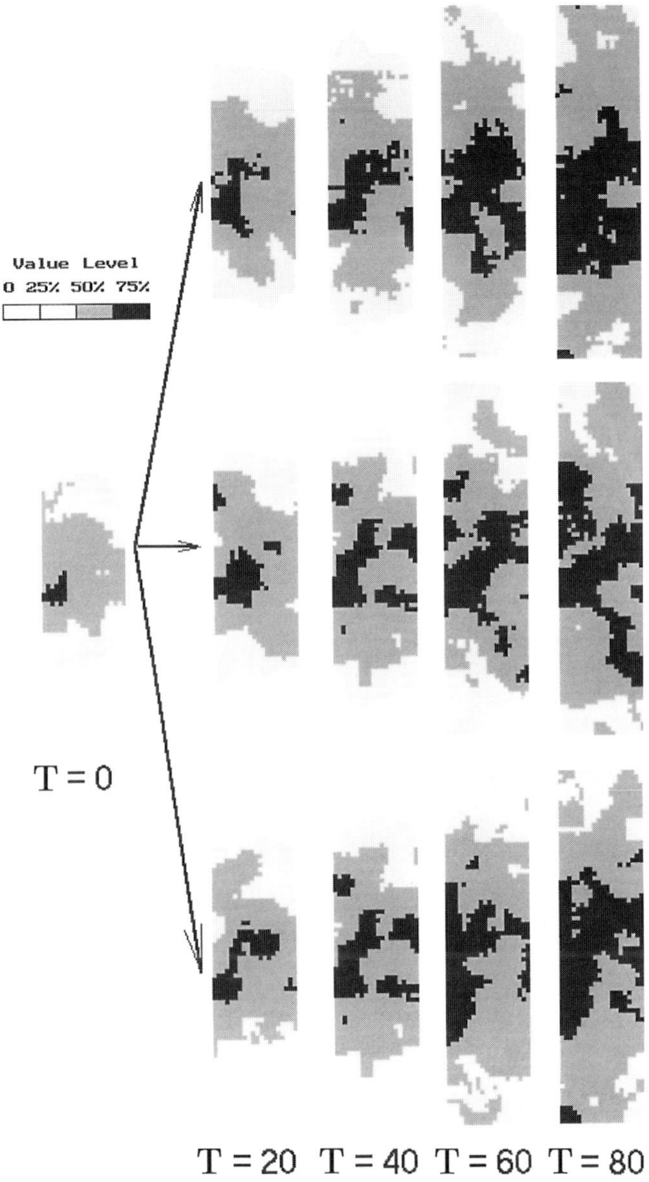

Fig. 12.3. Three time evolutions of housing value surface resulting from identical initial conditions

can be seen in Fig. 12.3, they resulted in three socio–economic spatial urban histories. From the point of view of artificial planning experience, these games illustrate the interplay between the unique and historical in urban evolution, versus the general and causal. It can be seen that the differences between the evolving three urban scenarios/spatial distributions are growing in time, with the implication that each process is unique and as such historical and non-causal. If at all, causal relations in such urban evolution might be meaningful only within a very limited time slot at the beginning of the process (from $t = 0$ to $t = 40$). Beyond that the complexity and self-organization properties of the system and its specific and unique spatial evolution become the more significant force. Having said this, however, some general structure can still be observed: in all maps and at all stages some form of core–periphery gradient can be observed in the urban land-value surface: the black, high-value area is always surrounded by a medium-value dark-gray area, itself surrounded by a light-gray low-value area.

12.4 Conclusions

As noted, there is no room here to further elaborate on the relevance and implications of our model to the Israeli reality. Instead we would like to conclude this discussion by emphasizing a point concerning the position of our model within the field of self-organization in regional analysis. It can be said that it stands somewhere in between a very abstract theorization of cities as self-organizing systems, and more operational models, which were built from the start as empirical tools. This midway position is the property of cell-space models in general which are, on the one hand, par-excellence theoretical devices, but on the other, are intuitively related to the physical structure of cities: cells as 'houses' and individuals which take residential location decisions in light of the properties of the cells/houses and the neighbors around them. This intuitive connection between the theoretical model and the actual structure of cities, makes cell-space models such as ours a rather convenient planning tool. It enables us to give planners a highly theoretical and abstract device on which, and with which, he or she can examine real situations in a general manner: the planner can define on the SPS situations which are structurally similar to the specific reality with which he or she is dealing, study them as if studying structurally similar scenarios in the past or in distant places, and thus gain the experience relevant to the situation at hand.

Part IV **Synergetic Cities**

13. Synergetic Cities I: The Pattern Recognition Approach*

13.1 Approaches of Synergetics

Synergetics – Haken's theory of self-organization – is now some 30 years old. Over the years four different approaches have been developed within this body of theory to cope with phenomena of complex systems: the microscopic, the macroscopic, the phenomenological (Haken 1993), and the conceptual or hermeneutic. The first is a "bottom-up" approach, starting with an interaction between the individual parts of the system which, beyond a certain threshold of a control parameter, gives rise to order parameters which then enslave the system. The second is a "top-down" approach typical of cases where a description of some macroscopic properties of the system can be performed, whereas information on the individual parts of the system is not sufficient or doesn't exist. The macroscopic approach enables one to guess the underlying dynamics by means of the entropy (or information) maximization principle, for example.

The third, phenomenological, approach begins with the finding that close to instability points the behavior of the system is governed by a few order parameters, the equations of which can often be written directly. The prominent example here is the Kelso's experiment that has become the finger movement paradigm, relevant to a whole class of phenomena related to human behavior. The fourth approach stems to the fact that beyond being a formal mathematical method to describe complex systems, synergetics is also a theory about self-organizing systems. That is to say, it is a conceptual framework, with a set of interrelated concepts and specific views about the various facets of the general phenomena of complexity and self-organization. As a general hermeneutic theory synergetics has proved itself a rather useful conceptual framework for the interpretation of "soft", non-quantifiable phenomena such as language, nationalism or planning, on the one hand, and of models constructed outside the mathematical formalism of synergetics, on the other. Parts I and III of this book are examples of the first case, and Part II of the second.

With respect to cities and regional systems, one can identify two lines of research that have developed within synergetics (Table 13.1). The first, the

* By Hermann Haken and the author

Table 13.1. A comparison between two lines of research developed within synergetics: the pattern recognition approach and the master equation approach

	The Pattern Formation → Pattern Recognition Approach	The Master Equation Approach
Stage I: Deterministic and stochastic processes	Evolution equations for state variables	Evolution equations for probability distributions
Stage II: A verge over fluctuations	Deterministic evolution equations	
Outcome	*Explicit* consideration of order parameter(s), slaving principle, circular causality, etc.	*Implicit* consideration of order parameter(s), slaving principle, circular causality, etc.
Fundament interest	Phase transition, qualitative change	
first applied to	Physics, chemistry, biology	Sociology as "sociodynamics"
More recent applications	Computers, cognition, brain functioning and very recently cities and urbanism	Cities, urban and regional systems, trafic systems

master-equation approach, starts from the formulation of a master equation with properly chosen transition probabilities. Such an equation determines the temporal evolution of the probability distribution of the socio–(economic) configuration ("state variables"). The solutions of the master equation can be obtained in a few special cases only. Therefore, in practice, a suitable average over the master equation is taken which leads to deterministic evolution equations which, in real cases, are solved numerically. In many cases, this approach treats the order parameters and the slaving principle in a rather implicit manner. This approach is characteristic, for example, of Weidlich and co-workers studies in sociology, economics and urban dynamics (Weidlich, 1987; Weidlich and Haag, 1983). For many years this was the main synergetic approach to cities, and most applications thus far have been within this conceptual frame (e.g. Haag et al. 1992, Pumain, Saint-Julien and Sanders 1987, Sanders 1992).

The second, the pattern recognition approach, has a long history, but in the domain of cities it is relatively new. In the frame of its microscopic description it also starts from state variables for which evolution equations are formulated directly. Chance events are taken care of by stochastic forces. In general, averaged deterministic equations are used, however. An explicit consideration of the order parameters, the slaving principle and the other tenets of synergetics is in the foreground of this approach. This second line of research typifies the synergetics' treatment of pattern formation, cognition, pattern recognition and brain activities, as developed in the last three decades by Haken and co-workers. Only very recently has this approach been applied to the study of cities as self-organizing systems (Haken and Portugali 1995). The present chapter is an elaborated version of this original paper, whereas Chap. 14 that follows further extends this approach to the issue of decision making in the context of urban dynamics and city planning.

The concepts of order parameters and slaving allow the formulation of phenomenological and of conceptual–hermeneutic synergetics. On the other hand, within macroscopic synergetics a close "relative" of the master equation, namely a Fokker–Planck equation, can be constructed by use of experimental data.

13.2 The Pattern Recognition Approach to Cities

Why study the self-organization of cities from the perspective of Haken's synergetics approach to pattern formation and pattern recognition? Much of the advance made in recent years in the theory of synergetics was connected to the issue of cognition in general, and to what is often termed "synergetics' paradigm of pattern recognition", in particular (Haken 1979, 1985a, 1988, 1990). Until very recently none of this new body of theory, whose most recent statement is Haken's (1996) *Principles of Brain Functioning*, has been related to the domain of regional science and the study of cities. One aim of this chapter is thus to show how the dynamics of cities can be conceptualized within this new theoretical framework.

As elaborated in Chap. 3, the synergetics approach to cognition started with the analogy made by Haken (1979) between self-organization processes of pattern formation in liquids, and the psychological processes of pattern recognition. We suggest that this analogy is specifically attractive to the study of cities, which, as graphically illustrated above in Fig. 3.8, can be seen as both physical (streets, buildings, residential areas, industrial zones and so on), and cognitive (i.e. cognitive maps of streets, cities etc.) self-organizing patterns. We thus suggest seeing the dynamics of cities in terms of a complex interplay between their physical structure, pattern, or map and their cognitive structure, pattern, or map.

The importance of the cognitive dimension of cities and regions has long been appreciated. In fact, it acted as the major impetus to the emergence in the early Sixties of mental maps studies and it is still an impetus to what is more recently termed cognitive maps studies (for an updated literature and recent advances see Portugali 1992, 1996a, and further bibliography there). While the research on cognitive maps has enriched our understanding about the mechanisms involved in shaping individual and collective cognitive maps and their relations to human behavior in space, in way-finding, navigation, shopping behavior, residential decision making and the like, this body of knowledge has not as yet been fully incorporated into urban and regional modeling. In the latter, researchers still tend to follow economists by assuming that individuals behave in space as "economic persons". The result is a rather problematic situation: on the one hand, there is a growing body of empirical and theoretical studies which reveal the various facets of the relations between the cognitive maps of individuals and their residential, migration, shopping, commuting etc., behavior in the urban environment, that is to say, in all those activities which make the variables of urban dynamics and modeling. On the other hand, most of this knowledge stands outside urban and regional modeling. The development of economically oriented urban and regional models and cognitively oriented behavioral studies run in parallel with no real connection.

The economic person assumption in regional analysis has recently been criticized and new directions have been suggested. In particular Fischer et al. (1990) and Sonis (1991) have suggested to replace the economic person by a 'social person' as the basic assumption about individuals' behavior in the city. In Chap. 7 we discussed these ideas at some length, and in Chap. 14 below we suggest that the notion of social person can be regarded as a synergetic decision-making heuristic. A further improvement is the psychological notion of spatial cognitive dissonance, which we proposed in Chap. 7 and then used in our City-2 simulation game of residential cultural segregation.

In this chapter we further elaborate on the relation between the material and cognitive aspects of the city. Our view is that the above noted analogy between pattern formation and pattern recognition creates a potential for such a connection. In fact a start toward the realization of this potential has already been made: in several previous studies (Portugali 1990, Portugali and Haken 1991) it has been suggested to approach the issue of cognitive mapping from the point of view of Haken's synergetics approach to pattern recognition and associative memory. The discussion below goes one step further and shows how the paradigm of pattern recognition can be used as a framework within which the interplay between the material pattern of cities and the cognitive pattern of cities can be conceptualized and subsequently analyzed. A rather detailed mathematical model that allows us to study and elucidate mechanisms that lead to self-organized structures in cities or settlements does this.

13.3 The Model

Let us consider the development of a city within a certain region which may be growing, however. Using Cartesian coordinates x, y, we may denote each individual location. But we may use any other numbering, say, by an index j as well. We then consider different types of inhabitants that live at that location j or work there. Introducing an appropriate coding, we may distinguish between their ages, professions, religion, and so on; for instance, we may indicate their age by an index α_1, their profession by an index α_2, etc. Similarly we may introduce the size of an enterprise, an office, and so on by a similar index. More precisely speaking, we denote by $q_{j,\alpha}$ the number of people at site j and of type $\alpha = (\alpha_1, \alpha_2 \ldots)$, or similarly the size of an enterprise, an office, and so on. In the following we shall talk explicitly of the number of people, but the approach may be generalized to that of enterprises, and so on. Because we are interested in the growth processes, we have to consider the rate of change of the population. We shall do this in two steps. In the next section we shall consider linear models which allow us incidentally to introduce some basic concepts of synergetics. The main emphasis, however, will then lie on the development of nonlinear models in Sect. 13.5, because only those are able to reproduce the formation of structures.

13.4 Linear Models

We study the rate of change of a population at site i of type α. This change is given by the equation

$$\dot{q}_{i,\alpha} = \sum_{j,\beta} \lambda_{i\alpha,j\beta} q_{j,\beta} . \tag{13.1}$$

According to it, the rate of change at site i of population of type α is the sum of the inputs from other populations at site j of type β and, with $j = i, \beta = \alpha$, the net growth (or decay) rate of the population of type α at site i. Equation (13.1) may describe, for instance, the movement of an inhabitant from one part of the town to another part, taking a new apartment there, or it may describe the rate of birth or death at the same site, or it may describe the change of a profession at the same site or being accompanied with the movement to another site. Quite clearly, (13.1) allows for a wide range of interpretations. Equation (13.1) does not include interregional or inter-city migration, that is, to say, the influx or outflux of populations. To this end, a further term $f_{i\alpha}$ must be added. In the present chapter we shall neglect this effect, however. $\lambda_{i\alpha,j\beta}$ will be denoted as the transition rate. If the transition takes place from one place to another one without any other change of the characteristics of the individual under consideration, we can split the transition rate into a product

$$\lambda_{i\alpha,j\beta} = \lambda_{ij}^{(1)} \delta_{\alpha\beta} . \tag{13.2}$$

In the following we shall use quite often two abbreviations, namely we replace the pair of indices i, α by a single index

$$i, \alpha \to I , \tag{13.3}$$

which may then by arrangement of numbering run from 1 till N. We may lump the quantities q together into the so-called state vector

$$\boldsymbol{q} = (q_1, q_2, \ldots, q_N) . \tag{13.4}$$

By means of (13.3) we may write (13.1) in the shorter form

$$\dot{q}_I = \sum_J \lambda_{I,J} q_J . \tag{13.5}$$

The state vector (13.4) describes a pattern in the following sense: to each index I, or specifically to each index j and α there is a specific number that denotes the size of the population q_I at that point. We thus expect a hilly landscape, which will be denoted "a pattern". We study the solutions of (13.1) or (13.5). This is a well-known procedure in mathematics, but here we want to give it a form that immediately allows us to discuss the growth of specific patterns. We introduce the matrix Λ by

$$\Lambda = (\lambda_{I,J}) . \tag{13.6}$$

We denote the right-hand eigenvectors belonging to (13.6) by $\boldsymbol{v}_k$ and the corresponding eigenvalues by λ_k, i.e. we consider the solutions of the equations

$$\Lambda \boldsymbol{v}_k = \lambda_k \boldsymbol{v}_k . \tag{13.7}$$

The eigenvectors may again be considered as describing patterns

$$\boldsymbol{v}_k = (v_{1k}, v_{2k}, \ldots, v_{N,k}) . \tag{13.8}$$

In the case of complex eigenvalues, oscillations may occur which we shall not consider in the present context, however. We may group the eigenvalues into three classes, namely those $> 0, = 0$, and < 0. Thus we obtain Table 13.2:

Table 13.2. Three classes of eigenvalues

$\lambda_1, \ldots, \lambda_M$	$>$	0	$\boldsymbol{v}_1, \boldsymbol{v}_2, \ldots, \boldsymbol{v}_M$
$\lambda_{M+1}, \ldots, \lambda_L$	$=$	0	$\boldsymbol{v}_{M+1}, \ldots, \mathbf{v}_L$
$\lambda_{L+1}, \ldots, \lambda_N$	$<$	0	$\boldsymbol{v}_{L+1}, \ldots, \mathbf{v}_N$

According to basic mathematics, the matrix Λ can be written in terms of the eigenvalues λ_k, the eigenvectors $\boldsymbol{v}_k$ and their adjoints $\boldsymbol{v}_k^+$

$$\Lambda = \sum_{k=1}^{N} \lambda_k \boldsymbol{v}_k \cdot \boldsymbol{v}_k^+ . \tag{13.9}$$

We shall assume that the eigenvectors are normalized

$$(\boldsymbol{v}_k \boldsymbol{v}_k) = 1 \tag{13.10}$$

and that the orthogonality relations

$$(\boldsymbol{v}_k^+ \boldsymbol{v}_{k'}) = \delta_{kk'} \tag{13.11}$$

hold. According to basic theorems of ordinary differential equations, the general solution of (13.5), when formulated by means of the state vector (13.4), can be written in the form

$$q = \sum_{k=1}^{N} c_k e^{\lambda_k t} \boldsymbol{v}_k. \tag{13.12}$$

The coefficients c_k are determined by the initially given distribution of $\boldsymbol{q}$ at time $t = 0$ according to

$$c_k = (\boldsymbol{v}_k^+ \boldsymbol{q}(0)). \tag{13.13}$$

The temporal change of the pattern described by $\boldsymbol{q}$ is described by (13.12). When we want to study the long time behavior, we may split (13.12) into three parts, according to Table 13.2,

$$\boldsymbol{q} = \sum_{k=1}^{M} c_k e^{\lambda_k t} \boldsymbol{v}_k + \sum_{k=M+1}^{L} c_k \boldsymbol{v}_k + \sum_{k=L+1}^{N} c_k e^{\lambda_k t} \boldsymbol{v}_k. \tag{13.14}$$

While the last term vanishes, the second term remains constant so that our interest may be focussed on the first term that describes the growing patterns. While (13.14) may give us a hints how patterns evolve and give us a basis to study different influences on the growth of patterns, linear models suffer from various serious drawbacks. First of all, as is evident from the first term in (13.14), there is unlimited growth. There is no selection of a definite final pattern, rather various patterns grow simultaneously and appear as a pure superposition without influencing each other. In addition, slight perturbations may change the growth parameters λ_k, for instance from neutral, to growing, or to decay. Thus the models bear an unrealistic instability in themselves. It is, therefore, desirable to develop nonlinear models. Before we do so, we cast our previous results into a form that allows us to make contact with basic notations of synergetics. One of these is the concept of an order parameter. In order to introduce it, we rewrite (13.14) in the form

$$\boldsymbol{q} = \sum_{k=1}^{L} \xi_k(t) v_k(t) + \sum_{k=L+1}^{N} s_k(t) v_k(t), \tag{13.15}$$

where the terms with a positive or neutral growth rate are lumped together in the first sum, those with negative growth rate in the last sum. Multiplying (13.15) with $\boldsymbol{v}_k^+$ from the left, and using the orthogonality relation (13.11), we obtain the equations

$$\dot{\xi}_k = \lambda_k \xi_k \qquad k = 1, \ldots, L \tag{13.16}$$

and

$$\dot{s}_k = \lambda_k s_k, \lambda_k < 0, \quad k = L+1, \ldots, N. \tag{13.17}$$

While the solution of (13.16) is given by

$$\xi_k(t) = c_k e^{\lambda_k t}, \quad \lambda_k \geq 0 \tag{13.18}$$

with exponential growth rates, (13.17) leads to solutions that vanish in the course of time. The amplitudes $\xi_k(t)$ are called order parameters for reasons that become obvious in the next section.

13.5 Nonlinear Models

From a formal point of view, nonlinear models may be defined by the equation

$$\dot{q}_I = f_I(\boldsymbol{q}), \tag{13.19}$$

where f_I is a nonlinear function of the state vector $\boldsymbol{q}$. One possibility to obtain such nonlinearities is to expand the right-hand side of (13.19) into a Taylor series

$$\dot{q}_I = \sum_J \lambda_{I,J} q_J + \sum_{JK} \lambda_{IJK} q_J q_K + \sum_{JKL} \lambda_{IJKL} q_J q_K q_L + \dots . \tag{13.20}$$

As is well-known from nonlinear dynamics or studies in synergetics, (13.20) allow for an enormous amount of different kinds of dynamic behavior. Therefore, to find equations that can be understood and handled, we shall restrict our study to special classes. One such class has been studied in synergetics with respect to pattern recognition by the synergetic computer (Haken 1991, see also Sect. 14.5). In this approach the mental process of pattern recognition is conceived as a process of pattern formation – similar to material processes of pattern formation in fluids. City growth is an interesting case because it is related to both material and cognitive pattern formation. On the one hand, the city is a material entity and the ordinary city map depicts this property (built up area, spatial distribution of population, and so on). On the other, the city is also a cognitive entity and the various cognitive city-maps, constructed by psychologists, geographers and architects, depict that property (mental distances, desirability areas, environmental spatial attention and the like). In fact, this interplay between the material pattern/map and the cognitive pattern/map is a major force behind the dynamics, growth and evolution of cities. Therefore, direct analogies with this work on the synergetic computer and models on city growth can be established. (A preliminary discussion and a graphical illustration of the city as an interaction between pattern formation and recognition, have already been presented in Chap. 3 and Fig. 3.7 above.)

We first cast (13.20) into a form that allows us to make contact with these former models and then we shall show how these models can be given a rather interesting interpretation with respect to city growth. In accordance with this model, we assume

$$\lambda_{IJK} = 0 \tag{13.21}$$

and a special form for λ_{IJKL}. The equations (13.20) then acquire the form

$$\dot{\boldsymbol{q}} = \sum_k \lambda_k \boldsymbol{v}_k \cdot \boldsymbol{v}_k^+ \boldsymbol{q} - B \sum_{k \neq k'} \sum (\boldsymbol{v}_{k'}^+ \boldsymbol{q})^2 (\boldsymbol{v}_k^+ \boldsymbol{q}) \boldsymbol{v}_k - C(\boldsymbol{q}^+ \boldsymbol{q}) \boldsymbol{q}, \tag{13.22}$$

where

$$B, C > 0. \tag{13.23}$$

We wish to convert the equations of the standard model of the synergetic computer into one which allows for a new interpretation. To this end we introduce the expansions

$$\boldsymbol{q} = \sum_k \boldsymbol{v}_k \cdot \boldsymbol{v}_k^+ \boldsymbol{q} \tag{13.24}$$

and

$$\boldsymbol{q}^+ = \sum_k \boldsymbol{q}^+ \boldsymbol{v}_k \cdot \boldsymbol{v}_k^+ . \tag{13.25}$$

We then form $\boldsymbol{q}^+\boldsymbol{q}$ by inserting (13.24) and (13.25) and making use of the orthogonality relation (13.11). We then obtain

$$\boldsymbol{q}^+\boldsymbol{q} = \sum_k (\boldsymbol{q}^+ \boldsymbol{v}_k) \cdot \boldsymbol{v}_k^+ \boldsymbol{q}. \tag{13.26}$$

One can easily show that

$$(\boldsymbol{q}^+ \boldsymbol{v}_k) = (\boldsymbol{v}_k^+ \boldsymbol{q}) \tag{13.27}$$

holds so that (13.26) acquires the form

$$\boldsymbol{q}^+\boldsymbol{q} = \sum_k (\boldsymbol{v}_k^+ \boldsymbol{q})^2 . \tag{13.28}$$

Using this result and (13.24) again, we obtain

$$(\boldsymbol{q}^+\boldsymbol{q})\boldsymbol{q} = \sum_{k'} (\boldsymbol{v}_{k'}^+ \boldsymbol{q})^2 \sum_k (\boldsymbol{v}_k^+ \boldsymbol{q}) \boldsymbol{v}_k . \tag{13.29}$$

Inserting (13.29) into (13.22) and making a rearrangement of the sums, we obtain

$$\dot{\boldsymbol{q}} = \sum_k \lambda_k \boldsymbol{v}_k (\boldsymbol{v}_k^+ \boldsymbol{q}) - \sum_{kk'} \underbrace{\{B(1-\delta_{kk'}) + C\}}_{D_{kk'}} (\boldsymbol{v}_k^+ \boldsymbol{q}) \boldsymbol{v}_k (\boldsymbol{v}_{k'}^+ \boldsymbol{q})^2 . \tag{13.30}$$

Equation (13.30) can again be cast into a new form by taking the double sum under the first sum so that we obtain

$$\dot{\boldsymbol{q}} = \sum_k \left\{ \lambda_k - \sum_{k'} D_{kk'} \left(\boldsymbol{v}_{k'}^+ \boldsymbol{q} \right)^2 \right\} \left(\boldsymbol{v}_k^+ \boldsymbol{q} \right) \boldsymbol{v}_k . \tag{13.31}$$

This equation allows for an interesting and surprising interpretation, namely the linear term that was determined by the eigenvalues λ_k is now replaced by the curly bracket, i.e. it is as if the eigenvalues λ_k depend now on the state vector $\boldsymbol{q}$ under the sum in the curly bracket in (13.31). At this time, a comparison with models for pattern recognition may be made, namely as was shown there, the parameters λ_k have the psychological meaning of attention parameters. When they are zero, no pattern can be recognized, when they are equal to each other, the originally expressed strongest pattern will be reinforced and survive, if they are different, then there is a bias towards recognition of patterns in which λ_k is bigger than the others. The psychological notion of attention parameter (λ_k) is directly relevant to urban dynamics. In the city one may speak of the cognitive maps of individuals, socio–cultural groups and whole collectivities. As internal representations of environments, such as cities, cognitive maps imply that the person (as an individual or a member of a group) is selectively attentive to certain features or situations in the environment. (For example, Figs. 1, 2 in Portugali 1990 and Figs. 1 in Portugali and Haken 1992).

One of the major importances of cognitive maps is that despite the fact that they are invisible, it is according to them that individuals, groups and collectivities behave in the environment. In the city game cognitive maps thus function as attention parameters: when they are zero", no new pattern would emerge in the city; when they are equal to each other, i.e. when two groups of individuals are equally attentive to a certain area in the city, the originally expressed strongest pattern will be reinforced and survive; if they are different, i.e. if one group draws stronger attention to its own aspirations, then there is a bias towards the realization of patterns in which λ_k is bigger than the others. As we have seen, λ_k regulates the growth. Because of the now appearing nonlinearity, this growth becomes limited and it becomes even in a way regulated, in which one pattern wins the competition, namely the one that is strongest expressed in the beginning. This seems to be an attractive model for a city growth, where there is a multiples of final possible states, but because of competition only one pattern wins, namely the one that was initially expressed most. This offers an interesting model for a planning of cities which leaves enough room for self-organization as we will discuss in the next chapter. Equations (13.31) are still rather complicated, but the order parameter concept allows us to reduce and simplify these equations appreciably. To this end, we introduce the order parameters by means of

$$\xi_k = (\boldsymbol{v}_k^+ \boldsymbol{q}) \quad k = 1, \ldots, L \tag{13.32}$$

and the so-called enslaved variables by

$$s_k = (\boldsymbol{v}_k^+ \boldsymbol{q}) \quad k = L+1, \ldots, N. \tag{13.33}$$

Multiplying (13.31) from the left with $\boldsymbol{v}_k^+$ leads us to

$$\dot{\xi}_k = \xi_k \{\lambda_k - \sum_{k'=1}^{L} D_{kk'} \xi_{k'}^2 - \sum_{k'=L+1}^{N} D_{kk'} s_{k'}^2\}, k = 1, \ldots, L \tag{13.34}$$

and

$$\dot{s}_k = s_k \{\lambda_k - \sum_{k'=1}^{L} D_{kk'} \xi_{k'}^2 - \sum_{k'=L+1}^{N} D_{kk'} s_{k'}^2\}, k = L+1, \ldots, N. \tag{13.35}$$

Let us first consider (13.35). Because the factor of s_k has only negative terms, $\dot{s}_k$ is negative, i.e. even if a nonvanishing s_k was present initially, it will vanish in the course of time. That means, when we look for longer time scales, then we may neglect (13.35) entirely. In addition, we may look at (13.34) for the case that $\lambda_k = 0$. By means of the same argument as the one just mentioned, we realize that ξ_k vanishes for sufficiently large times

$$\xi_k \to 0, \quad k = M+1, \ldots, L \quad (\lambda_k = 0). \tag{13.36}$$

Thus the whole dynamics is governed by equations of the form

$$\dot{\xi}_k = \xi_k \left\{ \lambda_k - \sum_{k'=1}^{M} D_{kk'} \xi_{k'}^2 \right\}. \tag{13.37}$$

Equation (13.37) allows for a simple interpretation: The pattern that is associated with the order parameter ξ_k grows by means of λ_k, but λ_k is limited by the sum accompanying it in the curly bracket. In psychological terms we might say that the attention in the growth of pattern k is diminished by the growing patterns which may be the same or different as the one under consideration. In "city terms" we would say that the *cognitive* pattern/map, and thus the attention in the growth of pattern k, is diminished by the growing *material* pattern/map which may be the same or different as the one under consideration. Let us consider some special cases. If only one $\lambda_k > 0$, (13.37) reduce to

$$\dot{\xi}_k = \xi_k(\lambda - C\xi_k^2). \tag{13.38}$$

In this case a stable pattern amplitude ξ_k is reached for a large time, namely

$$\xi_k \to \sqrt{\lambda/C} \quad \text{for} \quad t \to \infty. \tag{13.39}$$

Here the growth is limited and only one pattern, namely the one with the positive λ_k is developed. Let us consider the case of two positive λs. In this case, (13.37) acquire the form

$$\dot{\xi}_1 = \xi_1(\lambda_1 - (B+C)\left(\xi_1^2 + \xi_2^2\right) + B\xi_1^2) \tag{13.40}$$

$$\dot{\xi}_2 = \xi_2(\lambda_2 - (B+C)\left(\xi_1^2 + \xi_2^2\right) + B\xi_2^2). \tag{13.41}$$

The case in which the attention parameters λ_1 and λ_2 are the same size was discussed elsewhere (Haken 1991). In this case it turns out that only one pattern survives even if initially two patterns were present, namely the one which was initially more strongly expressed. The case that $\lambda_1 \neq \lambda_2$ will be discussed in detail elsewhere. Here we mention the essential result that if, for instance, $\lambda_1 > \lambda_2$ the pattern (1) will be expressed instead of (2), even if it was initially somewhat smaller than (2). However, if (2) was initially very strongly expressed then (1) has no chance; in that case pattern (2) wins. At any rate, there will be no coexistence between the two patterns, only one pattern will be finally expressed. An interesting case arises when no cross-attention is paid, i.e. if two communities, or two populations, do not pay attention to, or cognitively are not aware of, each other. (For an empirical study of no cross attention in cognitive maps of Jews and Arabs see Portugali 1990, 1993 Chap. 9.) In this case, we have

$$D_{kk'} = D_k \delta_{kk'} \tag{13.42}$$

and the two patterns that evolve are independent from each other. They are governed by order parameters obeying the equations

$$\dot{\xi}_k = \xi_k(\lambda_k - D_k \xi_k^2), \quad k = 1, 2 \tag{13.43}$$

with the steady-state solutions

$$\xi_k = \sqrt{\lambda_k / D_k}. \tag{13.44}$$

To study which kind of spatial patterns may be connected with this dynamics, we consider the special case in which we may decompose the matrix Λ into two parts

$$\Lambda = \Lambda^{(1)} + \Lambda^{(2)}, \tag{13.45}$$

where each part is a square matrix. This means that movements take only place within one part of the territory. Actually, our considerations hold for a much wider case in which the positions are quite arbitrary and the only condition is that the two parts of the population communicate only between parts belonging to the same kind of population. The eigenvectors of (13.45) are given in of the form

$$\boldsymbol{v}_1 = \begin{pmatrix} \vdots \\ 0 \\ 0 \end{pmatrix}, \tag{13.46}$$

$$\boldsymbol{v}_2 = \begin{pmatrix} 0 \\ 0 \\ \vdots \end{pmatrix}, \tag{13.47}$$

$$\Lambda^{(1)} \boldsymbol{v}_1 = \lambda_1 \boldsymbol{v}_1, \tag{13.48}$$

$$\Lambda^{(1)} \boldsymbol{v}_2 = 0, \tag{13.49}$$

$$\Lambda^{(2)} \boldsymbol{v}_1 = 0, \tag{13.50}$$

$$\Lambda^{(2)} \boldsymbol{v}_2 = \lambda_2 \boldsymbol{v}_2 \tag{13.51}$$

Growth and final pattern formation occur entirely independently in the two regions of populations.

The above case of "no cross (cognitive) attention" and the resultant material pattern formation, is of the utmost importance in the growth of cities. Cities are commonly described as huge mosaics of spatially segregated populations (national, ethnic, religious, socio–economic, ..., groups) and activity areas (industrial, business, recreation, ..., zones). No cross cognitive attention and the resultant material pattern as derived above, provide the rationale to the emergence and persistence of this urban mosaic: a few, coexisting, order parameters govern the dynamics. Two interesting and interrelated questions arise here: First, what happens on the boundaries between the otherwise non-attentive zones in the city? Second, is there a "super order parameter" which governs the relations between the non-attentive, spatially independent, zones? (For a preliminary discussion see Chap. 5 above, and Haken and Portugali, in preparation.)

13.6 Another Origin of the Saturation of Attention

The nonlinear terms arising in (13.37) can be thought of as stemming from a dynamic process in which the attention parameters λ_k saturate in the course of time. The corresponding equations are given by

$$\dot{\lambda}_k = \gamma(\lambda_k^0 - \lambda_k) - \gamma \sum_{k'} D_{kk'} \xi_{k'}^2 . \tag{13.52}$$

They can be interpreted in the following manner: The attention parameter λ_k changes in the course of time, because of the two terms on the right-hand side. The first term means that if the attention parameter λ_k is disturbed, it will relax in a time $\frac{1}{\gamma}$ towards a constant size λ_k^0. The second term means that the attention parameter is changed because of growing patterns. If patterns grow, the attention parameter λ_k for a specific pattern is decreased. However, when we assume the explicit form of $D_{kk'}$ according to (13.30), we may write

$$\sum_{k'} D_{kk'} \xi_{k'}^2 = (B + C) \sum_{k'} \xi_{k'}^2 - B\xi_k^2 . \tag{13.53}$$

This form reveals that the attention fades away by means of the growth of all patterns, but that it fades away much less for the growth of the eigen pattern. It is as if here collectively a mental map for the city as a whole is formed which favors the one pattern to which the attention is focussed. Here highly interesting questions arise, namely in how far can one speak of collective mental maps, in how far can here planning agencies be involved. In light of the interplay between the attention parameters λ_k and the order parameters as above, city plans might be regarded as yet additional attention parameters λ_k (or cognitive maps) which participate in the process of self-organization in the city. Planning as a whole might be seen as a conscious attempt to identify the various attention parameters in the city as they arise, and to affect their strength. Further on we note that the form of (13.52) as a function of the order parameters, can be different so that different saturation mechanisms of the attention parameters λ_k may be considered. We further remind the reader that by appropriately chosen $D_{kk'}$ also oscillations in attention and order parameters may occur as is known from the perception of ambiguous patterns, such as vase/face, and so on (Ditzinger and Haken 1990).

13.7 Concluding Remarks

In the present chapter we have tried to cast the problem of the growth of cities into a mathematical form that allows for explicit solutions. Depending on internal mechanisms, one or several patterns may be selected and we have studied what influences pattern selection. Clearly the next step must consist in filling this mathematical frame with life, for instance by studying concrete cases of the choice of the λs in (13.1). It must also be left to further studies to include the effect of external migration, that is, the influx or outflux of populations. This can be done rather easily in the case of linear models, but in the case of nonlinear models probably new ideas will be required. Yet another next step would be to examine the possibility to introduce planning activities into the model and thus transform it into an applicative tool at the hand of planners. Quite clearly, the growth of cities and the evolving structures of patterns is determined by the transition rates λ and then by the attention parameters. It will be therefore of interest to study the form of these transition rates more closely and how they can be influenced. They are surely influenced by means of traffic, but also mentally by the readiness of people to go from one place to another one, which may be determined by favorable positions and the new place, and so on. Quite evidently, it will also be necessary to identify several time scales, for instance for building roads, establishing mental maps, etc. In other words, one may also have to take into account temporal changes of the λs.

14. Synergetic Cities II: Pattern Recognition, Cognitive Mapping and Decision-Making*

14.1 Introduction

In this chapter we elaborate on Haken's recent demonstration that the synergetic paradigm of pattern recognition can be used as a conceptual and mathematical framework for the study of decision making in general and in the context of urban and regional planning in particular. The elaboration includes an extension concerning cognitive mapping, a reference to Tversky and co-workers' studies on the psychology of decision making, and a reformulation in terms of Portugali's notion of IRN and Haken and Portugali's model of SIRN. The discussion throughout the chapter follows the above description.

14.2 Pattern Recognition as Decision-Making

In two recent studies Haken (1996, 1998) has suggested that the process of pattern recognition as conceptualized by synergetics (Chap. 3 above) is, in principle, a decision making process: a person, or a computer, is offered part of a pattern which is stored in memory, together with many other patterns. The person/computer is then asked to decide to which of the stored patterns the offered part belongs. The principles here are, first, that the recognizer has to take a decision about a whole pattern on the basis of incomplete partial information on it, second, that this is implemented on the basis of the similarity of the offered part to known stored patterns. According to Haken, this situation is characteristic also of decision making in the context of urban and regional planning. In the latter every decision about a future planned situation is, by definition, taken on the basis of partial and incomplete information, and it is typical that the decision is taken on the basis of its similarity to known situations from the past.

In both cases, pattern recognition and decision-making, there is often a gap between the known data and the required data needed to decide upon a specific action (Fig. 14.1). In the ideal case the known data coincide with the required data. In general, however, the known data are insufficient, i.e., there are a certain number of unknown data. How do humans fill the gap

* By Hermann Haken and the author

Decision Making as Pattern Recognition

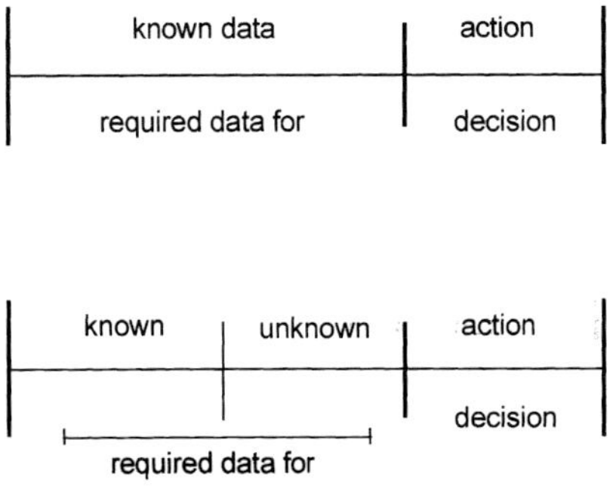

Fig. 14.1. In the *upper part*, the known data coincide with the required data for taking action and making a decision. In the *lower part* the known data are insufficient

of unknown data? This is what we want to analyze in the following. A simple though nontrivial example is provided by a tennis player. Some analysis shows that his or her time for the necessary reaction is too short to allow an analysis of the sensory input data before starting motor action. Thus tennis players have to act on specific clues that are based on their experience and training. But, in general, the problem is still more complicated. Consider to this end Fig. 14.2. It indicates that, at least in general, the known data can be complemented in a variety of ways to fill in the gap of the unknown data. Depending on how we fill in the unknown data, different decisions or actions may be taken. This figure is in a sense oversimplified, because even if all the data are known there may still be several decisions that are compatible with all the known data.

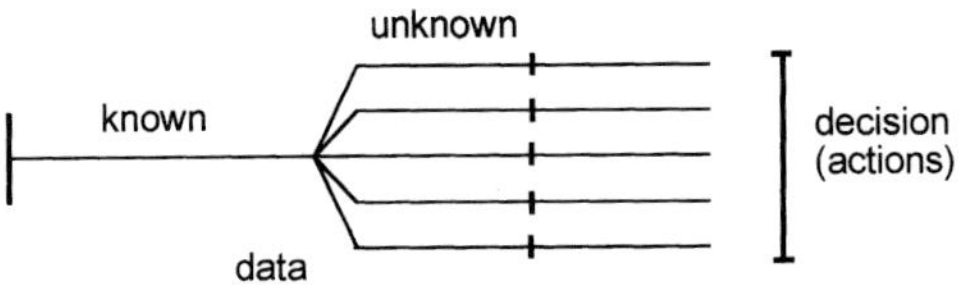

Fig. 14.2. The known data may be complemented in a variety of ways. Each of these ways might entail a different decision and action

How do we fill in the unknown data? Our main theme will be that we often rely on a similarity between a given situation and a previous situation. When we want to cast this similarity into a mathematical frame, we have to look for similarity measures (see Sect. 14.4 below). Of course, in a nonmathematical way we may rely on analogies or metaphors. A number of psychological factors are of importance, such as awareness, attention, bias, and beliefs. When we take seriously the analogy, which we shall discuss below, with pattern recognition, we can expect time-dependent choices as in the case of ambiguous figures. This implies that we make decisions that show oscillations, or in the course of time there may be random choices. In more detail,

Table 14.1. Correspondence between the elements and processes of pattern recognition and those of decision making

Pattern Recognition	Decision Making
patterns pictures arrangement of objects visual, acoustic signals movement patterns actions (often encoded as vectors)	data, quantitative, qualitative, yes/no rules, laws, regulations algorithms, programs flow charts, diagrams orders multi-dimensional in short: "data"
prototype patterns learned or given	sets of known complete "data" learned or given
test patterns	incomplete data in particular "action" lacking
similarity measure dynamics bias attention, awareness	 $\rightarrow$
unique identification or oscillations between two or more percepts hysteresis .. complex scenes saturation of attention	unique decision or oscillations between two or more decisions do what was done last time even under changed circumstances multiple choices failure, new attempt based on new decisions

we propose to draw the following analogies between pattern recognition and decision making (cf. Table 14.1). In decision making the data correspond to patterns treated in pattern recognition. The data may be quantitative or they may consist of specific rules, laws, or regulations. They may be in the form of algorithms, or when we think of computers, in the form of programs or flow charts. Diagrams may also be considered as constituting such data. In pattern recognition the patterns may consist of pictures or of the arrangements of objects. The patterns may be visual or acoustic signals. Quite often these patterns are encoded as vectors, which may be constant or time-dependent. Of course, in decision making the data may be multi-dimensional.

So far we have been discussing the analogy between the objects dealt with in decision making and in pattern recognition. In both cases the prototype patterns or the sets of known complete *data* may be learned or given. Incomplete data in decision making have their analog in pattern recognition in the form of incomplete test patterns. How can we exploit this analogy to study decision making? In analogy to pattern recognition we may introduce a similarity measure, for instance, the overlap of prototype patterns and the test pattern. We can then establish a dynamics that is based on the similarity measure and may also include bias, attention parameters, or awareness. So, from a formal point of view, the whole procedure that we encounter in pattern recognition may be transferred to a scheme describing decision making (for mathematical details cf. Sect. 14.4 below).

What will be the consequences? They are listed in Table 14.1. In pattern recognition and in decision making, we may find a unique identification and a unique decision, respectively. But in a number of cases we may be confronted with oscillations between two or more percepts, or between two or more decisions. These oscillations are not unusual in our daily life as everybody knows. Here we can trace them back to a fundamental mechanism of the human cognitive abilities. A very important analogy arises when we look at the hysteresis effect as analysed by the synergetic computer in the context of pattern recognition (Fig. 14.3). Translating this effect into decision making means the following: A person does what he or she did last time even under changed circumstances. The analogy between pattern recognition and decision making can be carried further. In pattern recognition we usually deal with complex scenes, where the synergetic computer and probably the human brain analyze such a scene by means of a saturation of attention. Once part of a scene has been recognized, we focus our attention on the other objects. Such a case is illustrated in Fig. 14.4. In our analysis of decision making, multiple choices correspond to complex scenes and the saturation of attention in pattern recognition, can now be translated as follows: Based on our attention

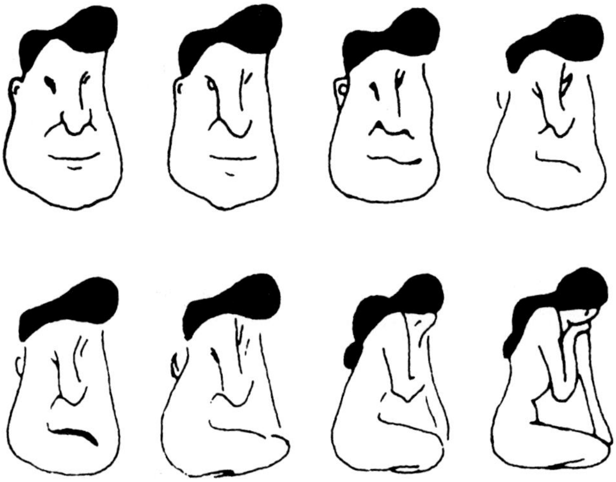

Fig. 14.3. Hysteresis in pattern recognition. When the sequence of figures is visually scanned from the upper left to the lower right, the switch from a face to a girl occurs in the lower row. When scanned in the reverse direction, the switch occurs in the upper row. As illustrated below, in Sect. 14.7, this might be interpreted as an illustration to the effect of the anchoring visual heuristic in decision-making

Fig. 14.4. A complex scene as recognized by the synergetic computer is analogous to a complex situation with multiple possible choices in decision making

we make a first choice. When we encounter a failure, the *attention* parameter for that endeavor is put equal to zero. We then make a new attempt based again on our attention for a new kind of endeavor, and so on. Depending on our previous experience there may be a hierarchy of attention parameters through which we work starting with the highest attention parameter. This interpretation is related to the notion of *heuristics* and its use in the process of decision making (*Wagenaar* (1993)). In Sect. 14.7 below we further discuss heuristics in relation to Tversky and Kahanman's psychological approach to intuitive judgment.

Summarizing these ideas we can state that the mechanisms discussed in the case of pattern recognition can be translated into those of decision making. This can be done not only at a qualitative level but also quantitatively at the level of computer algorithms in analogy to the synergetic computer. Quite obviously, our analysis is by no means complete and other strategies may be of equal importance. Artificial intelligence and here especially the approach by expert systems must be mentioned. A problem encountered here is that of branching, where the various branches become extremely numerous and decision making eventually becomes very difficult. We believe that this branching problem can be circumvented by the approach we outlined here, because, as in pattern recognition, the various possibilities are taken care of in a parallel fashion.

14.3 An Extension Concerning Cognitive Mapping

The above view can be extended by adding that the synergetic process of decision making is similar also to the synergetics conceptualization of cognitive mapping (Portugali 1990, 1996, Portugali and Haken 1992). In both decision-making and cognitive mapping, it is typical that a person makes a judgment (takes a decision or constructs a whole image/cognitive map of a city, for example) on the basis of only partial information on it (the consequence of the decision or the structure of the city). Furthermore, in pattern recognition the task of recognition is completed, usually, when the offered partial pattern fully resembles one of the stored complete patterns. In cognitive mapping this is rarely so. The usual case is that, because of size, there is no complete stored pattern, and as a consequence cognitive maps are usually incomplete and their structure and parts are often distorted, vague or unknown (Portugali, 1990). This is so also with decision making, for the simple reason noted above that the future at which the decision is aiming is by definition not fully known. A decision, in this respect, is an action based on a cognitive map of the future.

Cognitive mapping is related to decision making in the context of cities in yet another respect. As noted in the previous chapter, one of the most important aspects of cognitive maps is that it is mainly according to them that people navigate in the environment, choose routes for commuting, decide to leave their homes and migrate to another city, choose a new home, and so on (Portugali 1990, Portugali and Haken 1992). That is to say, cognitive maps provide the basis for decision making in space.

14.4 Optimization

As it will turn out in the mathematical formulation, there are two ways to steer the competition process between different kinds of decisions or, if one wishes, plans. The first approach, i.e. the foundation of pattern recognition, is based on the assumption that all stored patterns are basically equivalent to each other, but that there is an initial bias towards a specific similarity between a new situation and a formerly known one. Under these circumstances the pattern with the strongest similarity to the newly given pattern wins the competition and the decision making process. On the other hand, we may give all the patterns that are basically possible the same opportunity, but we pay to a specific pattern a higher attention than to the other one. Then the mathematics shows that, indeed, the pattern to which the highest attention is paid wins the competition in spite of the fact that all other patterns are equivalent. This is, of course, of psychological interest also. The question arises whether we can overcome these psychologically based decisions by specific optimization procedures. Astonishingly enough, the optimization procedure developed by Haken (1998) is based on precisely the same processes underlying decision making. It seems that here we are unearthing relationships that have not been explored yet, but that deserve further exploration because of the challenge of combining intuition and rigorous methods in regional planning.

In order to elucidate ideas and to be as simple and concrete as possible, we consider quite a specific problem in urban and regional planning, namely the spatial allocation problem. That is to say, the allocation of buildings to sites, of individual households to apartments, public parks to residential areas and so on. (For a general discussion and a survey of various approaches see Portugali 1980, Chap. 4.) Consider, for example, the allocation of buildings to sites. For this purpose let us label sites by the index i and specific buildings, such as homes, service centers, schools, factories, etc. by an index j. Then quite evidently there is a relationship between site i and building j with a specific efficiency κ_{ij}. The task is to allocate the buildings to the sites so that the total effficiency function $\sum \kappa_{ij}$ becomes a maximum with a specific choice of the assignments between j and i. In the next section we will cast this optimization problem into a mathematical form. As it will turn out, the solution of this allocation problem is closely related to both the pattern recognition and pattern formation problem. This may not be so surprising, because spatial allocation means that we have to find and determine a specific pattern. What is interesting, however, is that the formalism turns out to be practically identical in both cases. Furthermore, the individual efficiency factors turn out to be just the attention parameters in pattern recognition and urban dynamics that we have discussed above in Chap. 13. In terms of the attention parameter we thus look for such solutions of the optimization problem so that we pay the highest attention to the highest efficiency.

This is a surprising result, because it is not trivial at all and is, on the other hand, a direct consequence of this formalism. Quite evidently, these concepts can be pushed much further, because the issue of allocation can be given a much broader formulation, which includes traffic networks, interrelationships between different kinds of buildings, cultural and social groups and the like. While the formulation of this problem is, indeed, possible to a rather large extent, its explicit formulation would require too much space here and must be postponed to a later occasion.

14.5 Mathematical Formulation

We will present the mathematical approach in several steps. Basically we have to distinguish between the micro- and the macro-level. At the microlevel we describe the state of a system by the state vector $\boldsymbol{q} = (q_{1,\ldots}, q_N)$. In physics q_j may be, for instance, the density of a fluid in a cell labeled by the index j. In the context of cities q_j may denote the number of buildings at a site j. When we split the index j into further indices, we may also differentiate between different kinds of buildings at the same site j. Basic to synergetics is the idea that the state vector $\boldsymbol{q}$ changes in the course of time because of interactions between different sites. We may also include the effect of general external conditions that are described by so-called control parameters α, which in a fluid may be the amount of heating or which in cities may be the number or rate of immigrants, for example. Further we will take into account chance events by adding so-called fluctuating forces $\mathbf{F}$ that may depend on time. The general evolution equations are then given by

$$\dot{\boldsymbol{q}} = \mathbf{N}(\boldsymbol{q}, \alpha) + \mathbf{F}(t), \tag{14.1}$$

where $\mathbf{N}$ is a nonlinear function of the state vector $\boldsymbol{q}$. The basic assumption underlying the approach of synergetics is based on the role of the control parameters. As is witnessed by many processes in a variety of systems, the behavior of a system may change dramatically if a control parameter exceeds a certain critical value α_c. If $\alpha = \alpha_0$ is below such a control parameter value, for instance the rate of immigration, we may deal with a specific solution $\boldsymbol{q}_0$, for instance a homogeneous distribution of buildings or of a liquid

$$\alpha_0 : \boldsymbol{q}_0 . \tag{14.2}$$

But when the control parameter α exceeds a critical value, a totally new solution may set in, for instance a city may spontaneously grow and change its character dramatically. In order to take account of this effect, we make the hypothesis

$$\alpha : \boldsymbol{q} = \boldsymbol{q}_0 + \boldsymbol{w}(t), \tag{14.3}$$

where $\boldsymbol{w}$ is assumed to be initially small. This allows one to linearize (14.1). The solution of the linearization reads

$$\boldsymbol{w}(t) = \exp(\lambda_j t)\boldsymbol{v}_j , \tag{14.4}$$

where $\boldsymbol{v}_j$ is a time-independent vector that corresponds to a specific pattern. The exponent $\lambda_j \geq 0$ indicates growth and we put the index $j = u$ ("unstable") so that

$$\lambda_j \geq 0 \quad : \quad \lambda_u, \boldsymbol{v}_u . \tag{14.5}$$

Other configurations may decay, which is indicated by $j = s$ ("stable")

$$\lambda_j < 0 \quad : \quad \lambda_s, v_s . \tag{14.6}$$

The exact solution of (14.1) can then be represented by the superposition

$$\boldsymbol{q} = \boldsymbol{q}_0 + \sum_u \xi_u(t)\boldsymbol{v}_u + \sum_s \xi_s(t)\boldsymbol{v}_s , \tag{14.7}$$

where we sum up over the unstable modes with index u and the stable modes with index s. ξ_u and ξ_s are time-dependent amplitudes. The slaving principle of synergetics – that is quite fundamental – shows that the amplitudes ξ_s can be uniquely expressed by the amplitudes ξ_u

$$\xi_s = f_s(\xi_u) . \tag{14.8}$$

Since the number of all possible patterns $\boldsymbol{q}$ is much larger than the number of ξ_u, (14.8) implies an enormous information reduction. In a number of important cases, the behavior of the whole system is governed by one or a few amplitudes ξ_u that are called the order parameters. As it turns out, the whole dynamics of the complex system, described by (14.1), can be reduced to the dynamics of the order parameters ξ_u that again obey nonlinear equations that can be derived explicitly in each specific case

$$\dot{\xi}_u(t) = \tilde{N}_u(\xi_u, \alpha) + F_u(t). \tag{14.9}$$

The abstract formalism described in (14.1)–(14.9) allows us to calculate the evolving patterns in a great variety of systems, expecially in physics, chemistry and biology, but also in sociology, etc.

We now turn to pattern recognition. In this case we choose a specific order parameter equation, namely of the form

$$\dot{\xi}_u(t) = \xi_u \left(\lambda_u - \sum_{u'} B_{u,u'} \xi_{u'}^2 \right) + F_u(t). \tag{14.10}$$

Such an equation arises already in fluid dynamics and describes the competition between patterns. In pattern recognition, the λ_u's play the role of attention parameters. If the λ_u's are the same for all so-called modes u, and $B_{u,u'}$ is of the form

$$B_{u,u'} = A\delta_{u,u'} + B(1 - \delta_{u,u'}), B > A, \tag{14.11}$$

(14.10) have the property that the initially strongest amplitude ξ_u wins the competition so that, eventually, $\xi_u = 1$ for the initially biggest ξ_u and $\xi_u = 0$ for all others. As a side remark we mention that (14.10) can be conceived as a gradient dynamics

$$\dot{\xi}_u(t) = -\frac{\partial V}{\partial \xi_u} \tag{14.12}$$

where the potential V is given by

$$V = -\frac{1}{2} \sum_u \lambda_u \xi_u^2 + \frac{1}{2} \sum_{uu'} B_{u,u'} \xi_u^2 \xi_{u'}^2. \tag{14.13}$$

Before we exploit this property, we briefly discuss what determines the biggest ξ_u in pattern recognition. To this end, we form the so-called overlap function between the stored prototype patterns $\boldsymbol{v}_u$ (Fig. 3.6, *top*, in Chap. 3) and a test pattern vector $\boldsymbol{q}$ (Fig. 3.6, *bottom*, in Chap. 3) according to the scalar product

$$\xi_u(0) = (\boldsymbol{q}(0)\boldsymbol{v}_u) \, . \tag{14.14}$$

For mathematical simplicity we assume the pattern vectors $\boldsymbol{v}_u$ to be orthogonal which in the present context is not an important restriction. An example for stored prototype vectors $\boldsymbol{v}_u$ in the form of city maps and the recognition process is shown in Fig. 14.5. According to the dynamics described by (14.10), the final solutions are given by

$$\xi_u(t = \infty) = \begin{Bmatrix} 1 & u = u_c \\ 0 & u \neq u_c \end{Bmatrix} \tag{14.15}$$

where the order parameter wins with the biggest overlap (14.14), i.e. with the greatest similarity between the initially given vector $\boldsymbol{q}$ and the stored prototype patterns $\boldsymbol{v}_u$.

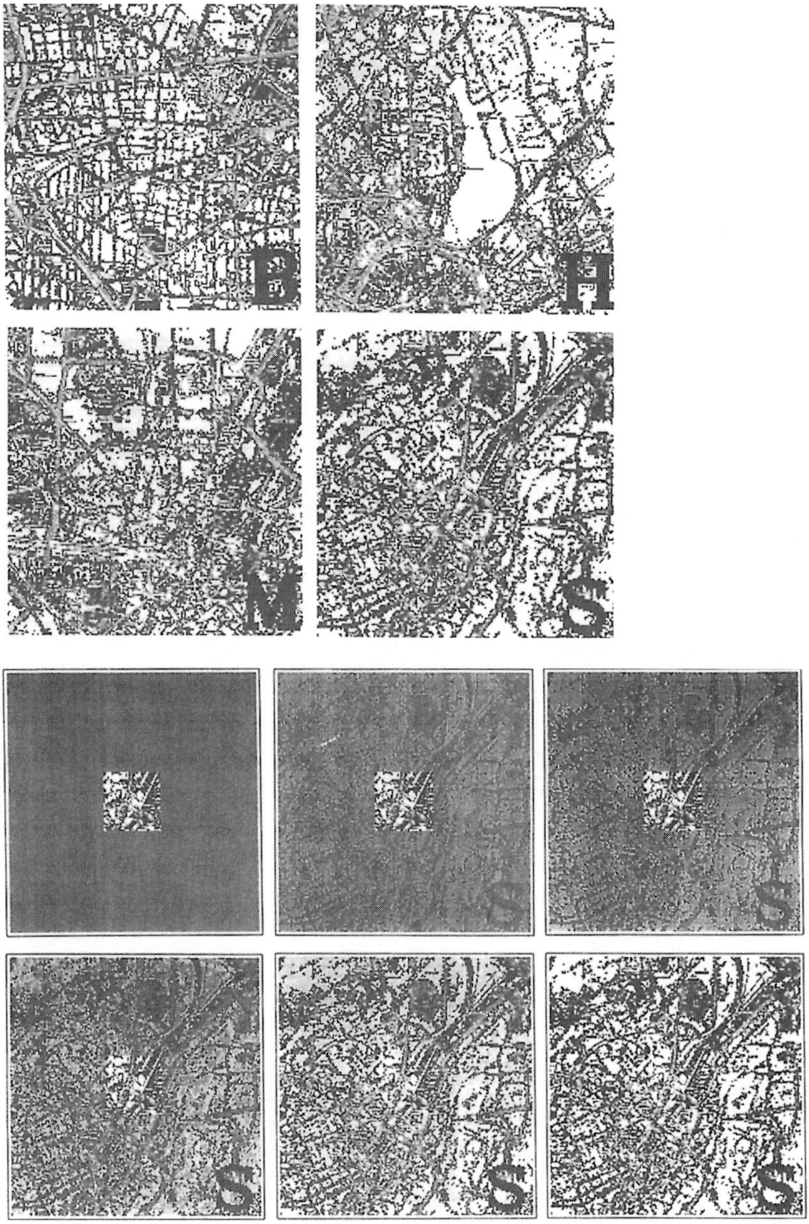

Fig. 14.5. City maps with their labels (Hamburg, Berlin, Stuttgart, Munich) stored in the synergetic computer as prototype patterns (*top*), are recognized on the basis of only a "window" of incomplete map offered to the computer (*bottom*)

In order to take into account the saturation effect of attention that occurs in ambiguous figures (as in Fig. 14.6, for example), we make the attention parameters time-dependent according to the differential equation

$$\dot{\lambda}_u = \lambda_{u,0} - \lambda_u - \xi_u^2. \tag{14.16}$$

In the example of two possible interpretations, (14.10) jointly with (14.16) yield the oscillatory process as given by Figs. 14.6–14.7.

Fig. 14.6. Example of an ambiguous pattern: vase or two faces?

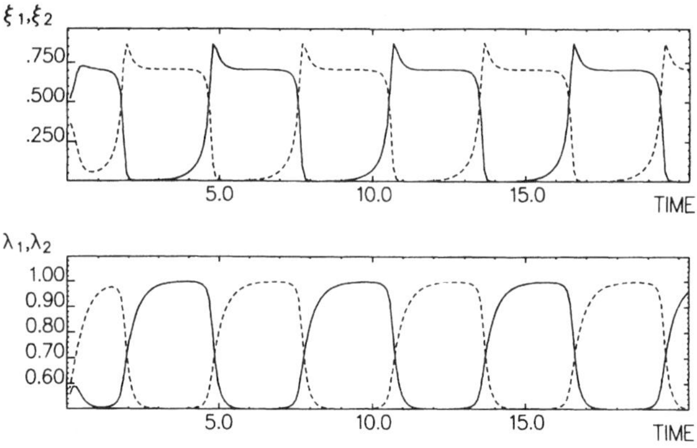

Fig. 14.7. Oscillations of the order parameters (*upper part*) and the corresponding attention parameters (*lower part*)

Finally, we mention that we can include also a bias in (14.10) and change it in such a way that the effect of hysteresis (above, Fig. 14.3) is taken into account.

So far we have given an outline of the process of pattern recognition and correspondingly of decision making. While in pattern recognition the prototype vectors $\boldsymbol{v}_u$ and the test pattern vectors $\boldsymbol{q}$ can be rather easily encoded by means of gray values of Fig. 14.5 (and of Fig. 3.6 in Chap. 3), decision making is a harder problem, because we have to encode abstract patterns in terms of vectors $\boldsymbol{v}$ and $\boldsymbol{q}$. For a comparison between pattern recognition and decision making consult Table 14.1 above.

In order to finish the sketch of the mathematical approach, let us now look at the process of spatial allocation, where the total efficiency has to be maximized. Let us denote the different building sites in the city by index i and the specific buildings by index j. The efficiency relating building j to a building site i is denoted by κ_{ij}. Then the aim is to maximize efficiency according to

$$\sum_{ij} \kappa_{ij} \xi_{ij}^2 = \text{Max!}, \tag{14.17}$$

where ξ_{ij} may take the values 1 or 0 only. Provided we have the same number of buildings as of building sites, and only one building can be built on a site, then ξ_{ij} must be a permutation matrix, for instance in the case of three buildings and three sites in the form (14.18). It is important to note that each row and column may contain only a single 1, and that all other matrix elements must be equal to 0.

$$\begin{pmatrix} 0\,1\,0 \\ 1\,0\,0 \\ 0\,0\,1 \end{pmatrix}. \tag{14.18}$$

In order to solve the optimization problem (14.17), we have to take into account the specific possible form of the permutation matrices which can be cast into the following constraints

$$\sum_{j} \xi_{ij}^2 = 1, \tag{14.19}$$

$$\sum_{i} \xi_{ij}^2 = 1, \tag{14.20}$$

$$\xi_{ij}^2 \xi_{kj}^2 = 0 \quad \text{for} \quad i \neq k, \tag{14.21}$$

$$\xi_{ij}^2 \xi_{ik}^2 = 0 \quad \text{for} \quad j \neq k. \tag{14.22}$$

We now take up an earlier idea (Haken 1990) on how to take the constraints (14.19)–(14.22) into account. We define a total potential in the form

$$V = -\sum_{ij} \kappa_{ij}\xi_{ij}^2 + \Lambda_1 \sum_i \left(\sum_j \xi_{ij}^2 - 1 \right)^2 + \Lambda_2 \sum_j \left(\sum_i \xi_{ij}^2 - 1 \right)^2$$
$$+\Lambda_3 \sum_j \left(\sum_{i \neq k} \xi_{ij}^2 \xi_{kj}^2 + \sum_{i \neq k} \xi_{ji}^2 \xi_{jk}^2 \right) \tag{14.23}$$

and postulate eqs. (14.1). Such a generalization has been formulated more recently by Starke (1997), however, with one fundamental difference, because he does not include the term (14.17). He assumes that the competition between the configurations ξ_j is provided by the initial values, where ε_i, ξ_{ij} with the largest initial value wins the competition under the constraints (14.19)–(14.22). Here our approach is basically different, because in the present case the configurations ξ_{ij} win the competition that maximize (14.17) under the constraints (14.19)–(14.22). One can show that this competition process can be won only if the coefficients of the Lagrange parameters $\Lambda_1, \Lambda_2, \Lambda_3$ are big enough. The question of distinguishing between different local minima may still be open in contrast to (14.12), (14.13), where a general theory of possible patterns and their minima of the potential function (14.13) exists.

14.6 Discussion

Thus far we have tried to show how a number of concepts that are applicable to the question of planning of settlements stem from a single root. The question, of course, arises in the present context, where self-organization comes in. In the approach outlined above, the original (14.1) contain, on the one hand, the internal interactions between the different components of a system and control parameters. Under the impact of control parameters describing specific sets of conditions that are based on the environment, like rivers, mountains, etc., or, in our case more realistically, on certain planning actions, the system (14.1) acquires a certain state. It is important to note that the number of control parameters α is much smaller than the number of the components of the state vector $\boldsymbol{q}$. In spite of this fact, because of the internal dynamics imposed on the system by the specific form of the function N, the system may acquire only a few stable states under favorable conditions imposed on the system by the control parameters. Thus only few specific patterns may evolve. The mathematical formulation allows one to determine these new evolving patterns. The evolving patterns are determined by a few order parameters ξ_u only. Then, in the next step, we have inverted the whole consideration, namely we have asked how we recognize patterns. This has led us to the concept of associative memory. The crucial role is played here by the attention parameters λ_u that may be considered as parameters controlling our attention. For instance, if we set attention parameters for specific patterns equal to zero, then we are not able to perceive these patterns. The attention parameters determine also the competition between different stored patterns, when a new pattern is given. If all patterns are given the same chance, the attention parameters decide which pattern is finally recognized.

Finally we formulated a maximization problem, which again could be cast into a form in complete analogy to the pattern recognition problem. Here the efficiencies κ_{ij} play the role of attention parameters. But because of the constraints not all configurations with the highest attention can be realized, but compromises must be found, which is expressed by a total potential function (14.23). Quite clearly, the mathematical approach can be extended in a variety of ways, for instance to more complicated maximization problems that take into account the interaction between different sites and also take into account still more complicated constraints than in (14.19)–(14.22).

Furthermore it may be stated that the minimization of the total potential function (14.23) need not be taken care of by a method of steepest descent as described by (14.12), but that other minimization procedures, such as genetic algorithms, and so on, may be employed here, too.

Finally we remark that (14.10) define the algorithm of a so-called synergetic computer, which is a fully parallel computer, and represents an alternative to neural networks (see Sect. 14.8 and Fig. 14.9, below). It can thus be stated that precisely the same formalism allows us to treat pattern formation by self-organization, pattern recognition, decision making and optimization from a unifying point of view.

14.7 Decision-Making Heuristics

As noted above, the correspondence between pattern recognition and decision making is related to the notion of heuristics. That is to say, to cognitive strategies, schemata or models that people tend to employ when making judgement in everyday life. This notion is also the cornerstone of Tversky and Kahneman's cognitive approach to decision-making. According to them, when facing complex decision situations with high degree of uncertainty, people tend to rely on a limited number of heuristic principles which reduce the complex tasks of assessing probabilities and predicting values to simpler judgemental operations. (Tversky and Kahneman 1974, p. 1124).

In a series of studies conducted during the 70s they have identified five such heuristics: representativeness, availability, anchoring, similarity, and decision frame (Tversky 1977, Tversky and Kahneman 1971, 1972, 1974, 1981, Tversky and Gati 1978). With respect to these heuristics they further say that in general,

> these heuristics are quite useful, but sometimes they lead to severe and systematic errors (Tversky and Kahneman 1974, p. 1124).

Smith and Lundberg (1984) have suggested that these heuristics are typical of individuals' decision making in the context of cities, whereas Krafta et al. (in preparation) have integrated them in a model of urban navigation and cognitive mapping. Table 14.2 gives some details of these heuristics and the way they are relevant to the context of urban dynamics.

While in both the synergetics and the TK (Tversky-Kahneman) approaches to decision making heuristics play an important role, the two approaches differ from each other in some important respects. In TK there is a separation between judgement, decision and action. In fact, they do not study decisions and actions at all. Apparently, their implicit assumption is that judgement entails a decision and action. In the synergetics approach to decision making, judgement, decision and action (or behavior) are inseparable – they are elements in a single decision-making system. This property is further elaborated below in Sect. 14.8.

TK is essentially an empirical approach followed by a verbal interpretation that makes use of elementary probability theory. Synergetics, as introduced above, is a theory about complex systems with its own mathematical formalism and modeling approaches. The synergetic analogy between pattern recognition and decision-making thus implies that the decision-making process is a complex, self-organizing, synergetic system. As a consequence, the aim of synergetics is not to study errors in order to identify judgmental heuristics, but rather to model what people really do when taking decisions. Furthermore, while TK considers a solitary decision making individual, synergetics looks also at collective effects. That is to say, the way individuals' decisions and actions influence each other. This theme, too, is further elaborated in Sect. 14.8 that follows.

Having stated the differences, it can now be said that some of the synergetic pattern recognition analyses of ambiguous patterns can be interpreted as associated with decision-making heuristics. For example, in the old/young woman pattern recognition case (Fig. 14.8), it was found that men tend to see the young woman first, and only then the old woman. This might be interpreted as a result of a 'visual availability heuristic', similar to a TK verbal heuristic.

Another case is the phenomenon of hysteresis as modeled by the synergetic computer (above, Fig. 14.3). In the context of the present discussion the interpretation might be that the form with which the scanning process starts has the effect of a visual anchoring heuristic. If the scanning of Fig. 14.3 starts from the top-left form, then the face acts as an anchor; if from bottom-right, the girl.

The above two examples will suffice to draw attention to the possibility that a given heuristic might be active in the context of both verbal intuitive judgement and visual pattern recognition. Further similarities with respect to other heuristics can be made but will not be pursued here as they deserve a separate study. In closing this section on heuristics, however, we would like to emphasize two aspects of heuristics that are specific to synergetics.

Table 14.2. Seven heuristics and their interpretation in the context of cities. The first five follow Krafta's et al. (in preparetion) elaboration on Tversky and Kahneman heuristics; the last two follow the discussion in this chapter

Heuristic	Description	Urban interpretation
Similarity	The similarity of two items is expressed as a function of their common and distinctive features.	Recognition of fundamental rules of urban composition, such as a grid layout, built form, or urban fabric, is performed through similarity.
representativeness	Probability that an object or event belongs to a particular class is judged by the degree to which a description is representative of a stereotype.	Urban categories are identified on the basis of architectural stereotypes, such as church, skyscraper, boulevard, tower, arch, etc.
Availability	Probability of an event, or frequency of a class, is assessed by the ease with which instances or occurrences can be brought to mind, or recalled.	Availability would make universal symbols more easily identified and recalled, such as Pizza Hat premises, Macdonald's signs, London pubs, etc.
Decision frame	The frame that a decision maker formulates of the problem (gain versus loss, etc.) is influenced by norms, habits, and personal characteristics of the decision maker.	Urban frames for decisions (congested versus free, public versus privet, etc.) depend on the cultural code of each agent (e. g. a tourist, a taxi driver, a policeman, etc.).
Anchoring	The tendency of people to make estimates by starting from an initial base value that is adjusted to yield the final answer.	City's internal representations can contain certain categories such as style, urban violence, town size, etc., which can be 'fired on' early in the process of cognition; once switched on, it stays on and is only eventually reassessed.
Synergetic I: Collective effects	When facing a complex decision situations people tend to rely on what other people are doing or saying.	Drivers, pedestrians, intra- and inter-urban immigrants, tend to 'follow the stream'. That is: to take decisions in line with what others are doing.
Synergetics II: Attention parameters effects	When facing a complex decision situations people often employ several heuristics in a sequence. First, the attention parameter calls into use a certain heuristic. Then, when exhausted, another attention parameter heuristic emerges and so on.	Intra- and inter-urban immigrants, for example, often start with a given location decision heuristic (say, synergetics I); if it doesn't work they switch to an alternative heuristic and so on.

Fig. 14.8. Young woman or old woman?: an ambiguous pattern in cognition as a 'visual availability heuristic' in decision making

First, that when facing a complex decision situation people tend to rely on what other people are doing. This synergetic heuristic refers to the fact that many decision situations are characterized by strong interaction and information exchange among the individual agents, and thus exhibit pronounced collective effects. TK do not consider such situations. In fact, we have already encountered this situation, in Chap. 7, in connection with the notion of "social person" suggested by Fischer et al. (1991) and Sonis (1994). In light of the above discussion we would like to suggest that the notion of social person, or the "synergetic heuristic" as we prefer to call it, is one among several heuristics that people might employ when taking decisions in the complexity of city dynamics.

Second, that the notion of a heuristic is intimately related to the synergetics concept of an attention parameter that we have discussed in the previous chapter, in connection with urban dynamics and cognitive mapping, and in the present chapter, in connection with the analogy between pattern recognition and decision-making. Employing a given heuristic implies activating an attention parameter in a specific way that corresponds to a given heuristic. Once activated, the emerging attention parameter controls the decision process that now proceeds in line with that given heuristic.

This perception has the advantage that it allows one to move from a decision-making by means of a single heuristic, to a complex decision situation that might involve the use of several heuristics (again, a situation that TK do not consider). As already noted, this situation is analogous to pattern recognition of a complex scene. Based on a certain heuristic, the decision maker becomes attentive to a certain aspect in a complex decision situation, and makes a first choice at the problem. If this attempt fails, saturated, or exhausted, a new attention parameter associated with another heuristic becomes active, and so on. What follows is a sequential process of decision-making. In the next section we further develop the sequential and collective aspects of decision making in the context of our model of SIRN (synergetic inter-representation networks).

14.8 SIRN (Synergetic Inter-representation Networks) and Decision-Making

The abbreviation SIRN is composed of two elements: Synergetics and IRN (Inter-Representation Networks). The notion of IRN was suggested by Portugali (1996) as a challenge to both behaviorism that considered the brain/mind as a black-box and cognition as an essentially external, stimulus–response, process, and the currently dominant computationism which likens the brain to a computer and insists that cognition is essentially an internal process involving internal representation only. IRN suggests viewing the cognitive system in general and the one associated with cognitive mapping and urbanism in particular, in terms of IRN. That is to say, that the individual's cognitive system is a network composed of internal elements representing the environment, and external elements representing the mind. From this follows that many cognitive processes, those connected with cognitive mapping and urbanism included, are implemented by a complex interaction between the internal and external elements of this network. In fact, this perception is the rationale implicit in the models developed in Part II of this book.

In a subsequent study, Haken and Portugali (1996) have cast this new perception into the formalism of synergetics by developing the model SIRN (synergetic inter-representation network). The building of this model starts with the algorithm of the so-called synergetic computer. The latter is a fully parallel computer that represents an alternative to the conventional neural network model (Haken, 1996). As noted above, both the decision-making model developed in this chapter and the city model of Chap. 13, were defined in terms of this algorithm. Figure 14.9 is a graphical exposition of this decision making model. Here, the input layer refers to the information available to the decision making actor, the inner layer of order parameters to the emerging alternative plans, and the output layer to the resultant decision and action. The derivation of the SIRN decision-making model begins by looking at Fig. 14.9 from the side, as indicated by the arrow. The result is Fig. 14.10. The SIRN model (Fig. 14.11) is constructed by adding to the internal input and output layers of Fig. 14.10, external input and output components, while leaving the layer of order parameters as before.

In this model every decision-making actor is subject to two types of inputs, internal and external. The internal input refers to information, knowledge and planning experience as it is internally represented in the memory (or memories) of the decision making actor(s); the external input, to the information enfolded in the externally represented environment. The decision-making

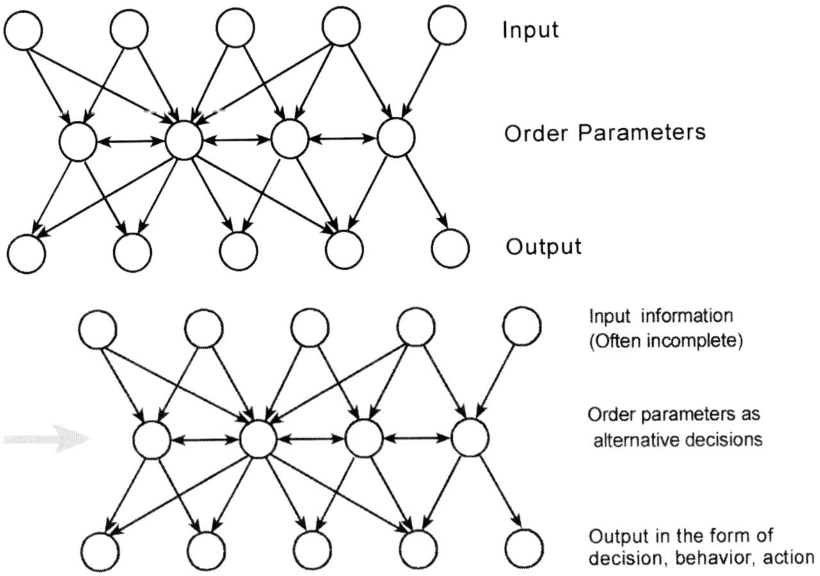

Fig. 14.9. A three-layer pattern recognition network of the synergetic computer (*top*), transformed into a decision making network (*bottom*). The first (*upper*) layer, previously modeling the neurons, now describes input information. This first layer projects on the second layer that represents the order parameters, here refering to alternative decisions and action rules. The third layer represents the output from the order parameter layer, which in the present context takes the form of decision, behavior or action. Though formally similar to a neural computer arrangement, the algorithm of the synergetic computer is quite different, e.g. the model neurons are interacting by means of soft nonlinearities.
Note that learned patterns are encoded in the connections between the first and second layers and those between the second and third layers. In the case of static patterns, the connections between the order parameters are of the same universal form, whereas in the case of dynamical patterns the order parameter connections may depend on the movement patterns to be generated. These remarks hold for all the following figures

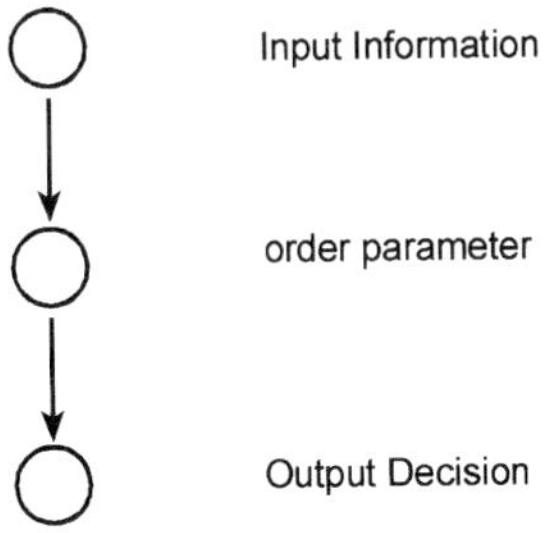

Fig. 14.10. Figure 14.9 seen from the side as indicated by the arrow

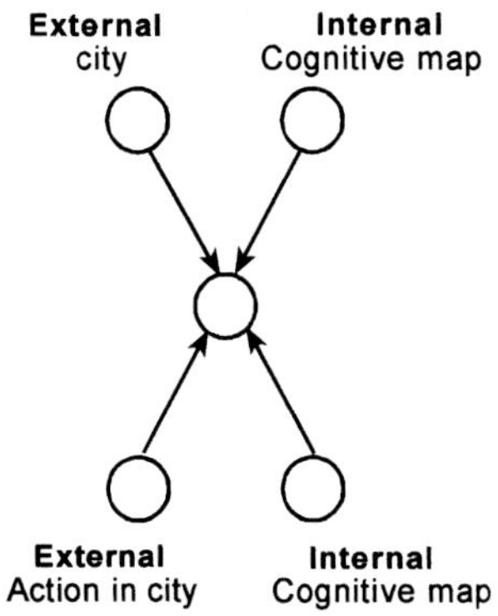

Fig. 14.11. The basic SIRN process representing a decision making agent. As can be seen, this decision maker is subject to two forms of input: externally represented input which comes from the city, and internally represented input which "comes" from the mind. The latter might be a cognitive map of the city, for example. This input information enters the middle layer that represents the order parameters as alternative decisions and actions. The competition and interaction between the latter give rise to two forms of output: externally represented output in the form of action in the city, and internally represented output which feeds back to the mind (cognitive map). Note that similarly to Fig. 14.10, the network corresponding to Fig. 14.11 is seen from the side so that each circle represents a whole set of model neurons

heuristics discussed above in Sect. 14.7 may appear here as either internal or external inputs. The "availability heuristic" might be an example of a typically internal input, whereas "anchoring", a typical external output. The complex and parallel interaction between these two sets of inputs enters the middle layer that might symbolize a brain, an individual, a city's planning committee, or the managerial board of a firm, in which one or several decision rules, in the form of order parameters, have been established. Note that the same order parameter(s) may govern quite different external outputs, such as a location decision in the city implemented by a resident or a firm, movement in the environment, and the like. The order parameters, decisions and plans of the middle layer produce two kinds of output, again internal and external. The external outputs are the plans in the form of ad-hoc and routinized patterns of action as described above, in Chap. 11, whereas the internal output is the new experience, information and knowledge which re-enter memory in a kind of feedback process, and become the internal input for the next step in the process.

An important property of this SIRN decision making and planning model is, first, that the process is sequential – it is an ongoing, space–time interaction between internal and external representations. Second, it evolves as a typical self-organizing system: at the start, the internal–external sequential process of interactions is characterized by strong or chaotic fluctuations, then a certain order parameter takes over, enslaves the internal and external elements of the system, and from that stage onward, the system evolves in a routinized steady state fashion. This scenario is typical as we illustrate below of decision making and planning at both individual and collective levels. In our original paper from 1996, we have developed three prototypes of the SIRN model: the intrapersonal, interpersonal and interpersonal with a common reservoir (Haken and Portugali 1996). The first and the last are relevant to the subject matter of this chapter and in the following we reformulate these prototypes as decision-making models.

14.8.1 Intra-personal Decision Making

The intra-personal process is typical of a decision making process of a solitary person. Solitary in the sense that no external representation created by another person, is involved in the process. An architect designing a building or a planner designing a city, or a part of it, is a case in point. Figure 14.12 graphically illustrates the process and will help to convey ideas. This designer may start, for example, with a vague idea internally represented in mind and an externally represented map of the plot or of the city. The first decision might be where in the plot to locate the building, or where in the city to locate a commercial center, for example. Based on the interaction between his or her internally represented information, and the map as an external representation, the designer takes a first decision by putting it on sketch paper. By externally representing this decision, that is, by drawing it, the person can see "how it looks like". The latter provides feedback information for approving the first decision, changing it, and/or moving to the next decision. The latter might be where to locate the living room, or where to locate a central public park in relation to the already "existing" (on paper) commercial center. Once this decision is internally taken and externally put on the paper, the designer can move to the next decision and so on in an on-going interplay between internal and external representations.

The principle in the above example (in fact, in many if not most of creative works) is that the process involves a sequence of decisions and actions at various scales, and that the play between internal and external representations is at the basis of that process.

14.8.2 Interpersonal and Collective Decision Making

One of the key features of the notion of inter-representation concerns the fact that once an external representation is produced, it becomes a public domain. For example whenever a person produces an external representation such as a talk, smile, ...etc., it becomes a public domain in the sense, first, that other people can hear, see ...what that person thinks, feels, and so on. That is to say, other people can know or sense some parts and aspects of what is internally represented in that person's head. Second and most importantly, other people can connect themselves to that person's thoughts, feelings and so on. External representations thus provide an important medium for interpersonal communication and to the emergence of collective phenomena such as language, culture, myths, style, fashion and what is often termed "collective memory".

In developing the notion of SIRN we've made a distinction between three forms of external representation: mimetic, lexical (or linguistic) and artificial. The first two have no existence independent of the biological memories of the people who produce them, whereas the third one has. Artificial external representations such as books, stone-tools, city-plans, buildings, cities and the like, have an existence independent of the biological memories that have originally produced them. Applying the above types of external representation to the domain of decision making and planning, one can make a distinction between three types of externally represented plans: mimetic plans which include all decisions and plans that make use of the mimetic capabilities of planners, ranging from a facial expression, through plans that mimic other cities or plans, to routinized patterns of space–time movement. Lexical plans would be all plans the external representation of which is lexical. The plan that emerges out of a planning team's "brainstorm" is a case in point here: all members of the team come out of this group dynamics with a shared set of ideas, concepts and categories to which non-members have no access. This is so since this lexical plan is fully dependent on the biological memories of the planning team members. Artificial plans are all plans that are artifacts in the sense that they have existence independent of their producers. A planning report that includes data, drawings, maps and text is an example (and also the structure of the city and the city's order-parameter plans noted in Chap. 11).

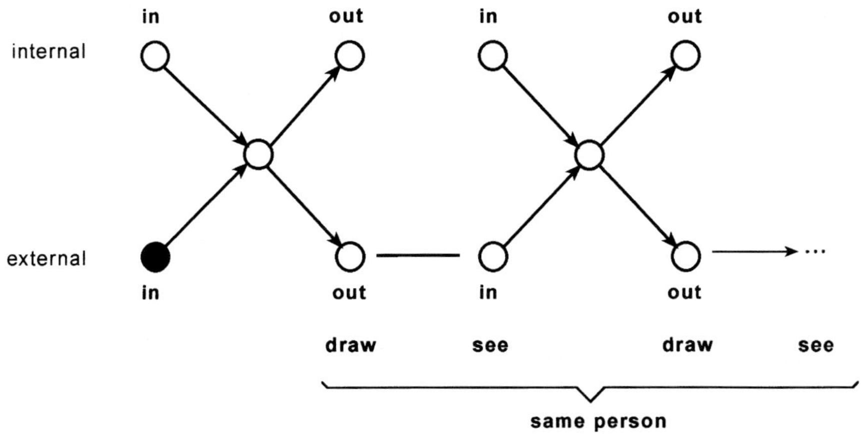

Fig. 14.12. Intra-personal decision making process of a person designing a building or a city

In our SIRN decision-making model this dynamics between the personal mimetic, lexical and artificial plans, and the collective artificial plan is captured by the diagram in Fig. 14.13. As can be seen, the SIRN process is sequential as before, and proceeds by an ongoing interplay between internal and external representations. The difference is that here the individual's externally represented output enters what Haken and Portugali (1996) have termed a *common reservoir*, when the latter is also the source from which the individual extracts the externally represented input for the next iteration in this sequential planning/decision making process. In the context of planning this common reservoir might be the plan and the planning report which emerge out of the SIRN group dynamics of the members of a planning team, and it might also be the structure of the city as a whole and its overall order-parameter plans. In the following we consider three examples that illustrate various aspects of the interpersonal and collective decision making process.

Consider, first, the decision making process of a person attempting to buy an apartment in the city. Let us say that this is person 1 in Fig. 14.13. At the beginning the decision maker, person 1, is subject to two flows of input: internal input in the form of a cognitive map representing some previous experience and knowledge about cities in general and about that city in particular, and, a flow of external input, in the form of apartments' architecture, location and price that the individual encounters. The latter, as can be seen, comes from the city as a common reservoir. Note, that the internal flow might often have the effect of the availability heuristic, for example, whereas the external input, of an anchoring heuristic. The on-going interaction between the external and internal flows of input, ordered and enslaved by the order parameters, which emerge in the process, entails also an interactive interplay between internal and external flows of output. In this sequential interplay between external and internal representations, a certain order parameter, referring to the price, architecture and location of that person's demanded apartment, is determined. That order parameter acts as an anchoring heuristic to a subsequent apartment considered by that person.

This example is interesting as it shows that the boundary between the intra- and inter-personal SIRN is not always clear-cut. On the one hand, the above scenario can be interpreted as an intra-personal process: a learning process of a solitary decision-maker. On the other, this is also an interpersonal SIRN process; first, since every external input in the process comes from the city as common reservoir, and second, since every act of learning (e.g. looking at an apartment for sale) participates in shaping the demand/supply relations in the city. Once the person buys an apartment, the SIRN process becomes fully collective.

Second, consider now Fig. 14.13 as the decision making process of a stranger who is coming to a new city, learns some routes, drives along them and after some time becomes an ordinary daily commuter in the city. At the

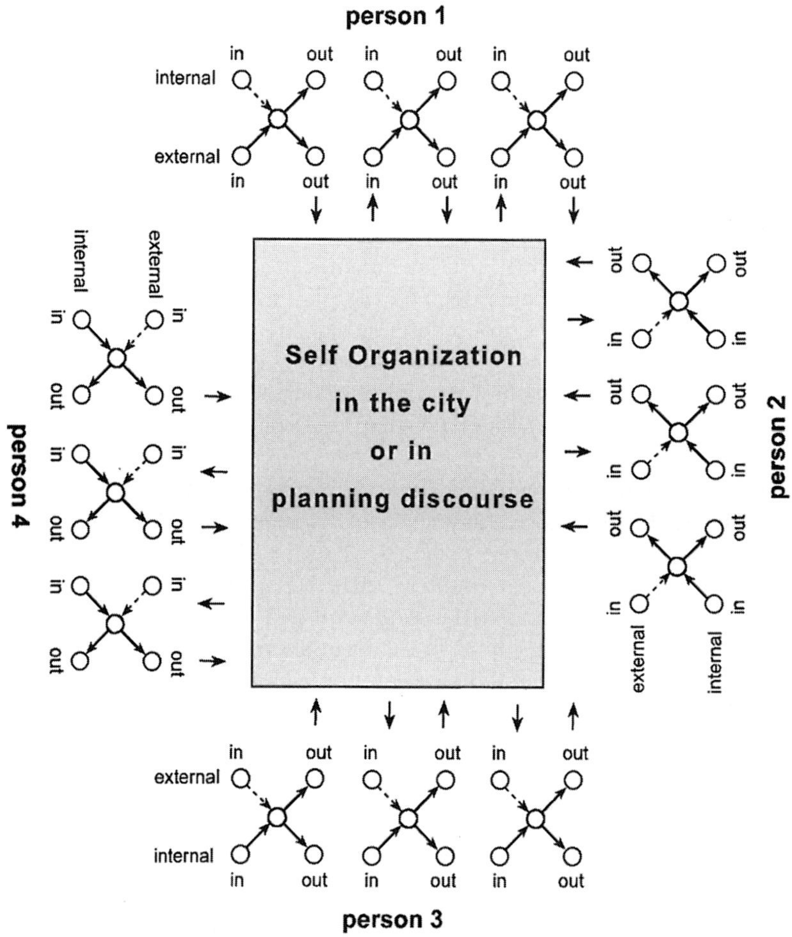

Fig. 14.13. An interpersonal and collective SIRN as a decision-making model. All communication and interaction between the agents involved in the process are made via a "common resevoir" which might be a planning team's discussion, the city as a whole or parts of it

beginning (Fig. 14.13, person 3, *left*, for example), the individual is subject to a flow of two forms of input. A cognitive map, internally representing some previous experience and knowledge about cities in general and about that city in particular, and a flow of external input which comes in, from the city, as the individual advances in the city's space. The on-going interaction between the external and internal flows of input, ordered and enslaved by the order parameters that emerge in the process, entails also an interactive interplay between internal and external flows of output. In this sequential interplay between external and internal representations, objects and patterns in the external environment are being determined, marked, and internally represented as the IRN of the emerging cognitive map. The output of the first excursion along the route provides part of the input for the second excursion, and so on in iterations.

At the beginning, the individual might try several excursions, or commuting configurations, or plans, which can differ markedly from one another, in terms of time allocation and routes chosen. This is a kind of a trial and error learning stage in the planning process, and it is thus often characterized by strong fluctuations from excursion to excursion. Eventually the decision-maker will settle on a certain configuration, which will then become the individual's routinized space–time order-parameter plan for commuting. Once this order parameter is established, it enslaves all other competing configurations and from now on the space–time commuting pattern becomes routinized.

What is typical of the above two examples and to urban dynamics in general, is that the individual decision makers (persons 1–4 in Fig. 14.13, for example) need not be aware of each other and of the fact that they participate in a collective decision making/action process that determines the city's land-use structure, its social and cultural structure, transportation structure, and the like. As we have already seen in Part II, individual decisions that are taken locally eventually affect the global structure of the city. This situation of "non awareness", or implicit collectivity, is thus a common property of decision making in the city context (See Chap. 11 for further discussion and elaboration). An example of a case were the collectivity of the process is apparent and explicit is the group dynamics of a planning team. For this purpose Fig. 14.13 can now be envisioned as a discussion table with four planners sitting around and engaged in a discourse. The aim of the latter is to decide on a given city plan, or urban policy. Each person comes to the table, so to say, with ideas, images, etc., internally represented in his or her memory. As the discourse develops, each person externally represents his or her ideas as an input to the discourse, takes other ideas inputted by other members of the team, internalizes them, outputs new, say integrative, ideas, and so on. The discourse thus becomes a common reservoir of possible decisions and actions to which the participants contribute and from which they extract new information as input and so an in an ongoing process.

A real-life case study of the above example is currently under investigation and research by Alfasi (forthcoming). Her empirical data consists of recorded protocols from the discussions of a certain planning team over a period of two years. Examining these protocols she shows how, by means of an interplay between internal and external representations as above, plans, policies and decision rules emerge as order parameters, enslave competing decisions and plans and thus bring the previously somewhat chaotic system of competing ideas into a steady state.

14.9 An Outline for a SIRN Decision-Making Model

Following the above graphical exposition of the SIRN decision-making process, the next step is to model and thus operationalize it. There are several ways to do so that we currently study and develop. The one that we outline below follows the suggestion of Portugali (1998). We start from City-4, the model developed in Chap. 8, in which each individual agent was defined by means of what we have termed as m-code. The interaction between agents with their m-codes and between them and the city's infrastructure gave rise to a few cultural groups and to the city's overall structure. From the perspective of the present discussion we can now add the following. First, that since in City-4 cultural groups emerge out of the interaction between individual m-codes, it is possible also to define each cultural group by means of an m-code. Second, that since the infrastructural elements of the city (the cells in City-4) and the city as a whole acquire their properties from the properties of the individuals who live in and around them, it is possible to define each infrastructural element by means of an m-code, too.

Now, the m-codes of agents refer to their intentions, wants, previous knowledge, information and so on, that is to say, to their internal representations. In the same manner one can regard the m-code of cultural groups that emerge in the city, the city infrastructure (the cells in City-4) and of the city as a whole, as external representations. In both internal and external representations, the m-code will be representing information (such as categories, intentions, plans, etc.) that is enfolded in material objects: in the brain of the agents, in the cultural areas of the city, in its elementary cells and in its overall material structure.

The above formulation by means of m-code, has the advantage that both the internal and external representations can be conceptually and formally described by means of the same units and can thus interact in a meaningful way. In terms of our graphical SIRN models (Figs. 14.9–14.13) this implies describing $q(\mathrm{i})$ and $q(\mathrm{e})$, the terms for internal and external representations, by means of m-codes. This is shown in Fig. 14.14 that describes a typical decision maker participating in the inter-personal, collective, decision-making process described in Sect. 14.8.2 and Fig. 14.13. As can be seen, this agent is subject to two flows of information: internal, which comes from internal representations in the mind, and external which comes from external representations in the city. Both forms of representation, or sources of information, are defined by means of m-codes: the whole city as an externally represented "common reservoir" or "collective memory", $q(\mathrm{i})$ – the term for internal representations, and $q(\mathrm{e})$ – the term for external representations. As before, $q(\mathrm{i})$ and $q(\mathrm{e})$ refer to both input and output. Note, however, that as an input $q(\mathrm{e})$ is termed also "legibility window". That is to say, it is that part of the city as a collective memory and a common information reservoir that is legible to the decision-making agent. In the original formulation (Haken and Portugali

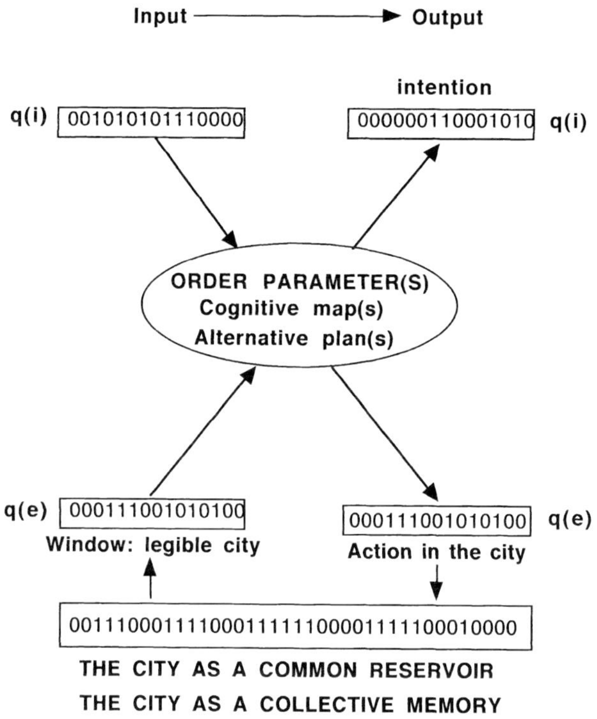

Fig. 14.14. A SIRN model of a typical decision-maker participating in the interpersonal, collective, decision-making process, when all representations, internal, external, private and collective, are described by means of m-codes

1996) we have termed it a "personal window" and symbolized it by $W(\mathrm{k})$; the notion "legibility" is borrowed from Lynch's (1960) *The Image of the City.*

As illustrated in Fig. 14.14, the interaction between internal and external representations gives rise to order parameters which as suggested in Chap. 13, can be interpreted as cognitive maps. In the context of the present discussion, the order parameters can be interpreted as alternative decision rules, or plans, and as such as cognitive maps of the future. As can be further seen in Fig. 14.14, the order parameters as plans imply two forms of output, again internal and external. An internal output that feeds back to newly shaped information, intentions, etc., stored in memory and an external output which is behavior and action in the city. The latter might be a move to a new apartment, or office and the like. "Action in the city" participates in the city's dynamics, that is to say, in the interaction between the many other such actions taken in the city, and between them and the city's physical infrastructure. These synergistic interactions give rise to order parameters at the scale of the whole city which then enslave the various elements of the city and their m-code, from which individuals extract their legibility window and so on. The result is a complex process of circular causality between self-organizing systems at two scales: the scale of the individual decision-making agent as a self-organizing system, and the city as a whole as a self-organizing system. In Part II of this book we have simulated and studied some aspects of this complex circular causality by means of our FACS City models, and have used synergetics as a conceptual–hermeneutic framework. In Haken and Portugali (1996) we have indicated how SIRN can be cast into the formalism of synergetics. In this chapter we have re-interpreted SIRN as a decision-making model and suggested how, by the use of m-codes, it can be integrated with our FACS models. What remains to be done is to implement the outline suggested here as a full-scale formal decision-making model. This is a task we intend to accomplish in the future.

Part V Self-Organization and Urban Revolutions

15. Self-Organization and Urban Revolutions

15.1 From the Urban Revolution to La Révolution Urbaine

In 1950 Gordon Childe published a paper entitled 'The urban revolution'. Twenty years later, in 1970, Henri Lefebvre published a monograph with the very same name *La Révolution Urbaine*. Childe's revolution has since become a seminal paper in the history, anthropology and archaeology of urbanism; Lefebvre's revolution is currently a most influential work among geographers and urbanists commenting on the transition from 20th to 21st century urbanism. Both men write from the vantage point of the present – from the great interest of 20th century science and society in the phenomenon of urbanism which in the last 100 years has reached dimensions and scale never known before in human history; yet, both men do not write about the present. Childe's work is part of an intellectual effort to look back to the past with the hope to find there insight into the phenomenon that presently shapes our life; Lefebvre's is an attempt to assess the potential and implications of today's urbanism – his thesis is prescient (Gregory 1996, p. 379), it has a futuristic flavor about it.

With the 'urban revolution' Childe refers to the rise of an urban society in Mesopotamia, some 5500 years ago. According to him this event should be seen as one of the most significant revolutions in human history. The urban revolution, according to Childe, is of the magnitude of the very few grand events in human history: the hominid revolution which marks the emergence of human beings, the Neolithic revolution which has transformed human society from hunting and gathering to food production, that is, agriculture; the urban revolution, and more recently the industrial revolution.

Lefebvre, per contra, claims that while cities are indeed with us for some 5500 years, urbanism, that is, a genuine urban society, has yet to arrive; in fact only now, in the 20th century and in particular in the second half of it, we can observe its emergence in front of our very eyes.

The data, that is to say, the archaeological record, upon which Childe's revolution is based is generally agreed upon. This record indicates that at the middle of the 4th millennium B.C. one observes an abrupt and sudden (in historical terms) eruption in the form and structure of settlement and society: A new form of settlement appears, first in Uruk in Mesopotamia and very soon it diffuses in space and time all over the ancient Near East. But the eruption, i.e. the revolution, is not only in the form of settlement but in all aspects of life, culture and society: the emergence of a highly specialized, class divided, society with an articulated division of labor, long-distance trade, new forms of politics and government, and above all the appearance of writing and as a consequence a transition from pre-history to history. The urban revolution of 5500 years ago is at the very same time *the rise of civilization.*

While there is a general consensus among scientists about the overall picture of Childe's revolution as portrayed above, there is no consensus about his theory, that is, his interpretation concerning the causes of the revolution, the definition of a city (i.e. of 'what is a city'?), and about the distinction between the appearance of cities and the emergence of urbanism – a distinction which stands at the core of Lefebvre's revolution.

Similar things can be said on Lefebvre. There is a strong 'feeling' among scientists that something very dramatic is happening right now, at the end of the 20th century and on the verge of the 21st. The rapid demographic growth of human population, the associated rapid urban growth – from 16 cities of over one million people at the turn of the 20th century to over 500 such cities today, the seemingly contradictory conjunction between the emerging global economy and global village, on the one hand, and the parallel emergence of cultural pluralism and localism, on the other, the internet, the information highway, the notions of 'network society', 'space of flow', 'information society' and 'information city', suggested by Castells (1996), the emergence of the environmental problem, the green movements with no specific inclination to the traditional 'right' versus 'left' dichotomy that has dominated society, social philosophy and politics for over a century, and the very recent realization that the environmental dilemma is essentially a dilemma of cities and urbanism.

This sensation of a changing world, cities and urbanism was the ground for our discussion in Chap. 2 of the changing image of the city from the tamed city of modernity to the untamed city of postmodernity – a discussion which has paved the way for our suggestion that the city be viewed in terms of self-organization. In the remainder of this section we shall have a closer look at, and compare, the urban revolutions of Childe and Lefebvre, as well as at some other interpretations and alternative viewpoints.

15.1.1 Gordon V. Childe: The Urban Revolution

The first urban revolution, according to Childe, was a genuine Marxist revolution, the logical dialectical consequence of the Neolithic revolution that preceded it by some 5000 years. The Neolithic revolution was a revolution in the economic base of society. The latter was transformed from hunting

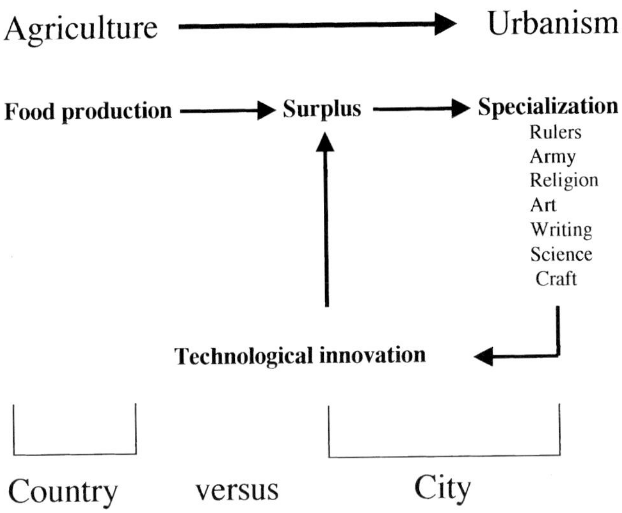

Fig. 15.1. Childe's urban revolution as the logical conclusion of agricultural, food-producing society

and gathering into food production, that is to say, to agriculture. Food production entailed the emergence of surplus, which in its turn has allowed the emergence of ruling classes, full-time specialists in short- and long-distance trade, art, science, and above all writing, and so on in a snow-ball feedback effect. The cities were the locus, the spatial outcome and material expression of all those changes (Fig. 15.1).

Childe's urban revolution had two components in it. One, was the above chain of events that led from the agricultural revolution to the urban revolution. The Neolithic revolution and the agricultural society are seen here as the pre-conditions for the subsequent emergence of an urban society. Two, a definition of a city – an answer to the question of 'what is a city?' A city, suggested Childe, is a settlement which has the following necessary and sufficient characteristics: (1) large 'urban size' settlement area (compared to agricultural settlements), and by implication population; (2) centralized capital accumulation resulting from the imposition of taxation and tribute; (3) monumental public works; (4) writing; (5) science; (6) long-distance trade; (7) a stratified class-society; (8) full-time craft specialists (9) a territorially-based political society – the state – in place of the agricultural one which was based on kin ties; (10) representational art.

15.1.2 A Spencerian Interpretation

In the 18th century Herbert Spencer published an essay on population in which he suggests, contra Malthus, that the exponentially growing population pressure on the linearly growing environmental carrying capacity, is not going to lead society to starvation, conflicts, wars and disasters, as Malthus predicted, but rather to social, cultural and technological progress. According to Spencer, the conflicts that entail population pressure upon the environment, are among the key factors that force society to improve itself, to invent new technologies and more advanced and efficient socio–political forms. In short, the conjunction between population pressure and conflict is the engine of progress (Spencer, in Carneiro, 1967, pp. 10–33). This view which for years was largely forgotten, became popular again in the mid Sixties as a result of two studies. The first and more dominant one was Boserup's (1965) work in which, independently from Spencer, she suggested the very same mechanism, this time in connection with *The Conditions of Agricultural Growth*. The second, which was directly based on Spencer, was Carneiro (1970) in his studies on the origin of the state.

In an article in *Science* from 1970, Carneiro criticizes Childe's theory for being *voluntaristic* and *automatic*. Voluntaristic, in the sense that it assumes that in order to increase collective productivity people will voluntarily give up their freedom and shift from an equal agricultural society to an unequal state-society. Automatic, in the sense that it assumes that agriculture automatically brings surplus. Both assumptions are wrong, says Carneiro and supports his saying by archaeological evidence from central and south America. In their stead he suggests a Spencerian view according to which population growth and pressure associated with wars and conflicts are the genuine explanatory variables which have pushed society up the progress ladder to a state society. Other scholars influenced mainly by Boserup's (1965) model have reiterated a similar scenario both with respect to the origin of the state and the rise of urbanism (Adams 1972).

A typical population pressure scenario of the urban revolution would proceed as follows (Fig. 15.2): Several, relatively small, agricultural communities are dispersed in a circumscribed area. These communities undergo an exponential population growth. The latter might be the result of natural increase, new immigrating communities who enter the area, or both. Since the living area is circumscribed and limited, population growth implies an increasing population pressure upon the environment – more and more territories are being occupied and cultivated. This, in turn, eventually leads to conflicts, struggles and wars over the land among the various communities. Wars end with winners and losers and this is the source for the class-divided society: The losing communities become full-time food producing ruled classes, whereas the winning communities, in order to preserve the fruits of their victory, become full-time rulers, with armies and other ruling classes. They fortify themselves in fortified settlements which are called 'fortified cities', or in short 'cities', so that the division of labor in this emerging class-divided society takes also a spatial dimension in the form of 'town–country antagonism', and so on.

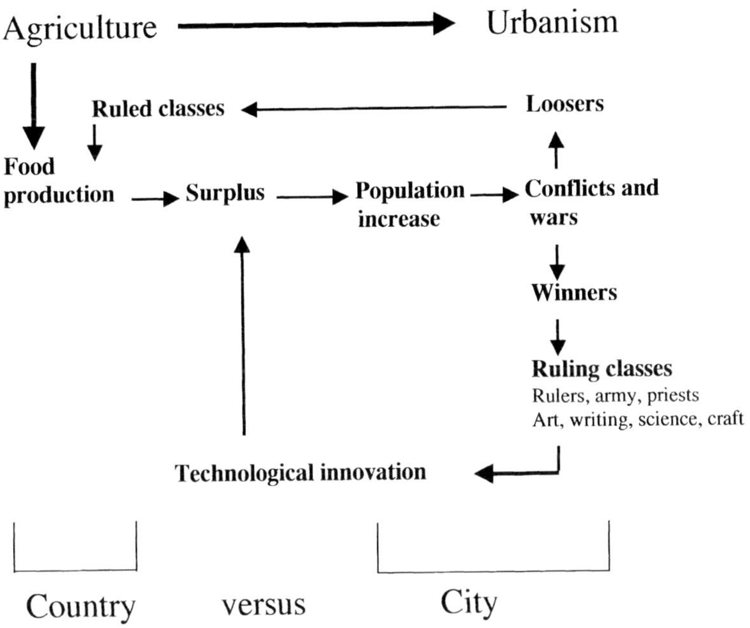

Fig. 15.2. A Spencerian, population pressure scenario of the urban revolution

15.1.3 The First City

The general consensus among students of urbanism, especially since Childe's *revolution*, is, first, that it happened in the middle of the 4th millennium B.C., in Mesopotamia. Second, that Uruk is the first known city. Third, that the Neolithic–agricultural revolution with its subsequent agricultural society, created the necessary and sufficient conditions for its eruption. This view fits well with the archaeological record which shows that suddenly, since the middle of the 4th millennium B.C., we find cities 'everywhere': first at the core area of Mesopotamia, and then, apparently as a consequence of a space–time diffusion process, in the rest of the Near East and far beyond. This view also fits well with the transition from a rural to an urban society and economy which characterized the urban revolution of modernity – the one which followed the industrial revolution. In fact, there is ground to claim that scholars' view about the urban revolution of 4th millennium B.C. is strongly influenced and even biased by the rural–urban transition, or revolution, of modernity.

The above picture is considered sound, because the data on the modern urban revolution is well documented and because from the point of view of field archaeology, the cities of antiquity are easy to identify (though not to define). This is so mainly because of their physical appearance: they are relatively bigger, characterized by rectangular architecture, road networks, public buildings and most importantly, fortifications.

The only problem with this consensus is that there are a few 'Popperian' empirical exceptions which falsify the above general impression. The most prominent and famous one is the site of Çatal Hüyük in Eastern Anatolia. Judged by its physical appearance, it looks like a city, and had it not been dated (by means of pottery and C14) to the 8th millennium B.C. it would no doubt be classified as a typical city. Another, somewhat controversial, such site, from the 6th millennium B.C., is Jericho, yet another is Tell Ghassul from the 5th millennium, and there are more.

Childe's revolution with its 10-criteria definition of a city, which pinpoint the first cities to 4th millennium B.C. Mesopotamia, was thus followed by heated discussions about the status of these premature cities. Thus Kenyon (1957) and Wheeler (1956) claimed that 6th millennium B.C. Jericho is the first urban settlement, to which Childe (1957), in a paper on 'Civilizations, cities and towns', responded that the key definitive criterion is 'writing', which does not exist in Jericho and other sedentary–fortified agricultural settlements. These controversies as well as similar controversies about the definition of a city today in the 20th century, eventually came to a frustrating dead-end. The main reason being, as noted in the first chapter of this book, that 'a city' is not an Aristotelian category, but a Wittgensteinian family-resemblance one. We shall come back to this issue in the next section of this chapter.

What Çatal Hüyük, Jerico, Ghassul and other such sites teach us is that the 'patent' of a town and a city was known and used more than 3000 years before the first urban revolution. The implications are, first, that the cities of the urban revolution were not a new invention, but rather a re-configuration, or a re-ordering, of known socio–spatial forms. Second, that one has to distinguish between cities and urbanism: cities are the sedentary settlements which form the family resemblance network described above and in Chaps. 1, 2. Urbanism is a family resemblance network, the elements of it are cities. Third, that agriculture was not the necessary pre-condition for the emergence of cities – a point elaborated in some length by Jean Jacobs (1969) in the first chapter of her book on *The Economy of Cities.*

In that chapter Jacobs challenges the conventional wisdom that the hunting–gathering–nomadic economy had no surplus, that the Neolithic revolution and the emergence of agriculture enabled humans, for the first time in their long evolution, to produce food and to create surplus, that this surplus enabled the emergence of non-food producing classes which were the necessary and sufficient condition for the urban revolution with its stratified society, etc. She challenges this conventional wisdom by speculating on an alternative scenario which shows, first, how hunters and gatherers can develop surplus, non-food producing classes, long distance trade, ... and cities. Second, how the existence of sedentary cities provided the conditions for the development of food production, that is, agriculture. The agricultural revolution, according to Jacobs, was not the pre-condition for, but the result of, sedentary urban society.

The attraction of Jacobs' speculation is the demonstration that exactly as agriculture and domestication might trigger a chain of events which end up with urbanization, so can urbanism and sedentarization trigger a chain reaction the outcome of which is agriculture. On the other hand Jacobs' claim that her sequence is better than the other sequence is problematic exactly as the rival claim is: both are, in fact, speculations that attempt to pinpoint the ultimate cause for their socio–cultural change and in this respect they are similar to many other attempts to causally explain cultural change. The problem with these attempts is that because of the openness and complexity of the systems concerned, there is no way to determine causal relations among their many components. As we have seen above, the theory of self-organization suggests several ways to study such open and complex systems in a non-causal manner.

15.1.4 Henri Lefebvre: La Révolution Urbaine

Cities and towns, according to Lefebvre, have been significant socio–spatial products for more then 5500 years, but the urban revolution, that is, a transition to a genuine urban society, is only now emerging. With the 'urban revolution' he means

> the total ensemble of transformations which run throughout contemporary society and which serve to bring about the change from a period in which economic growth and industrialization predominate to the period in which the urban problematic becomes decisive (Lefebvre 1970, in Harvey 1985b, p. 62).

As portrayed by Lefebvre this contemporary total ensemble of transformations, from the industrialized to the urban, is the last in a much larger-scale ensemble of transformations which run throughout human society from the rise of civilization to date. This perspective of Lefebvre's was recently discussed by Gregory (1994) and interpreted by him by reference to two diagrams (Fig. 15.3). The first follows Lefebvre's *La Révolution* from 1970, and the second its extension in light of his *The Production of Space* from 1974.

In Lefebvre's view the cities of the past were essentially urban forms – the spatial representations and products of their respective societies. Accordingly, the history of urbanism is seen as a sequential succession of city forms: the *political city*, the *merchant city*, the *industrial city*, and finally, the *urban society*, that I suggest to call the '*urban city*' – the only city in the sequence which is a genuine representation of an urban society. The sequence of urban forms is paralleled by an associated sequence of *fields*, *principles* and *space–time* conceptions. The *fields* (*rural*, *industrial*, *urban*) include, in addition to morphological properties, "experiences, and perceptions, times and

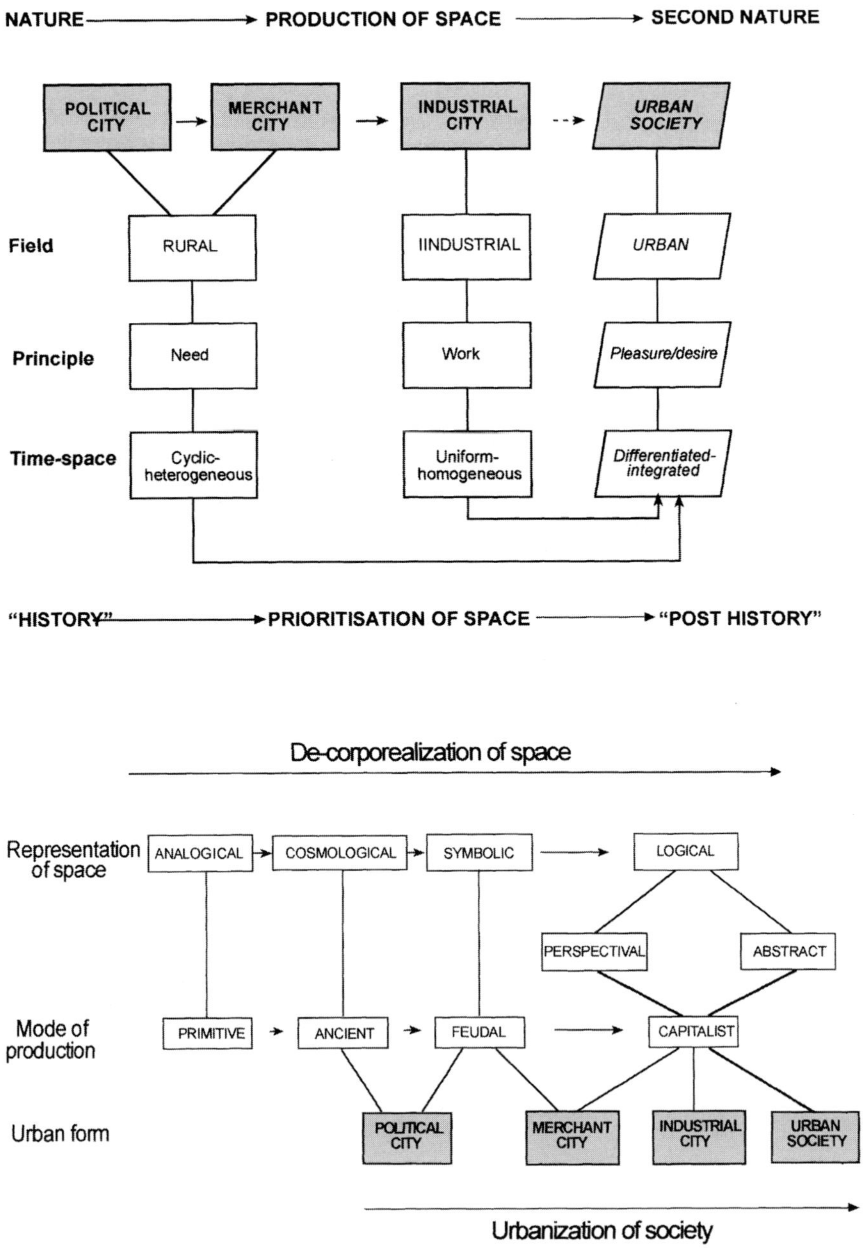

Fig. 15.3. Gregory's diagrammatic representation of Lefebvre's perception regarding the emergence of an urban society. *Top*: according to his *La Révolution Urbaine*. *Bottom*: according to his *The Production of Space*

spaces, images and concepts, languages and rationalities, theories and practices" (Gregory 1994, p. 375). The urban forms and their fields are dominated by *principles*: the rural is dominated by the imperative to satisfy *needs*, the industrial by the imperative of *work*, and the urban, representing as it is the 'post scarcity' society, transcends beyond both to satisfy what Lefebvre terms *jouissance*, that is, pleasure, desire, creativity, leisure, etc. And these progressions are further paralleled by associated conceptions of time and space. The political city with its rural fields and imperative to satisfy needs is associated with a cyclical–heterogeneous conception of time and space, the industrial city with its parallels is identified with a uniform-homogeneous-rational space–time, while the urban revolution brings the *differential* – a conception of time and space that according to Lefebvre incorporates and transcends the previous representations by producing a "time-space that is *simultaneously* differentiated and integrated" (Gregory ibid, p. 376).

In *La révolution*, Lefebvre focuses mainly on the city and its transformations: from the political, to the merchant, to the industrial and to the urban. The relations between the form of the city to the form of society are somewhat implicit. These connections are being explicated in the *production*, in which, as illustrated in Fig. 15.3 bottom, each city form is derived from, or represents, a more basic mode of production: the political is derived from/representing the ancient and feudal, the merchant the feudal and early capitalist, and the industrial and urban the capitalist mode of production. These relations are not deterministic, however, and Lefebvre emphasizes that while "each mode of production has its space; ... the characteristics of space do not amount to the general characteristics of the mode of production" (ibid, p. 382).

The above two sequences echo, on the one hand, as Gregory (1994) comments, Mumford's (1961, p. 7) *The City in History* which "opens with a city that was, symbolically, a world", and "closes with a world that has become, in many practical aspects, a city", and on the other, Marx and Engels' sequence of social forms: the *tribal* (or *primitive*), the *ancient*, the *feudal*, and the *capitalist*. The Eurocentrism of Marx and Engels' view and the fact that their sequence is based on the fragmentary archaeological and historical information available at mid 19th century, make Lefebvre's usage of the notions of the *ancient* mode of production and its associated political city a not very useful construct. In particular his political city lumps together, under a single title, some 4000 years of urbanism during which, as we now know, there were many forms and variations of city states, urbanized chiefdoms, states and empires.

There is also a methodological problem with Lefebvre's account: the basic (Marxist) methodology is to interpret urban changes as reflections of social changes (apart from the urban revolution). This methodology is justified and is similar, in fact, to the practice in self organization studies to distinguish between global and local variables, and to treat the former ones as order parameters of the latter. This approach is not sufficient because it leaves open and often obscures the question of urban continuity: how one city form, say the merchant, participated together with industrialization, in shaping the subsequent industrial city. We shall further elaborate this issue below.

15.1.5 Manuel Castells: Network Society with its Information Megacity

As noted above, while there is no consensus among urbanists about the details of the Lefebvre thesis, there is nevertheless a strong 'feeling' among scientists that something very dramatic is happening to cities and urbanism right now, at the end of the 20th century and at the gate of the 21st. Thus Castells in his recent book on *The Rise of the Network Society* writes:

> The development of electronic communication and information systems allows for an increasing disassociation between spatial proximity and the performance of everyday life's functions: work, shopping, entertainment, healthcare, education, public services, governance, and the like. Accordingly, futurologists often predict the demise of the city ... (Castells 1996, p. 394).

Are we envisaging the end of cities? asks Castells, and his answer is that while 'home centeredness' is an important trend of society at the end of the 20th century, "it does not mean the end of the city" (ibid, p. 398). It does mean, however, the emergence of a new *information age* which is "ushering a new urban form, the *information city*" (ibid, italics added).

His analogical model in developing the new information city is the old/previous industrial city. Like the old one, the new city, too, is an expression of society – "space is an expression of society", writes Castells (ibid, p. 410) when elaborating his notion of the *space of flow*. As in the industrial city here, too, the new city is not a replica of a single model, but it takes variable forms in different places and contexts. Despite this variability and diversity there are "some fundamental common features" (ibid), namely, that "the information city is not a form but a process ... characterized by the structural domination of the space of flow" (ibid). These common features are in their turn an expression of the newly emerging information/network society which is "based upon knowledge, organized around networks, and partly made up of flows" (ibid).

In order to illustrate the diversity of the newly emerging information city Castells discusses Mike Davis' (1990) *City of Quartz* and Joel Garreau's (1991) *Edge City*, as "representatives" of new urban forms in the USA. He then examines some changes in the more traditional urban centers of Europe where the transition from the industrial age and city to the network society and its information city is much more difficult, especially so in the smaller cities and metropoles. All these changes are just various aspects of the really significant urban revolution of end 20th century: the rise of *megacities*.

Tokyo, Sao Paulo, New York, Ciudad de Mexico, Shanghai, Bombay, Los Angeles, Buenos Aires, Seoul, Beijing, Rio de Janeiro, Calcutta, Osaka, are the main members in the club of world's megacities, to which one might probably add Moscow, Jakarta, Cairo, New Delhi, London, Paris, Lagos, Dacca, Karachi, and possibly a few others. They are big – with population beyond 10 million, some over 20, they are magnets to their hinterlands, they are the nodes of the new global economy and they form huge configurations of urban areas. One such megacity in the making is presented in Fig. 15.4. It is not yet on the map, but, promises Castells, it is going to be a preeminent megacity of the 21st century. Already in 1995 this network of metropoles and cities extended over an area of 50000 square km, with a population of over 40 million, connected by a network of highways, railways, air transport, boats and an ever growing electronic communication system. But the most distinctive feature of the megacity is that it is "globally connected and locally disconnected, physically and socially..." (ibid, p. 404).

15.1.6 The Modern, the Postmodern and the Hyper-modern City

Views such as Lefebvre's and Castells' as elaborated above, take us back to the observation made in chap. 2 of the recent changing image of a 20th century city from the modern to the post modern city. That is to say, from an ordered, controllable, tamable and plannable entity, just waiting for us to scientifically analyze it and reveal its regularities and laws, to an image of a shrew environment untamed and untamable. As noted above, today's image of the city is the city of postmodernity: untamed, shrew, capricious, ever-changing, chaotic, pluralistic. In this postmodern city nothing is stable, nothing is true nor matters for more than a second; all must go, must move, clear the way to the new next whatever it is.

This postmodern picture of cities has several important similarities to what one might call the *hyper-modern city* – the image of the city as portrayed by the theory of self-organization. Like postmodernism and its post modern urban phenomena, self- organization and its hypermodern self-organized city, come with an air of recentness, of the newly acquired qualities of our global village and city at the end of the 20th century and on the verge of the 21st, of hypermodernity. But is it really so? Is self-organization a property of cities and urbanism which is only now emerging in front of our very eyes and is going to shape our future? The answer is no (Portugali 1997). The reason being that while theories of self-organization and of self-organized cities are the product of the second half of the 20th century, processes of self-organization, like cities, are not – exactly as chaotic river streams are not, nor butterfly effects, chemical reactions, pattern formations in liquid and the like. What is new, and this is the thing that creates around the notion of self-organization an air of futurism, is the theory itself and the fact that it is intimately related to the sensation of life and urbanism at the end of the 20th century.

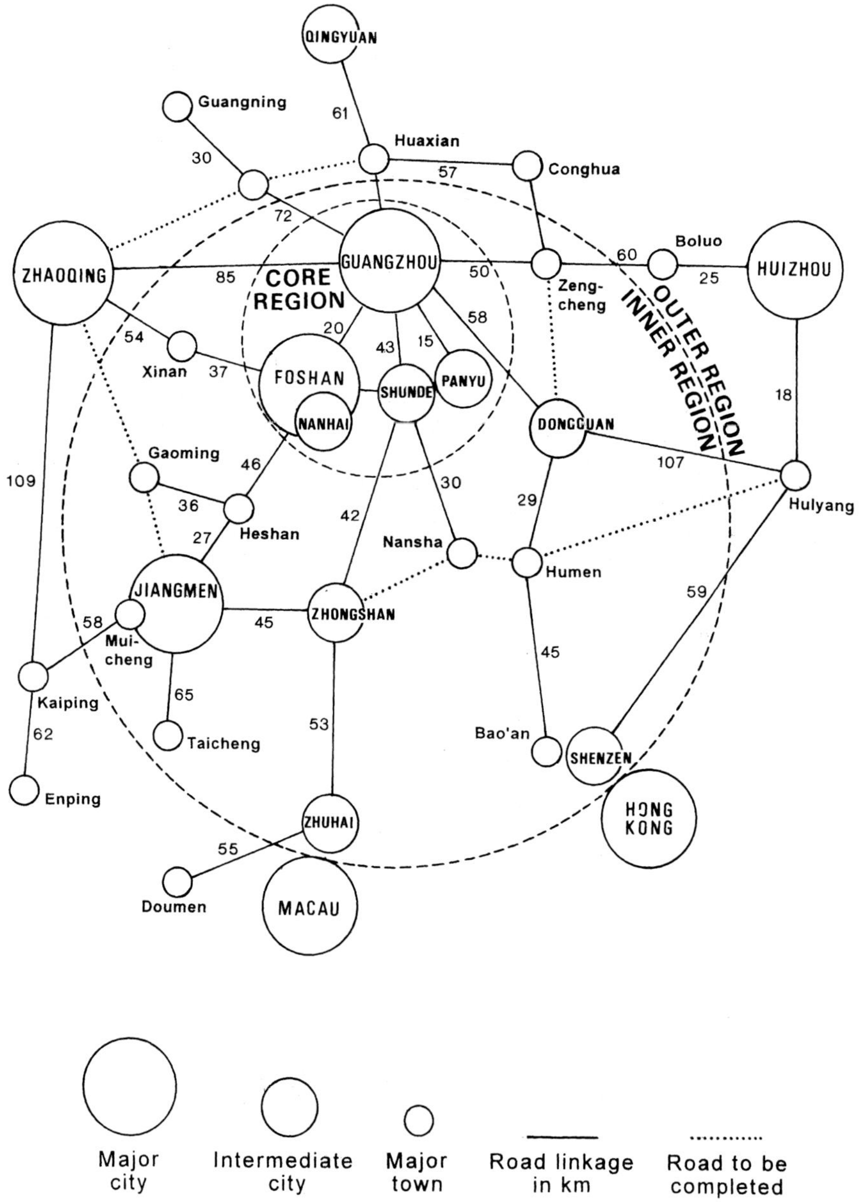

Fig. 15.4. Woo's (1994) diagrammatic representation of the urban complex (megacity) that is growing in the urban region of the Pearl River Delta

15.2 Self-Organization, Social Theory and Socio-spatial Revolutions

15.2.1 On Punctuated Equilibria and Social Revolutions

> Gradualism, the idea that all change must be smooth, slow and steady, was never read from the rocks. It represented a common cultural bias, in part a response of nineteenth century liberalism to a world in revolution. But it continues to color our supposedly objective reading of life and history ... The history of life, as I read it, is a series of stable states, punctuated at rare intervals by major events that occur with great rapidity and help to establish the next stable era.

So writes Gould (1980, p. 226) in *The Panda's Thumb* were he reiterates his and Eldrege's notion of *punctuated equilibria* and reflects on natural history. But he is only partly right. Gradualism indeed represents a cultural bias strongly associated with nineteenth century liberalism, but specifically so with respect to the life sciences which have evolved with Darwin's gradualist theory of evolution as their basic paradigm. This was not the case, however, with the social sciences and humanities. In the latter, punctuation was never an exception but a very dominant and central view. It shows up, implicitly, in the common tendency to describe human history in terms of *long durée* ages (stone-age, bronze-age, iron-age, etc.) or periods (Helenistic, Roman, Medieval ...), and it appears explicitly in theories of social evolution, conflict theories of social change, and above all in Marxism. While the various theories of social evolution emphasise evolutionary stages, conflict theory and in particularly Marxism put specific emphasis and attention to revolutions. Indeed, the whole of human history is often perceived here as a sequence of revolutions; ranging from small-scale revolutions such as the French, the Bolshevic, etc., to large-scale grand human revolutions such as the hominid, the Neolithic–agricultural, the urban, the industrial revolutions and today, some say, the information revolution. As we have seen in the previous section, the urban revolutions of the kind discussed by Childe, Lefebvre, and Castells, belong to the family of grand socio–spatial revolutions.

Revolutions, according to this view, are relatively short social events which, on the one hand, disintegrate and thus bring to an end old, relatively long, periods during which society was dominated by a certain mode of production, and on the other, open the way for the emergence of new social forces with a new mode of production (or, in the language of synergetics, a new order parameter). According to Marxism, and this is a commonly held view in social theory, this revolutionary form of social change is one of the characteristics which distinguish the evolutionary dynamics of human society from that of the non-human biological kingdom; the general consensus is that the latter evolves in line with Darwin's theory of biological evolution. Other important characteristics are historical irreversibility, and the view that human development and evolution is historically specific.

15.2.2 Similarities Between Self-Organization and Social Theory

One of the more interesting aspects of the theory of self-organization, and one which has made this theory specifically attractive as a potential bridge between the natural and the social sciences, is that it has found in the dynamics of inanimate matter properties hitherto assigned only to changes in the social and cultural human domain (Portugali 1985, 1993). One such property concerns revolutionary social change; another, irreversibility, and a third one the relations between the local and the global.

The term *revolution* in social theory is, in fact, *phase transition* in the language of self-organization. Or more specifically, as we have seen in previous chapters, self-organization theory suggests that the evolution of open and complex systems can generally be described by the following sequence:

$$\textit{steady state} \rightarrow \boldsymbol{chaos} \rightarrow \boldsymbol{bifurcation} \rightarrow \boldsymbol{phase\ transition} \rightarrow \textit{steady state}$$

Social theory perceives the dynamics of change in culture and society in a similar way, and in fact the bold-lettered components in this sequence describe in some details what in social theory is often termed a *revolution*. In both social theory and self-organization theory, the process of change starts when an old regime disintegrates and enters into strong fluctuations and chaos, both theories claim that this chaotic stage is necessary to enable new forces and orders to emerge, assume power and bring the system into a new steady state or a mode of production.

The same goes for the other common property. The notion *historical process* often comes to indicate that "history never repeats itself", that all historical events and processes are historically specific or, in short, are irreversible. In self-organization irreversibility is an inherent property, reflecting the nonlinear and chaotic nature of the process, and the fact that phase transitions often proceed via bifurcations.

The third similarity is that in both domains there is a tendency to interpret the behavior of the local parts by reference to the behavior or evolution of the global system. In social theory this shows up, for example, in notions such as *culture*, *environment*, or *mode of production* in Marxist thinking, which are often assumed to provide the boundaries, or to determine, the action, behavior and even images and inner world, of the 'local' human individuals. In self-organization theory this shows up for example in the theory of synergetics and its notions of the order parameter and the slaving principle. As we have seen in previous chapters, it is a common practice in synergetics to treat the slow-moving variables of the global system as order parameters that enslave the fast-moving local variables.

15.2.3 The Differences

On the other hand, however, the above comparison exposes also the differences between the two schools of thought. In social theory the morphology of the process is just a preparatory stage to the real stuff which is a causal explanation of the change – in our case the urban transitions, transformations and revolutions. As we have just seen, the various social theories of urbanism are methodologically identical in that they try to define an Archimedian point from which to derive the city and its dynamics. They differ among themselves in their suggestion as to what exactly is this Archimedian point. This insistence of social theory characterizes many interpretations of specific events (e.g. the cause for the First World War), as well as many general meta-historical interpretations of change. Thus historical materialists define the Archimedian point in the material conditions of existence, whereas for idealists it is the mind and its imaginative power, and so on. In fact, the various social theories differ among themselves on this very issue.

In self-organization theory, per contra, the morphology of the process is *the* focus of interest, while causal explanation takes a secondary role. Self-organization is, in a way, a morphological theory of systemic change; it says that change has a specific pattern: it starts with strong fluctuations and/or chaos, which lead to a bifurcation and phase transition from which the system then lands on a new steady state, and so on. The question of what was the cause of this chain of events is interesting, but secondary. This is so since self-organizing systems are in a far from equilibrium condition and as a consequence characterized by nonlinearities. In such a situation minor and marginal causes might trigger a global change.

The result is that from the perspective of self-organization there is no single Archimedian cause. In fact, there are no causes in the proper sense, but *triggers*. That is to say, no 'hard causes' that determine, and thus predict, the outcome of the process, but rather 'soft causes' that can trigger the system to self-organize itself but cannot determine and predict its fate. All types of such soft causes can trigger a social system into a revolutionary trajectory of chaos, bifurcation, phase transition and a new steady state. They might be materialist causes, idealist causes, internal–dialectical causes, external causes, socio–cultural or natural-environmental causes; they can also be fully random such as a random coincidence of small-scale random fluctuations which cause a global change otherwise absorbed by the system. In the language of synergetics these causes are termed 'control parameters' (see Chaps. 3, 13, 14). With the notion of control parameter, synergetics adds further insight into the interplay between "chance and causality". When a given control parameter grows beyond a certain threshold, the system, hitherto in a steady state, enters a stage of strong fluctuations and chaos. At this stage of multistability small chance events might cause major phase-transitions. Control parameters can thus trigger (and in this respect "cause") systemic change, but quite often cannot determine its fate. The theory of self-organization is, thus, first and foremost morphological. And this, I think, might be one of its most important contributions to the understanding of change in the human domain.

15.2.4 Self-Organization, Social Theory and Urban Revolutions

The above Archimedian social theory typifies also the various approaches to urbanism discussed above. They show up in the various attempts to see cities and urbanism as a logical outcome of some larger forces: the accumulation of surplus, population pressure, the invention of writing, long-distance trade, and combinations thereof. They show up in Castells' contention that his information megacity is 'an expression of society', it characterizes Lefebvre's *La Révolution*, in particular his second version as developed by him in *The Production of Space* and exemplified in Fig. 15.3 *bottom*, and it typifies also some recent self-organization studies on 'discontinuous evolution of urban historical forms' (see, for example, Rosser 1991, Chap. 9, and further bibliography there). In all these cases the city, its evolution and development, are seen as derivations and representations of larger social structures, or of larger socio–economic processes, inventions or events.

To theorize about cities and urbanism in relation to some other grand processes is, of course, important and illuminating, but to build the whole theory exclusively on external forces is to miss the most significant property of cities and urbanism that we have discussed in the first two chapters of the book: that cities and urbanism are with us for more than 5000 years. That is to say, they existed in an implicate form already in the Neolithic period, they had an explicitly central role in the rise of civilization with its newly invented writing, administration and long-distance trade, they were important with the rise of state societies, empires, Feudalism, mercantilism, capitalism, modernism, post-modernism ...

The real challenge is, therefore, to theorize about cities in their own sake, to regard their multiple appearances and to consider urbanism as a single evolving self-organizing system, to show how the various forms of cities emerge, not only as derivation from external forces, but from within and from each other. In Lefebvre's *La revolution*, for moments, this indeed seems to be the case: one gets the impression that the sequence of urban forms is being examined in its own sake, relatively independent of the sequence of their modes of production; especially so with respect to the outcome of the urban revolution which is at once an urban society and an urban city. This option with its implied bold proposal that the spatial, the territorial and the urban might become a social form independent from a specific mode of production (like industrial society), or even a mode of production, is being restrained in the sequence of Fig. 15.3 *bottom*. The latter implies that even though 'the characteristics of space do not amount to the general characteristics of the mode of production', in the last analysis, urban society is the spatial product of late capitalism.

But is there ground to the suggestion to view the spatial, the territorial or the urban as a mode of production, that is to say, to give them a status similar to 'capital' in capitalism? My answer is that if you look at society, city and urbanism from the perspective of self-organization theory the answer is yes! If you look from the perspective of self-organization you see that 'the urban revolution' was not a single event in human history: Childe's revolution was one, Lefebvre's révolution another, and there were several more in between. In these events a certain society underwent its own urban revolution in Lefebvre's sense that 'the urban problematic becomes decisive', and in the sense that society was perceived in terms of an urban cognitive map (a concept I'll elaborate below).

If you look at society as a self-organization system you see also other cases in which 'the spatial' dominated society. The most recent one is *nationalism* which, as I have shown in some detail elsewhere (Portugali 1993), has become the order parameter of modern society with the nation-state as its material form and the ideology of nationalism as its informational form. And if I am right in this suggestion, then an interesting question emerges: is it possible that the dramatic changes in 20th century cities and urbanism, that were discussed by many writers and above, indicate an urban revolution not just from the industrial- to the 'urban-city', but from nationalism as the generative order of modern society, to urbanism as the generative social order of postmodern or hypermodern society?

15.3 Self-Organized Urban Revolutions

My aim in this section is to elaborate on the above suggestions. That is to say, first, to show that by looking at society, city and urbanism from the perspective of self-organization theory one can theorize about cities in their own sake and regard their multiple appearances as a single evolving self-organizing system, so that the various forms of cities emerge, not only as derivation from external forces, but from within and from each other. Second, to regard the spatial, the territorial and the urban as a social form or even as an urban mode of production. I'll elaborate on the above two aims by illustrating empirically that it has happened more than once in the history of urbanism. In particular I'll show that Childe's revolution was the first that we know about, that there were several subsequent such revolutions, with the implication that Lefebvre's révolution might be the very recent one. Finally, in the light of the above, I shall examine the possibility that the dramatic changes in 20th century cities and urbanism, indicate an urban revolution not just from the industrial- to the 'urban-city' or the megacity, as suggested by Lefebvre and Castells, for example, but from nationalism as the generative order of modern society, to urbanism as the generative social order of postmodern or hypermodern society. I'll start with the empirical illustrations.

15.3.1 The First Urban Revolution from the Perspective of Self-Organization

In his neo-Darwinian theory on the origins of agriculture, Rindos (1984) has suggested that agriculture originated not from the invention of domestication technologies, but out of a slow evolutionary process by which societies have protected their food resources – plants and animals. This process of food protection, which at a later stage implied domestication, increased food productivity, pushed upward environmental carrying capacity, and as a result allowed population growth. On the other hand, it entailed also a higher and higher environmental vulnerability of the protected and domesticated plants and animals. As a consequence, Neolithic agriculture as a socio–spatial system was highly productive but at the same time highly unstable, characterized by frequent environmental crises of sedentarization, nomadization, and migration. These frequent migration processes were the mechanism behind the space–time diffusion of agriculture. The time development of this unstable agricultural system is illustrated in Fig. 15.5.

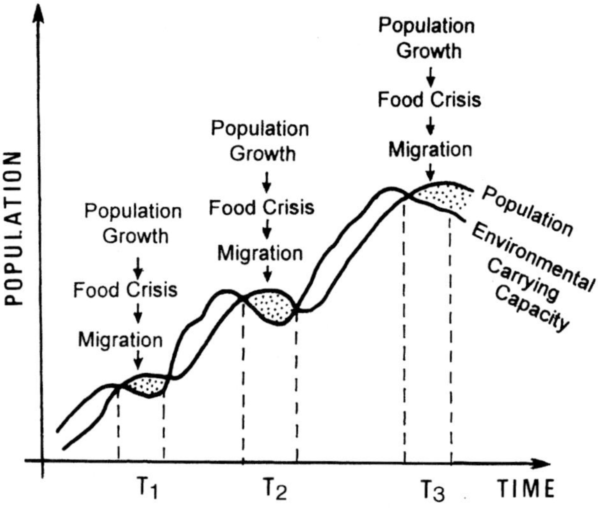

Fig. 15.5. The dynamics of an environmentally unstable agricultural system according to Rindos

Rindos' theory is attractive to our understanding of urbanism, first, because of the observation that agriculture as domestication was not a new technological invention, but rather a well known practice among hunters and gatherers. In a similar way we suggest below that urbanism was not a socio–spatial invention of the city, but rather the application and use of an existing and known form of settlement in a new and innovative way. Second, that the large-scale application of domestication entailed the evolution of what one might call *artificial environments*, that is to say, environmental niches with species which would not have survived the selection mechanisms of an otherwise natural environment. From this perspective, the transition from agriculture to urbanism was a revolution in the nature of society's artificial environment.

In addition to being a highly vulnerable and thus environmentally unstable system, agriculture was also characterized by relatively small, loosely connected, socio–spatial communities. The combined effect of these two properties was that the frequent environmental crises and migration processes that have characterized agriculture as a socio–spatial system, were socially and spatially of a local and limited scale. From this point of view, I suggest that urbanization has stabilized agriculture by revolutionizing the socio–spatial organization of society and by implication of agricultural production: from relatively small, loosely connected communities, to relatively large urban communities, strongly connected and interrelated by means of the systems of cities that have developed.

The 'urban revolution' has not changed the economic base of ancient societies – in the ancient world the economies of nomadic and sedentary non-urban societies and the economies of urban societies, were essentially agricultural. Nor was the urban revolution the consequence of a new invention – the city. We have seen above that the city as a settlement form existed long before urbanism. As in the case of agriculture, here, too, urbanism was not the invention of the city as a new socio–spatial form (as implied by Childe, for example) but rather the application and the use of cities in a new and innovative way. Cities and systems of cities have become the order parameters of society – the medium, and organizing elements, of society. Using a Marxist terminology, one can say that the urban revolution was not a revolution in the *means of production*, but in the *social relations of production*; the latter became dependent on cities. In particular it brought a major change in the socio–spatial organization of agriculture. It has revolutionized agriculture by a spatial division of agricultural production which implied a highly articulated and hierarchical system of socio–economic interdependencies, over vast territories (e.g. between Egypt, Anatolia and Mesopotamia).

The spatial dimension of the emerging urban order parameter was thus of crucial significance. The urban system, that is to say, a system of cities over vast territories, had 'globalized' agricultural production. This globalization became possible by the double role played by the cities. On the one hand, they have become local integrative centers enabling symbiotic relations among the previously, relatively independent, sedentary and nomadic agricultural communities. Within this spatial division of labor the cities acted as the mediators between the local regional system and the global system of cities.

On the other hand, cities were the elements of, and belonged to, the global system of cities, and as such alien and often antagonistic to their local region. Their interaction with other cities of the global system of cities gave them the military, economic, and political power to enforce stability and symbiotic relations in their local complementary regions. The internal stability of local regional systems and the stability of the global, interregional, system, thus became interdependent, and the system as a whole could absorb local instabilities hitherto fatal in the pre-urban era. This was achieved by means of accumulation, planning, long distance trade and other features of urbanism.

Relative to the agricultural, pre-urban system, the urban system was highly stable. However, once the global urban system as a whole, or major links of it, destabilized, the result might have been the collapse of the entire system or substantial sections of it: urbanism has thus reduced the frequency of socio–environmental crises, but has increased their intensity. Figure 15.6 describes this process within the framework of the Rindos model.

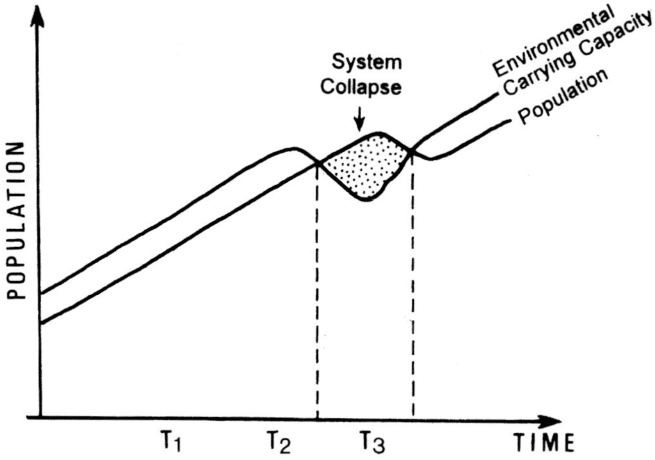

Fig. 15.6. The dynamics of an environmentally stable agriculture within an urban system according to Portugali (1994a)

15.3.2 A Sequence of Urban Revolutions

From the perspective of self-organization the difference between agriculture and urbanism can be described as in Fig. 15.7: Agriculture is here characterized by the co-existence of many, relatively small and autonomous, socio–spatial systems (systems A,B,C in the example of Fig. 15.7a). Each system undergoes its own space–time routine between socio–spatial stability, instability and migration, with results as described above. The effect of urbanism is shown in Fig. 15.7b: the hitherto autonomous agricultural systems have become local sub-systems in a larger global urban system. As illustrated in

Fig. 15.7b, this urban system enables the agricultural sub-systems to absorb, or overcome local instabilities previously fatal during the agricultural era.

While urbanism as a socio–spatial system was much more stable a system than agriculture, it could still collapse and undergo structural changes. The latter might be the result of some external or internal forces (i.e. control parameters) that trigger a process of self-organization, or the result of several local instabilities which randomly coincide, as exemplified in Fig. 15.7b. In the latter cases the urban system as a whole, or major parts in it, might collapse and undergo a major crisis. Thus, relative to the agricultural era as described in Fig. 15.7a, the urban system in Figs. 15.7b,c is characterized by relatively long periods of socio–spatial stability, followed by short, but very intensive environmental crises.

The theoretical pattern described in Fig. 15.7 can be nicely illustrated by the available archaeological and historical data on the space–time history of Palestine in the third and second millennia BCE, which in its turn reflects, in fact, the settlement evolution of the entire ancient Near East. As presented in Fig. 15.8, the process starts with a relatively short, unstable, pre-urban, agricultural period, at the end of the 4th millennium BCE, during which relatively small and independent groups of immigrants enter the country and interact here among themselves and with local socio–spatial segments. The socio–spatial reality of that period in Palestine, is characterized by the coexistence of many, relatively small and relatively autonomous agricultural communities at various degrees of sedentarization and nomadism.

This short unstable period ends with a revolution – the first urban revolution in the region. The population of the country undergoes a major transition and in a very short time becomes an urban society with properties as described above. This urban transformation marks the beginning of a relatively long period of urban society in Palestine, during which the cities become sub-systems in a 'global' urban system, one of whose focal points was Egypt and the other Mesopotamia.

With the collapse of this first urban system, the entire region between Mesopotamia in the north and Egypt in the south, enters a period of socio–spatial instabilities. In Palestine this period is characterized by the coexistence of small and relatively autonomous socio–spatial agricultural communities, not dissimilar to those of the first pre-urban period. Once again these sedentary and nomadic agricultural communities undergo their own urban revolution – the second in the region of Palestine, and become components in a global urban system.

A new urban culture rises in the region, and Palestine becomes a subsystem within a large scale Near-Eastern urban system. This urban system continues throughout most of the second millennium BCE. At the end of this urban steady state, the global urban system in the entire Aegean, Anatolian, Syrian and Palestinian regions collapses, and large scale processes of nomadization and migration of entire population groups take place. This global

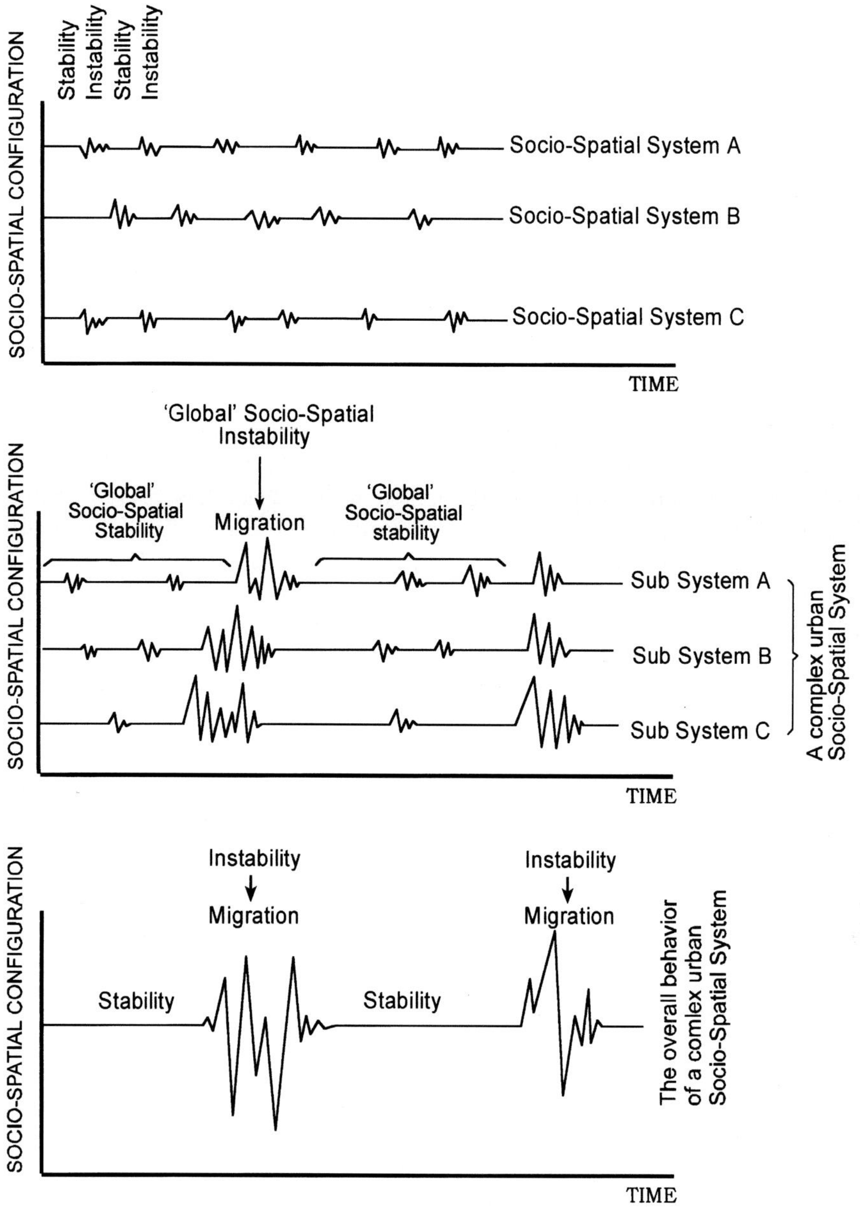

Fig. 15.7. The transition from agriculture to urbanism (source: Portugali, 1994a). (**a**) *Top*: Relatively small agricultural socio–spatial systems are environmentally vulnerable and subject to frequent instabilities. (**b**) *Middle*: The local agricultural communities noted above, become subsystems within a global urban system, thus capable of withstanding local perturbations hitherto fatal to the communities in the top diagram. (**c**) *Bottom*: The overall behavior of an urban system

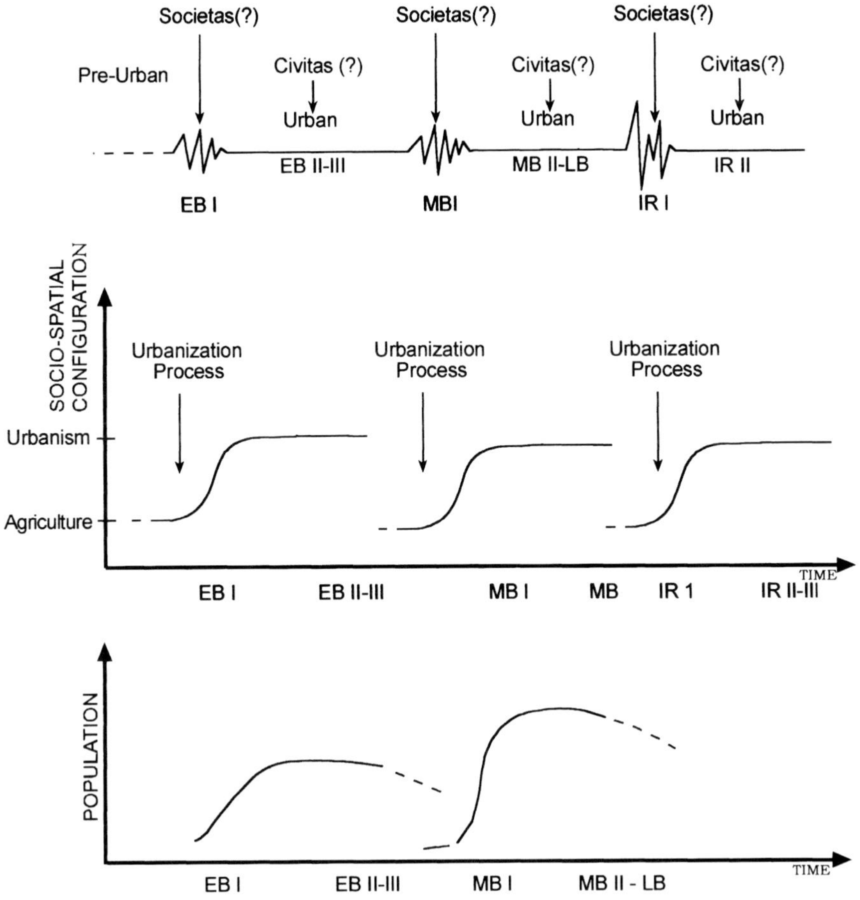

Fig. 15.8. The evolution of the settlement system in Palestine from the Early Bronze Period to the Iron Age. Source: Portugali (1994a). (**a**) *Top*: A description in terms of Fig. 15.7. (**b**) *Middle*: A description of the process as a rhythm between agriculture and urbanism, interrupted by global collapses of the urban system. (**c**) *Bottom*: Calculated population changes in the Early Bronze and Middle Bronze periods according to Gophna and Portugali (1988)

scenario is reflected in the local region of Palestine: relatively autonomous socio–spatial groups of various sizes, at various degrees of sedentarization and nomandism and at various forms of socio–spatial organization, coexist in the country in complex relations of interaction and conflict, and once again part of these groups undergoes an urban revolution – the third in the sequence.

The information about the events associated with the third urban revolution at the end of the 2nd millennium BCE, is much richer than that of the first and the second revolutions. In addition to archaeological data we now have corresponding historical data of written documents. From both we know that the urban system that collapsed toward the end of the second millennium BCE was a system of relatively small city states, each with its king, army, religious leadership, as well as dependent cities and farms. We further know that parallel to, and in between, the city states there existed a non-urban society of pastoralists and semi-nomadic agriculturalists that lived in relations of symbiosis and conflict with the urban society.

The collapse of this urban system entailed a large scale process of nomadization and migration of both the previously urban dwellers and the non-urban segments of that society. This socio–spatial reality is partly reflected also in the Biblical story which corresponds to that period. Of specific interest here are two lexical schemata (in Hebrew): 'Nahala ve ariah', that is, a tribal territory and its cities or towns, and 'Arim, banot and hatzerim', that is, a city, its daughters and farms. The first refers to a non-urban society (e.g. the Israelite tribes), whereas the second to an urban society (e.g. the Cenaanites and Philistines). This is presented graphically in Fig. 15.9. The third urban revolution in Palestine, at the end of the 2nd millennium BCE, was the revolution of the Israelites who underwent a transformation from a non-urban to an urban society.

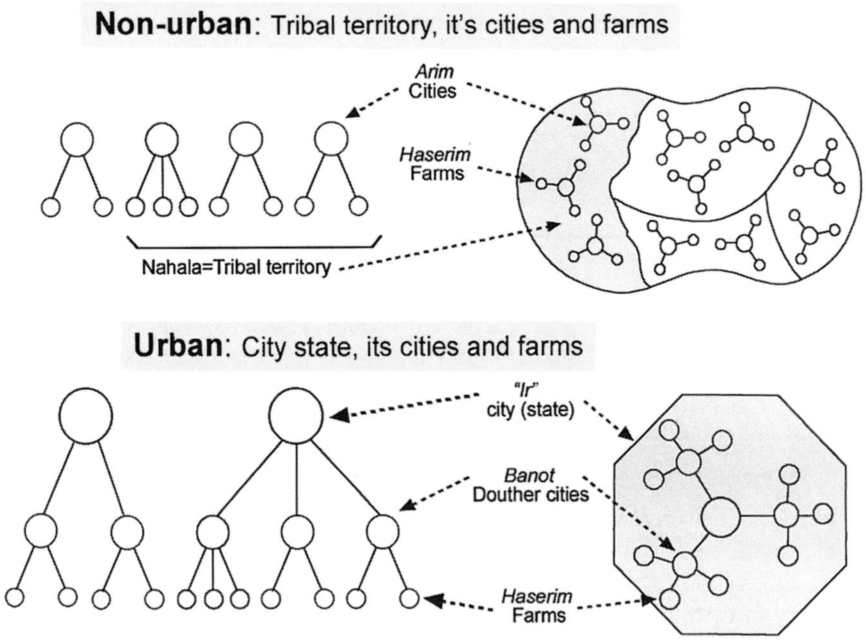

Fig. 15.9. *Top*: The Biblical lexical schemata '*Nahala ve ariah*', that is, a tribal territory and its cities or towns, versus, *Bottom*: '*Arim, banot and hatzerim*', that is, a city, its daughters and farms. The first refers to a non-urban society (e.g. the Israelite tribes), whereas the second to an urban society (e.g. the Cenaanites and Philistines)

15.4 Self-Organized Urban Revolutions at the End of the Second Millennium

15.4.1 Cities as a Self-Organized Family Resemblance Category

We have opened this book with an observation and a question. The observation was that cities and urbanism are with us for more then 5000 years and that despite their archaic origin they have survived and often dominated many of the subsequent cultural changes and societies. So much so that today, on the verge of the 21st century, cities and urbanism are still at the center of our life.

This observation was followed by a question: how is it possible that we employ the notion *city* to refer to a tiny little place such as 10th century B.C. Beer-Sheva with its 10 (!!!) dunams and few hundreds inhabitants, and to the 45 million people in the megacity that according to Woo (1994) and Castells (1996) is currently building up in the Pearl River Delta with Guangzhou at its center?

We have answered this question by saying that in order to understand this extraordinary process of change and continuity, cities and urbanism must be interpreted as categories with the following characteristics:

1. All cities, from the first city to the cities of today, must be looked at as a huge Wittgensteinian family resemblance network, in which, in every period, one can observe ideal type and basic-level cities.
2. This family resemblance network with its prototypes and basic levels has evolved and diffused in space and time, and in the process underwent evolutionary changes.
3. The network has evolved and diffused in space and time by means of SIRN (synergetic inter-representation networks). That is to say, as a self-organizing synergetic process, and by an interplay between internal and external representations.
4. The huge family resemblance network that forms the category 'city', with its changing prototypes and so on, is a single, open, complex, and diffusing self-organizing network with characteristics described at length in the above chapters.

15.4.2 Urbanism as a Generative Order Parameter

In this closing chapter of the book we have once again reiterated the discussion on the nature of cities and urbanism – this time from the perspective of recent discourse in social theory on the nature of cities and urbanism in general and at end 20th century in particular. As in Chaps. 1, 2, here too, one could identify the tension between change and continuity: between cities and urbanism as old, long-lasting phenomena, and cities and urbanism as markers of the dramatic changes that typify the 20th century on the verge of the

21st. That is, the strong notion that something very dramatic is happening to cities and urbanism today, in front of our eyes. "The twentieth century has been, then, *the* century of urbanization" writes Harvey (1996, p. 403) in his recent book, and as we have elaborated above, according to Lefebvre we are at the midst of *La révolution urbaine*, while Castells speaks of the new information city and the coming network-megacity.

Examining the discourse in social theory on society, city and urbanism we have suggested that by looking at these issues from the perspective of self-organization theory one can theorize about cities in their own sake and regard their multiple appearances as a single evolving self-organizing system. In this way the various forms of cities emerge, not only as derivations from external forces, but from within and from each other. We have further elaborated this point by reference to the first and subsequent urban revolutions.

This interplay between change and continuity and the implied possibility that the city emerges 'from within' was prominent already in the first urban revolution. The latter was not, as we have seen, an outcome of a new invention – the first city. Sedentary settlements with a city form were probably known and experienced long before the urban revolution. Rather, it was the emergence of urbanism as a generative socio–spatial order and of specific urban order parameters. The first urban revolution was a revolution in the sense that the spatial and territorial, in this case the city, emerged to a status of a social form – a kind of an urban mode of production.

In terms of Chaps. 1, 2 this implies that the urban revolution marks the start of a new family resemblance network with cities as its prototypical-basic level elements. 'Non-urban cities' such as Çatal Hüyük, Jericho or Rasul, which predated the urban revolution were isolated events in this respect, not-yet-urban cities with a status similar to that of domestication in hunting and gathering societies before the neolithic revolution – cities which were not connected to, and thus members of, the urban family resemblance.

The urban revolution was the eruption, so to say, of a new category – of a new family resemblance network. From now on it has evolved and diffused in space and time by means of the process of a synergetic inter-representation network (SIRN) as described above. That is to say, by means of an interplay between internal and external representations which took the form of an evolving self-organizing system. Thus, in the first urban revolution people took from their past knowledge – their collective memory – the city as a sedentary Neolithic settlement and externalized it in a new and innovative way – in the form that we now call an urban society. This was the case in subsequent revolutions too where non-urban, semi-nomadic, people took a known urban form, internalized it, and then externalized it in their own specific way. By so doing they not only extended the family resemblance category 'city', but also often modified it, installed in its center new prototypical cities and pushed to the periphery other non-central exemplars. In the third urban revolution discussed above the result was a new social form altogether:

the attempt to revolutionize the non-urban Israelites into an urban society resulted in a 'higher' order socio–spatial form – the state.

This internal–external interplay thus took the form of an evolving self-organizing system as conceptualized by the synergetics approach, with relatively long periods of steady state, during which a certain configuration of central prototype cities dominating and enslaving the categories 'city' and 'urban', followed by relatively short periods where the urban system disintegrated, collapsed, entered into strong fluctuations and chaos, from which it re-emerged into a new steady state with its often new prototypes, and so on. This space–time self-organization process took two forms. One, which was a consequence of the nature of urbanism as elaborated above and illustrated in Fig. 15.8, and second, as a space–time evolutionary process.

As we have seen above, the nature of urbanism is such that it would be better to think of it not in terms of *the, single, genuine*, urban revolution, but in terms of a sequence of urban revolutions. This point was elaborated above theoretically and empirically. As in the first revolution, so in subsequent ones, the fact that the various socio–cultural groups knew about, memorized, or even experienced, cities does not reduce the requirement that in order to become urban they had to undergo their own socio–spatial revolution. The result is that the first urban revolution and the subsequent sequence of two more urban revolutions described above, give a strong impression of repetitive cyclical processes which may justify considering the whole sequence as a single, large, socio–spatial system. The other result is a distinction between cities and urbanism.

15.4.3 Two Cognitive Maps: a Territory and Its Cities Versus a City and Its Territory

The first urban revolution was thus the emergence of a new family resemblance network, a new socio–spatial category – the urban – in which cities form the prototypical elements at the center of the category; each of the subsequent urban revolutions described was thus a revolution of a specific people, and a hermeneutic change (see Chap. 4 and Fig. 4.5 above) in the family resemblance category city and urban. In the language of social theory this implied that in each of these revolutions the spatial and territorial, in this case the city, emerged to a status of a social form – a new form of an urban mode of production. Before the first urban revolution cities did exist, but as peripheral elements in pre-urban, or non-urban, societies. This was the case with subsequent urban revolutions too. In each of the revolutions described above a non-urban society underwent its own urban revolution. Each of these societies, as we have emphasized, knew about cities and in certain cases had their own cities. But the latter were peripheral in their spatio–social organization. They were *societies with cities* but not *urban societies*.

This distinction between cities and urbanism, between a society with cities and an urban society, was nicely illustrated by the third urban revolution in

Palestine at the end of the second millennium BCE. As elaborated above, from the archaeological and historical record of that time there emerge two contrasting socio–spatial categories or cognitive maps. One which perceives space, society, and culture from the urban center outward – in terms of a central city, its dependent cities and their dependent farms, and the other from the perimeter inward, in terms of a (tribal) territory and its cities. As we have seen above, the third urban revolution in Palestine, at the end of the 2nd millennium BCE, started as a transformation of a non-urban society with a cognitive map of a *territory and cities*, to an urban society with a cognitive map of a *city and its territory*.

This tension between the two cognitive maps, the two forms of socio–spatial organization – the territory with its cities versus the city, its daughter cities and farms – is not confined to the urbanization of the Israelites. It showed up throughout history in a variety of contexts and configurations. In fact, the history of cities and urbanism, that is to say, the evolutionary process of the space–time diffusion of the family resemblance categories city, urban society, is characterized by this tension between the territorial and the urban cognitive maps, between periods and societies in which cities have occupied the prototypical center of society, and periods and societies in which the cognitive map of territories and their cities dominated. Thus Rome was a city with a territory – the Roman empire, whereas during the middle ages the feudal Europe was a society of territories and their cities. This tension is once again relevant today at the end of the second millennium A.D. Not between a tribal society and an urban society, as at the end of the 2nd millennium BCE, but between nationalism as the socio–spatial generative order which dominated 20th century society, and urbanism as an emerging socio–spatial order parameter which on the verge of the third millennium threatens the dominancy of nationalism.

15.4.4 From Nationalism to Urbanism: A New Urban Revolution at the Gate of the 21st Century?

In a recent study I have suggested that one examine nationalism from the theoretical standpoint of *Implicate Relations* – a conceptual framework developed from the conjunction of Bohm's notion of 'implicate order', and Haken's theory of synergetics (Portugali 1993). The suggestion was that nationalism has become the generative order parameter of modern society with the nation state as its material external representation, and the ideology of nationalism as its cognitive internal representation – as a collective cognitive map which resides in the minds of individuals. In the 20th century nationalism has become a political doctrine and ideology shared by people of otherwise opposing convictions (liberals, socialists, Marxists, etc.), often regarded by them as the precondition to achieve their grand social goals: Marxists–Leninists see it as a precondition for a genuine socialist society which is supposed to emerge when the nation(alist)-state will 'wither away' by itself; liberals interpret the nationalist imperative for national(list) self-determination as a basic and natural human right, and so on. Using Haken's synergetics the suggestion was that nationalism has emerged as a socio–spatial order parameter and that in the process it has enslaved the other competing political ideologies, such as liberalism, Marxism, socialism and even religious fundamentalism.

The essence of nationalism is a unity of people, territory and polity. People are naturally divided into, and belong to, nations, each nation has its own national territory, and must have its own self-government. In order to fulfill themselves as individuals, culture and society, people must self-determine themselves as a nation with their own national government on their own national territory. A person's duty to his or her national territory overrides that person's duty to any other territorial entity such as a city. A person is first French, Israeli, Italian ... and only then Parisian, Tel-Avivian, Roman, New-Yorker ... etc. (or European, Asian...). In the 20th century the world is perceived as divided into national territories, each with its own cities.

The 20th century has been the century of nationalism, there can be little doubt about that. The two world wars, the many smaller-scale wars, the rapid space–time diffusion of nationalism in the post colonial world, the recent disintegration of the ex-USSR and Yugoslavia into their past nationalist components are some of the most dramatic events in the process.

But the 20th century has also been, as we have elaborated above, a century of urbanism, and as we could see above there can be no doubt about that, too. The numbers and observations presented above speak very clearly about that. In fact, as noted by many scholars (e.g. Gelner 1983; Reisman 1970 and Chap. 11, above), nationalism and urbanization have developed hand in hand in symbiotic relations. Up till now, however, this tremendous urban growth took place within the confines of national territories, enslaved by the nationalist order parameter. At the end of the 20th century and on the verge of the 21st, nationalism is still the order parameter; spatially, politically and cognitively, the world is still a world of nations: rich, poor, capitalist, socialists, authoritarian, democratic.

But can we say the same about the big cities of the world? about Mexico–city with its almost 20 million people and the Megacity in the Pearl River Delta, with its 45 million people? Are these and similar urban giants, too, cities of their national territories? what is the cognitive map of the people living in these giants? do they perceive themselves first as members of their nations and second as citizens of their cities? or do the cities as material and mental entities start to overflow and transcend their national boundaries? are these emerging urban giants an indication for a new urban order parameter which is going to enslave nationalism and become the generative order parameter of the 21st century?

Concluding Notes: Self-Organizing Cities at the Gate of the 21st Century

The self-organizing cities we have discussed enable us to theorize about the very first city, its subsequent evolutionary forms, as well as about its current appearance as the hypermodern city of today. We see them all from the perspective of end 20th century. The self-organizing cities that we have discussed and simulated in this book are the theories of cities with which we are going to close the 20th century and enter the 21st century. This is so by definition. But the self-organized city is also an image, a cognitive map, a general cultural atmosphere about the nature of the cities and the urbanized world we live in. And this, as far as I can judge, is something new. In the very ancient past and in antiquity, cities always came with a strong feeling of stability, civilization, law and order. Think of *the urban revolution*, which is but a synonym to *the rise of civilization*, think of the classical Greek city (from which the word 'police' is taken), think of the cities of the Occident as portrayed, for example, by Max Weber or Mumford. Think also of the image of our cities as we used to look at them just a few decades ago – as they appeared in our urban theories. Todays images are very different and the various self-organization cities are part of this new image of the complex, chaotic, bifurcative, far-from-equilibrium urbanized world we live in. (This by itself is not surprising if you realize that Darwin's theory of evolution was the product of a wider cultural atmosphere of individualism, competition and economic survival of the fittest.)

But can our self-organized theoretical cities be of any use in our self-organized urban reality (imagined and/or real)? The general answer among practitioners of self-organization would be Yes! Self-organization is a theory about complexity and in an age of high complexity it can help us guide our actions. But how? Here we can identify two basic answers, two basic approaches.

The first approach would claim that self-organization is a conceptual and mathematical theory about complexity and consequently it will teach us how to control the complexity of our chaotic self-organizing cities, how to predict their behavior, plan their future and thus tame them. The whole thing is a matter of time. In time we will learn more about the control parameters of our cities, the way they give rise to self-organized criticalities, to various forms of structural stability and steady states governed by order parameters which have enslaved the many parts of the city. Self-organization is a new quantitative-mathematical theory which can revive the status of quantitative urban geography and regional science after more than two decades during which they were subject to strong criticism by structuralists, humanists and postmodernists. Self-organized cities are to give the quantitative urbanists an edge over the qualitativists, and thus once again widening the gap between the geographical appearances of Snow's two cultures.

The second approach claims that control parameters are controls only post-factum, or in models and laboratories. In real cities the control parameters themselves are self-organizing systems and as such uncontrollable. Self-organization here is essentially a new way of seeing, of realizing that self-organizing systems, be they humans or cities, are essentially and profoundly unstable, chaotic, far-from-equilibrium, unpredictable, and that therefore we have to find ways to live with their complexity. From this perspective follows, for example, a new type of action in the city, a new type of city planning, the aim of which is not to control, but to participate; a new type of urban theory which will be able to integrate Snow's two cultures as they show up in urban theory and the study of cities. This view and optimism stems from the assessment that the potential is here – that, as noted time and again throughout this book, many of the notions of self-organization are more than similar to those of the social theory of the city: the notion of an order-parameter of a city is very similar to the notion of *mode of production* of a city, and a process such as circular causality is not very different from the notions of *reproduction* and *structuration* in social theory and philosophy, and so on. The potential is here – it has just to be realized.

Which of the two approaches will dominate the future self-organizing cities, is hard to say; my personal inclination is, as you can guess, toward the second approach. But this might equally well be my personal wishful thinking and probably, in order to know better, we will have to wait for another book on *Self-Organization and the City* which will appear, say, toward the end of the first decade of the 21st century.

Bibliography

AAAG. (1979). Annals of the Association of American Geographers 69.

Adams, R. M. (1960). The origin of cities. *Scientific American* 203(3), 153–172.

Adams, R. M. (1966). *The Evolution of Urban Society*. Chicago: Aldine.

Adams, R. M. (1972). Demography and the "Urban Revolution" in Lowland Mesopotamia. In *Population Growth: Anthropological Implications* (B. Spooner, ed.). Cambridge, MA: MIT Press.

Ajzen, I. (1991). The theory of planned behavior. *Organizational Behavior and Human Decision Processes* 50, 1–33.

Albin, P. S. (1975). *The Analysis of Complex Socioeconomic Systems*. Lexington: Lexington Books.

Alexander, C. (1965). A city is not a tree. *Architectural Forum*, April–May, 58–62.

Alexander, C. (1979). *The Timeless Way of Building*. New York: Oxford University Press.

Alexander, C., Ishikawa, S. and Silvestein, M. with M. Jacobson, I. Fiksdahl-King and S. Angel (1977). *A Pattern Language*. New York: Oxford University Press.

Allen, P. A. (1981). The evolutionary paradigm of dissipative structures. In *The Evolutionary Vision* (E. Jantsch, ed.), pp. 25–71. Boulder: Westview Press.

Allen, P. and Sanglier, M. (1981). Urban evolution, self-organization and decision making. *Environment and Planning*, A, 13, 169–183.

Allen, P., Sanglier, M., Engelen, G. and Boon, F. (1985). Towards a new synthesis in the modeling of evolving complex systems. *Environment and Planning, B: Planning and Design* 12, 65–84.

Alonso, W. (1965). *Location and Land Use*. Cambridge, MA.: Harvard University Press.

Alstrom, P. and Leao, J. (1994). Self-organized criticality in the Game of Life. *Physical Review*, E, 49, R2507–R2508.

Angel, S. and Hyman, G. M. (1972). *Urban Fields and Geometry of Movement*. London: Pion.

Ashby, R. (1947). Principles of self-organizing dynamic system. *Journal of General Psychology* 37, 125.

Atkin, R. H. (1974). *Mathematical Structure in Human Affairs*. London: Heinemann Educational Books.

Atkin, R. H. (1977). *Combinatorial Connectivities in Social Systems*. Basel: Birkhouser.

Atkin, R. H. (1978). Q-analysis: a hard language for the soft science. *Futures* 10, 492–499.

Atlan, H. (1985). Two instances of self organization in probabilistic automata networks: epigenesis of cellular networks and self'-generated criteria for pattern discrimination. In *Dynamical Systems and Cellular Automata* (J. Demongeot, E. Goles and M. Tchuente, eds.). London: Academic Press.

Bak, P., Chen, K. and Creutz, M. (1989). Self-organized criticality in the Game of Life. *Nature* 342, 780–782.

Bak, P. and Chen, K. (1991). Self-organized criticality. *Scientific American* 28, 26–33.

Bak, P., Chen, K., Scheinkman. J. A. and Woodford, M. (1993). Aggregate fluctuations from independent shocks: self-organized criticality in a model of production and inventory dynamics. *Ricerche Economiche* 47(3).

Banzhaf, W. (1994). Self-organization in a system of binary strings. In *Artificial life IV* (R. A. Brooks and P. Maes, eds.), pp. 109–118. Cambridge, MA: MIT Press.

Barkley, R. J. (1991). *From Catastrophe to Chaos: A General Theory of Economic Discontinuities.* Boston, Dordrecht, London: Kluwer Academic Publishers.

Bartlett, F. C. (1932/1961). *Remembering: A Study in Experimental and Social Psychology.* Cambridge: Cambridge University Press.

Batten, D. and Johansson, B. (1987). The dynamics of metropolitan change. *Geographical Analysis* 19(3) 189–199.

Batty, M. (1996). *Urban Evolution on the Desktop: Simulations Using Extended Cellular Automata.* Unpublished paper, Centre for Academic Spatial Analysis, UCL.

Batty, M. (1997). Cellular automata and urban form: a primer. *Journal of the American Planning Association* 63(2), 266–274.

Batty, M and Xie, Y. (1994). From cells to cities. *Environment and Planning, B: Planning and Design* 21, 531-548.

Batty, M. and Xie, Y. (1999). Self-organized criticality and urban development. In *Population, Environment and Society on the Verge of the 21st Century* (J. Portugali, ed.).

Batty, M. and Longley, P. (1994). *Fractal Cities.* London: Academic Press.

Beauregard, R. A. (1989). Between modernity and postmodernity: the ambiguous position of US planning. *Society and Space* 7, 381–395.

Benati, S. (1997). A cellular automata for simulation of competitive location. *Environment and Planning, B: Planning and Design* 24, 205–218.

Benenson, I. and Portugali, J. (1995). Internal vs. external spatial information and cultural emergence in a self-organizing city. In *Spatial Information Theory: A Theoretical Basis for GIS.* (A. U. Frank and W. Kuhn, eds.), pp. 431–441. Berlin: Springer.

Bennett, R. J. and Chorley, R. J. (1978). *Environmental Systems: Philosophy, Analysis and Control.* London: Methuen.

Benovolo, L. (1971). *The Origins of Modern Town Planning.* Cambridge, MA: MIT Press.

Berry, B. B. and Horton, F. E. (1970). *Geographic Perspectives on Urban Systems.* Englewook Cliffs: Prentice-Hall.

Berry, B. J. L. and Kasarda, J. D. (1977). *Contemporary Urban Sociology.* New York: Macmillan.

Bertalanffy, L. von (1968). *General System Theory.* New York: George Braziller.

Bertuglia, C. S., Occelli, S., Rabino, G. A., Tadei, R. and Wilson, A. G., eds. (1987). *Urban Systems: Contemporary Approaches to Modelling.* Kent: Croom Helm.

Bertuglia, C. S., Clarke, G. P. and Wilson, A. G. (1994). *Modelling the City.* London: Routledge.

Bleicher, J. (1980). *Contemporary Hermeneutics: Hermeneutics as Method, Philosophy and Critique.* London: Routledge & Kegan Paul.
Boal, F. W. (1987). Segregation. In Social *Geography: Progress and Prospect* (M. Pasion, ed.). Beckenham: Croom Helm.
Bohm, D. (1957). *Causality and Chance in Modern Physics.* London: Routledge & Kegan Paul.
Bohm, D. (1980). *Wholeness and the Implicate Order.* London: Routledge & Kegan Paul.
Bohm, D. and Peat, F. D. (1987). *Science, Order and Creativity.* New York: Bantam.
Bohm, D. and Hiley, B. J. (1993). *The Undivided Universe.* London, New York: Routledge.
Boserup, E. (1965). *The Conditions of Agricultural Growth: The Economics of Agrarian Change Under Population Pressure.* Chicago: Aldine Publishing Company.
Bouligand, Y. (1985). *Symmetry Principle and Morphogentic.*
Breheny, M. and Hooper, A. (1985). *Rationality in Planning.* London: Pion.
Breitbart, M. M. (1981). Peter Kropotkin, the anarchist geographer. In *Geography, Ideology & Social Concern* (D. R. Stoddart, ed.), pp. 134–153. England: Basil Blackwell Publishers.
Burgess, B. (1925). The growth of the city : An introduction to a research project. In *The City* (R. Park, E. W. Burgess and McKenzie, eds.), pp. 47–62. Chicago: Chicago University Press.
Burgess, E. W. (1926/1968). *The Urban Community.* New York: Greenwood Press.
Burgess, E. W. (1927). The determination of gradients in the growth of the city. *American Sociological Society Publications* 21, 178–184.
Camhis, M. (1979). *Planning Theory and Philosophy.* London: Tavistock Publications.
Carneiro, R. L. (1967). On the relationships between size of population and complexity of social organization. *South-Western Journal of Anthropology* 23, 234–243.
Carneiro, R. L. (1970). A theory of the origin of state. *Science* 169(3947), 733–738.
Carneiro, R. L. (1972). From antonomous villages to the state: a numerical estimation. In *Population, Resources, and Technology* (B. Spooner, ed.). Philadelphia: University of Pennsylvania Press.
Castells, M. (1977). *The Urban Question.* Cambridge, MA: MIT Press.
Castells, M. (1996). *The Rise of the Network Society.* Malden, MA: Blackwell Publishers.
Chadwick, G. (1971). *A System View of Planning.* Oxford: Pergamon Press.
Chapman, G. P. (1977). *Human and Environmental Systems.* London: Academic Press.
Childe, V. G. (1950). The urban revolution. *Town Planning Review* 21, 3–17.
Childe, V. G. (1957). Civilizations, cities and towns. *Antiquity* 31, 36–37.
Christaller, W. (1933/1966). *Central Places in Southern Germany.* Englewood Cliffs, NJ: Prentice Hall.
Cooke, P. (1990). Modern urban theory in question. *Transactions, Institute of British Geographers* 15, 3.
Couclelis, H. (1985). Cellular worlds: a framework for modeling micro–macro dynamics. *Environment and Planning, A*, 17, 585–596.
Couclelis, H. (1988). Of mice and men: what rodent population can teach us about complex spatial dynamics. *Environment and Planning, A*, 20, 99–109.
Cressey, P. (1938). Population succession in Chicago; 1898–1938. *The American Journal of Sociology* 44, 59–69.

Davis, M. (1990). *City of Quartz.* London: Verso.
Dawkins, R. (1982). *The Extended Phenotype.* Oxford: Freeman.
Dawkins, R. (1986). *The Selfish Gene.* London: Granada Pub. (first published 1976).
Dear, M. and Scott, A., eds. (1981) *Urbanization and Urban Planning in Capitalist Society.* London, New York: Methuen.
Demongeot, J., Goles, E. and Tchuente, M., eds. (1985) Introduction: dynamic behaviour of automata. In *Dynamical Systems and Cellular Automata* (J. Demongeot, E. Goles and M. Tchuente, eds.). London: Academic Press.
Dendrinos, D. S. and Sonis, M. (1990). *Chaos and Socio–Spatial Dynamics.* New York, Berlin: Springer-Verlag.
Dennett, D. C. (1991). *Conciousness Explained.* Boston: Little, Brown.
Dennett, D. C. (1995). *Darwin's Dangerous Ideas.* New York: Simon & Schuster.
Desbarts, J. (1983). Spatial choice and constraints on behavior. *Annals of the Association of American Geographers* 73, 340–357.
Ditzinger, T. and Haken, H. (1990). The impact of fluctuations on the recognition of ambiguous patterns. *Biological Cybernetics* 63, 453–456.
Donald, M. (1991). *Origins of the Modern Mind: Three Stages in the Evolution of Culture and Cognition.* Cambridge, MA, London: Harward University Press.
Dunbar, G. S. (1981). Elisee Recules, an anarchist in geography. In *Geography, Ideology and Social Concern* (D. R. Stoddart, ed.), pp. 154–164. England: Basil Blackwell Publishers.
Duncan, O. and Duncan, B. (1955). A methodological analysis of segregation indices. *American Sociological Review* 20, 210–217.
Edelman, G. M. (1987). *Neural Darwinism: The Theory of Neural Group Selection.* New York: Basic Books.
Edelman, G. M. (1992). *Bright Air Brilliant Fire: on the Matter of the Mind.* London: Penguin Books.
Eigen, M. (1971). Molecular self-organization and the early stages of evolution. *Quarterly Review of Biophysics* 4(2 & 3), 149.
Eigen, M. and Winkler, R. (1983). *Laws of the Game.* New York: Alfred Knopf.
Eldredge, H. W., ed. (1967). *Taming Megalopolis.* New York, Washington, London: Frederick A. Preager.
Faludi, A. (1973). *A Reader in Planning Theory.* Oxford: Pergamon Press.
Festinger, L. (1957). *A Theory of Cognitive Dissonance.* Stanford: Stanford University Press.
Fischer, M. M., Nijkamp, P. and Papageorgiou, Y. Y., eds. (1990). *Spatial Choices and Processes.* New York: Elsevier Science.
Fischer, M. M., Haag, G., Sonis, M. and Weidlich, W. (1990). Accounts of different views in dynamic choice processes. In *Spatial Choices and Processes* (M. M. Fischer, P. Nijkamp and Y. Y. Papageorgiou, eds.), pp. 17–47. New York: Elsevier Science.
Fishbein, M. and Ajzen, I. (1975). *Belief, Attitude, Intention, and Behavior: An Introduction to Theory and Research.* Reading, MA: Addison-Wesley.
Foerster, H. von and Zopf, G. W., eds. (1962). *Principles of Self-Organization.* New York: Pergamon.
Forester, J. (1985). Practical Rationality in Planmaking. In *Rationality in Planning* (M. Breheny and A. Hooper, eds.). pp. 48–58. London: Pion.
Foucault, M. (1967). *Madness and Civilization.* London: Tavistock.
Galbraith, J. K. (1967). *The New Industrial State.* Boston: Houghton Mifflin.
Gardner, M. (1971). On cellular automata, self-reproduction, the Garden of Eden and the Game of Life. *Scientific American* 224, 112–117.
Gardner, H. (1987). *The Mind's New Science.* New York: Basic Books.

Garling, T. and Golledge, R. G., eds. (1993). *Behavior and Environment.* Amsterdam, London, New York, Tokyo: North-Holland.
Garreau, J. (1991). *Edge City: Life on the New Frontier.* New York: Doubleday.
Gazzaniga, M. S. (1985). *The Social Brain.* New York: Basic Books.
Geddes, P. (1915/1968). *Cities in Evolution.* London: Ernest Benn.
Gelner, E. (1983). *Nations and Nationalism.* Oxford: Basil Blackwell.
Gibson, J. J. (1979). *The Ecological Approach to Visual Perception.* Boston: Houghton-Mifflin.
Giddens, A. (1976). *New Rules of Sociological Method: a Positive Critique of Intepretive Sociology.* London: Hutchinson.
Giddens, A. (1984). *The Constitution of Society: Outline of the Theory of Structuration.* Berkeley: University of California Press.
Goldberg, M. A. (1985). Flexibility and adaptation: some cues for social systems from nature. In *Links Between Natural and Social Sciences* (J. Portugali, ed.), pp. 179–190. A special theme issue, *Geoforum* 16(2).
Golledge, R. G. and Timmermans, H., eds. (1988). *Behavioral Modeling in Geography and Planning.* London, New York, Sydney: Croom Helm.
Goodchild, B. (1990). Planning and the modern/postmodern debate. *Town Planning Review* 61(2), 119–137.
Gophna, R. and Portugali, J. (1988). Settlements and demographic processes in Israel's coastal plain from the Chalcholithic to the Middle Bronze Age. *BASOR* 269, 11–28.
Gould, S. J. (1980). *The Panda's Thumb.* New York: Norton.
Gould, P. (1981). Letting the data speak for themselves. *Annals of the Association of American Geographers* 77, 166–176.
Grabow, S. (1983). *Christopher Alexander: The search for a New Paradigm in Architecture.* London: Oriel Press Stockfield, Boston Henley.
Greene, J. C. (1981). *Science, Ideology and World View.* Berkeley, Los Angeles, London: University of California Press.
Gregory, D. (1978). *Ideology, Science and Human Geography.* London: Hutchinson.
Gregory, D. (1981). Human agency and human geography. *Transactions of the Institute of British Geographers, New Series,* 6, 1–18.
Gregory, D. (1989). Presences and absences: time–space relations and structuration theory. In *Social Theory of Modern Societies: Anthony Giddens and his Critics* (D. Held and J. Thompson, eds.), pp. 183–214. Cambridge: Cambridge University Press.
Gregory, D. (1994). *Geographical Imaginations.* Cambridge, MA: Blackwell.
Gregory, D. and Urry, J. eds. (1985). *Social Relations and Spatial Structures.* London: Macmillan.
Haag, G., Muntz, M., Pumain, D., Saint-Julien, T. and Sanders, L. (1992). Interurban migration and the dynamics of a system of cities. *Environment and Planning,* A, 24. 181-198.
Hagerstrand, T. (1967). *Innovation Diffusion as a Spatial Process.* Chicago: Chicago University Press.
Hagerstrand, T. (1970). What about people in regional science? *Papers and Proceedings of the Regional Science Association* 24, 7–21.
Haggett, P. (1972). *Geography: A Modern Synthesis.* New York: Harper and Row.
Haigh, M. J. (1985). Geography and general system theory, philosophical homologies and current practice. In*Links Between the Natural and Social Sciences.* (J. Portugali, ed.), pp. 191–203. A special theme issue, *Geoforum* 16(2).
Haken, H. (1979). Pattern formation and pattern recognition – an attempt at a synthesis. In *Pattern Formation by Dynamical Systems and Pattern Recognition* (H. Haken, ed.), pp. 2–13. Berlin: Springer.

Haken, H. (1983a). *Synergetics, an Introduction* (3rd Edn). Berlin: Springer-Verlag.
Haken, H. (1983b). *Advanced Synergetics.* Berlin: Springer-Verlag.
Haken, H., ed. (1985a). *Complex Systems – Operational Approaches.* Berlin: Springer-Verlag.
Haken, H. (1985b). Synergetics – an interdisciplinary approach to phenomena of self-organization. In *Links Between the Natural and Social Sciences.* (J. Portugali, ed.), pp. 205–212. A special theme issue, *Geoforum* 16(2).
Haken, H. (1987). *Advanced Synergetics: An Introduction* (2nd print). Berlin, New York: Springer.
Haken, H. (1988a). *Information and Self-Organization: A Macroscopic Approach to Complex Systems.* Berlin: Springer-Verlag.
Haken, H. (1988b). Synergetics in pattern recognition and associative action. In *Neural and Synergetic Computers* (H. Haken, ed.). Berlin: Springer-Verlag.
Haken, H. (1990). *Synergetics of Cognition.* Berlin: Springer-Verlag.
Haken, H. (1991). *Synergetic Computers and Cognition.* Berlin: Springer-Verlag.
Haken, H. (1993). Synergetics as a strategy to cope with complex systems. In *Interdisciplinary Approaches to Non-Linear Complex Systems* (H. Haken and A. Mikhailov, eds.). Berlin: Springer-Verlag.
Haken, H. (1996). *Principles of Brain Functioning: A Synergetic Approach to Brain Activity, Behavior and Cognition.* Berlin: Springer-Verlag.
Haken, H. (1998). Decision making and optimization in regional planning. In *Knowledge and Networks in a Dynamic Economy* (M. J. Beckmann, B. Johansson, F. Snickars and R. Thord, eds.). Berlin: Springer-Verlag.
Haken, H., Kelso, J. A. S. and Bunz, H. (1985). A theoretical model of phaze transition in human hand movement. *Biological Cybernetics* 51, 347–356.
Haken, H. and Portugali, J. (1995). A synergetic approach to the self-organization of cities. *Environment and Planning, B: Planning and Design* 22, 35–46.
Haken, H. and Portugali, J. (1996). Synergetics, inter-representation networks and cognitive maps. In *The Construction of Cognitive Maps* (J. Portugali, ed.), pp. 45–67. Dordrecht: Kluwer Academic Publishers.
Haken, H. and Portugali, J. (in preparation). Synergetics and captivity in a city.
Hall, P. (1982). *Great Planning Disasters.* Middlesex, England: Penguin Books.
Hall, P. (1992). *Urban and Regional Planning* (3rd Edn). London: Routledge.
Harris, C. D. and Ullmann, E. L. (1945). The nature of cities. *Annals of the American Academy of Political and Social Science* 242, 7–17.
Harvey, D. (1969). *Explanations in Geography.* London: Arnold.
Harvey, D. (1970). Social processes and spatial form: an analysis of conceptual problems of urban planning. *Papers of the Regional Science Association* 25, 47-69.
Harvey, D. (1973). *Social Justice and the City.* London: Edward Arnold.
Harvey, D. (1982). *The Limits of Capital.* Oxford: Blackwell.
Harvey, D. (1985a). *Consciousness and the Urban Experience.* Oxford: Blackwell.
Harvey, D. (1985b). *The Urbanization of Capital.* Oxford: Basil Blackwell.
Harvey, D. (1989). *The Condition of Postmodernity.* Cambridge: Blackwell.
Harvey, D. (1996). *Justice, Nature and Geography of differences*, Oxford: basil Blackwell.
Hay, E. J. (1988). *The Just-in-Time Breakthrough.* New York: John Willey & sons.
Howard, E. (1902/1965). *Garden Cities of To-Morrow.* London: Faber and Faber.
Hoyt, H. (1939). *The Structure and Growth of Residential Neighborhoods in American Cities.* Washington, DC: Federal Housing Administration.
Isard, W. (1956). *Location and Space Economy.* Published jointly by the Technology Press of Massachusetts Institute of Technology and Willey, New York.

Itami, R. (1994). Simulating spatial dynamics: cellular automata theory. *Landscape and Urban Planning* 30, 27–47.

Jackson, P. and Smith, S. J., eds. (1981). *Social Interaction and Ethnic Segregation.* London: Academic Press.

Jackson, P. and Smith, S. J. (1984). *Exploring Social Geography.* New York: George Allen & Unwin.

Jacobs, J. (1961). *The Death and Life of Great American Cities.* England: Penguin Books.

Jacobs, J. (1969). *The Economy of Cities.* New York: Random House.

James, D. and Taueber, K. E. (1985). Measures of segregation. In *Sociological Methodology* (N. Tuma, ed.). Jossey-Bass.

Jantsch, E. (1975). *Design for Evolution.* New York: George Braziller.

Jantsch, E., ed. (1981). *The Evolutionary Vision.* Boulder, CO: Westview Press.

Johnston, R. J. (1988). Exploring the role of geography. *Horizons* 23–24, 13–28 (in Hebrew).

Kelso, J. A. S. (1984). Phase transitions and critical behavior in human bimanual coordination. *American Journal of Physiology* 246, R1000–R1004.

Kelso, J. A. S. (1990). Phase transitions: foundation of behavior. In *Synergetics of Cognition* (H. Haken and M. Stadler, eds.), pp. 240–268. Berlin: Springer.

Kenyon, K. M. (1957). *Digging up Jericho.* New York: Praeger.

Kirk, G. (1980). *Urban Planning in a Capitalist Society.* London: Croom Helm.

Krafta, R., Portugali, J. amd Lemos, J. C. (in preparation). Cognition, automata and urban symbolic order.

Lakoff, G. (1987). *Women, Fire and Dangerous Things: What Categories Reveal About the Mind.* Chicago, London: The University of Chicago Press.

Langacker, R. W. (1987). *Foundations of Cognitive Grammar vol. 1: Theoretical Prerequisites.* Stanford: Stanford University Press.

Langton, G. (1986). Studying artifical life with cellular automata. *Physica* 22D, 120–149.

Larrain, J. (1982). *The Concept of Ideology.* London: Hutchinson.

Laszlo, E. (1987). *Evolution: The Grand Synthesis.* Shambhala, London: New Science Library.

Le Corbusier (1967). *The Radiant City.* London: Faber and Faber.

Lefebvre, H. (1970). *La Révolution Urbaine.* Paris: Gallimard.

Lefebvre, H. (1974). *The Production of Space.* English translation, 1995. Oxford: Blackwell.

Levi, Z. (1986). *Hermeneutics.* Tel-Aviv: Sifriat Hapoalim Hakibbutz Hameuchad (in Hebrew).

Ley, D. (1983). *A Social Geography of the City.* New York: Harper & Row.

Ley, D. and Samuels, M. (1978). *Humanistic Geography.* Chicago: Maaroufa.

Lieberson, S. (1981). An asymmetrical approach to segregation. In *Ethnic Segregation in Cities* (C. Peach, V. Robinson and S. Smith, eds.). London: Croom Helm.

Lierop, W. V. (1986). *Spatial Interaction Modeling and Residential Choice Analysis.* Vermont: Gower.

Lindblom, C. E. (1959). The Science of Muddling Through. In *A Reader in Planning Theory*, 1973, (A. Faludi, ed.). Oxford: Pergamon Press.

Lösch, A. (1954). *The Economics of Location.* New Haven: Yale University Press.

Lovelock, J. (1979). *Gaia.* New York: Oxford University Press.

Lynch, K. (1960). *The Image of the City.* Cambridge, MA: MIT Press.

Mainzer, K. (1996). *Thinking in Complexity.* Berlin: Springer.

Mandelbrot, B. B. (1983). *The Fractal Geometry of Nature.* San Francisco: Freeman.

Manneville, P., Boccara, N., Vichniac, G. Y. and Bidaux, R., eds. (1989). *Cellular Automata and Modeling of Complex Physical System.* Berlin: Springer-Verlag.

Marfoe, L. (1987). Cedar forest to silver mountain: social change and the development of long-distance trade in early Near Eastern societies. In *Center and Periphery in the Ancient World* (M. Rowlands, M. Larsen and K. Kristiansen, eds.). Cambridge: Cambridge University Press.

Margulis, L. (1995). Gaia is a tough bitch. In *The Third Culture* (J. Brockman, ed.). New York: Simon & Schuster.

Massey, D. and Denton, N. (1988). The dimensions of residential segregation. *Social Forces* 76, 281–315.

Maturana, H. and Varela, F. (1980). *Autopoiesis and Cognition.* Dordrecht, Holland: Reidel.

McCulloch, W. S. and Pitts, W. S. (1943). A logical calculus of the ideas immanent in nervous activity. *Bulletin of Mathematical Biophysics* 5, 115.

McKenzie, R. (1929). Ecological succession in Puget Sound region. *Publications of the American Sociological Society* 23, 60–80.

McLoughlin, J. B. (1969). *Urban and Regional Planning: A System Approach.* London: Faber and Faber.

McShane, J. (1991). *Cognitive Development.* Oxford: Basil Blackwell.

Meyerson, M. (1956). Building the Middle Range Bridge. In *A Reader in Planning Theory,* 1973, (A. Faludi, ed.). Oxford: Pergamon Press.

Mikhailov, A. S. (1990). *Foundations of Synergetics I.* Berlin: Springer-Verlag.

Mikhailov, A. S. and Loskutov, A. Y. (1991). *Foundations of Synergetics II.* Berlin: Springer.

Morrill, R. L. (1965). The negro ghetto: problems and alternatives. *Geographical Review* 55, 339–369.

Mumford, L. (1961). *The city in History.* London: Pelican Books.

Nicolis, G. and Prigogine, I. (1977). *Self-Organization in Nonequilibrium Systems: From Dissipative Structures to Order Through Fluctuations.* New York: Wiley-Interscience.

Ohno, T. and Mito, S. (1986). *Just-in-Time for Today and Tomorrow.* Cambridge, MA: Productivity Press.

Omer, I. (1996). *Ethnic Residential Segregation as a Structuration Process.* Unpublished Ph.D Thesis, Tel-Aviv University, Tel-Aviv.

Packard, N. H. (1985). Complexity of growing patterns in cellular automata. In *Dynamical Systems and Cellular Automata* (J. Demongeot, E. Goles and M. Tchuente, eds.). London: Academic Press.

Palma, A. and Lefevre, C. L. (1987). The theory of deterministic and stochastic compartmental models and its applications. In *Urban Systems: Contemporary Approaches to Modelling* (C. S. Bertuglia, S. Rabino, R. Tadei and A. G. Wilson, Eds.). Kent: Croom Helm.

Park, R. E. (1925/1967). *The City.* Chicago: Chicago University Press.

Pasty, H. (1992). Planning through debate – the communicative turn in planning theory. *TRP* 63(2), 143–162.

Peach, C. (1975). *Urban Social Segregation.* London: Longman.

Peach, C., Robinson, V. and Smith, S., eds. (1981). *Ethnic Segregation in Cities.* London: Croom Helm.

Peet, R., ed. (1977). *Radical Geography.* London: Methuen.

Phipps, M. (1989). Dynamical behavior of cellular automata under the constraint of neighborhood coherence. *Geographical Analysis* 21(3), 197–215.

Popper, K. R. (1959). *The Logic of Scientific Discovery.* London: Hutchinson.

Portugali, J. (1980). Allocation, distribution, social structure and spatial form: elements of planning theory. *Progress in Planning* 14(3), 227–310.

Portugali, J. (1984). Arim, Banot, Migrashim and Haserim: the spatial organization of Palestine in the 12–10 centuries B.C. according to the Biblical source. *Eretz Israel* 17, 282–290 (in Hebrew).
Portugali, J., ed. (1985a). *Links Between the Natural and Social Sciences.* A special theme issue, *Geoforum* 16(2).
Portugali, J. (1985b). Parallel currents in the natural and social Sciences. In *Links Between Natural and Social Sciences* (J. Portugali, ed.), pp. 227–238. A special theme issue, *Geoforum* 16(2).
Portugali, J. (1988a). Notes on socio–spatial change in the region of Israel: Chalcolithic to Iron Age. *Orient* XXIV, 55–70.
Portugali, J. (1988b). Theories of settlement and population and their importance to the demographic research in the region of Israel. In *Settlement, Population and Economy in Ancient Eretz Israel* (S. Bonimovich and M. Kohavi, eds.). Tel-Aviv: Tel-Aviv University (in Hebrew).
Portugali, J. (1989). Demographic speculations on evolution, settlement and urbanization. *Archaeologya 2* (in Hebrew).
Portugali, J. (1990). Preliminary notes on social synergetics, cognitive maps and environmental recognition. In *Synergetics of Cognition* (H. Haken and M. Stadler, eds.), pp. 379–392. Berlin: Springer.
Portugali, J. (1991). An Arab segregated neighborhood in Tel-Aviv: the case of Adjami. *Geography Research Forum* 11, 37– 50.
Portugali, J., ed. (1992). *Geography, Environment and Cognition.* A special theme issue, *Geoforum* 23(2).
Portugali, J. (1993). *Implicate Relations: Society and Space in the Israeli–Palestinian conflict.* Dordrecht, Boston, London: Kluwer Academic Publishers.
Portugali, J. (1994a). Theoretical speculations on the transition from nomadism to monarchy. In *From Nomadism to Monarchy* (I. Finkelstein and N. Na'aman, eds.), pp. 203–217. Jerusalem: Yad Izhak Ben-zvi.
Portugali, J. (1994b). The taming of the shrew environment. *Science in Context* 7(2), 307–326. Portugali, J. ed. (1996a). *The Construction of Cognitive Maps.* Dordrecht: Kluwer Academic Publishers.
Portugali, J. (1996b). Inter-representation networks and cognitive maps. In *The Construction of Cognitive Maps* (J. Portugali, ed.), pp. 11–43. Dordrecht, Boston, London: Kluwer Academic Publishers.
Portugali, J. (1996c). Notions concerning world urbanization. In *Contemporary Perspectives on Urbanization*, by Amos, F. J. C., Bourne, L. S. and Portugali J., *Progress in Planning* 46(3), 145–162.
Portugali, J. (1997). Self-organizing cities. *Futures* 29(4/5), 353–380.
Portugali, J. (1998). IRN and m-codes. Paris: Maison Suger.
Portugali, J. and Benenson, I. (1994). Competing order parameters in a self-organizing city. In *Managing and Marketing of Urban Development and Urban Life* (G. O. Braun, ed.), pp. 669–681. Berlin: Dietrich Reimer.
Portugali, J. and Benenson, I. (1995). Artificial planning experience by means of a heuristic cell-space model: simulating international migration in the urban process. *Environment and Planning*, A, 27, 1647–1665.
Portugali, J. and Benenson, I. (1997). Human agents between local and global forces in a self-organizing city. In *Self-Organization of Complex Structures: from Individual to Collective Dynamics* (F. Schweitzer, ed.), pp. 537–546. London: Gordon and Breach.
Portugali, J., Benenson, I. and Omer, I. (1994). Socio–spatial residential dynamics: stability and instability within a self-organizing city. *Geographical Analysis* 26(4), 321–340.

Portugali, J., Benenson, I. and Omer, I. (1997). Spatial cognitive dissonance and sociospatial emergence in a self-organizing city. *Environment and Planning, B: Planning and Design* 27, 263–285.

Portugali, J. and Haken, H. (1992). Synergetics and cognitive maps, In *Geography, Environment and Cognition* (J. Portugali, ed.), pp. 111–130. A special issue, *Geoforum* 23(2).

Portugali, J. and Gophna, R. (1993). Crisis, progress and urbanization: the transition from Early Bronze I to Early Bronze II in Palestine. Tel-Aviv 20(2), 164–186.

Prigogine, I. (1980). *From Being to Becoming.* San Francisco, CA: Freeman & Co.

Prigogine, I. and Stengers, I. (1984). *Order Out of Chaos.* New York: Bantam.

Pumain, D., Saint-Julien, T. and Sanders, L. (1987). Application of a dynamic urban model. *Geographical Analysis* 19(2), 152–166.

Redman, C. L. (1978). *The Rise of Civilization.* San Francisco, CA: Freeman & Co.

Reissman, L. (1970). *The Urban Process: Cities in Industrial Societies.* New York, London: The Free Press.

Relph, E. C. (1976). *Place and Placelessness.* London: Pion.

Rindos, D. (1984). *The Origins of Agriculture, an Evolutionary Perspective.* London: Academic Press.

Rosch, E., Mervis, C., Gray, W., Johnson, D. and Boyes-Braem, P. (1976). Basic objects in natural categories. *Cognitive Psychology* 8, 382–439.

Rosser, J. B. (1991). *From Catastrophe to Chaos: A General Theory of Economic Discontinuities.* Boston: Kluwer Academic Publishers.

Rumelhart, D. E., Smolensky, P., McClelland, J. L. and Hinton, G. E. (1986). Schemata and sequential thought processes in PDP models. In *Parallel Distributed Processing, Explorations in the Microstructure of Cognition. Vol. 2: Psychological and Biological Models* (J. L. McClelland, D. E. Rumelhart, and the PDP Research Group), pp. 7–57. Cambridge MA: MIT Press.

Ruthen, R. (1993). Adapting to complexity. *Scientific American* 268, 110–117.

Sanders, L., Pumain, D., Mathian, H., Guerin-Pace, F. and Bura, S. (1997). SIMPOP: a multiagent system for the study of urbanism. *Environment and Planning, B: Planning and Design* 24, 287–306.

Sanders. L. (1992). *Systems de Villes et Synergetique.* Paris: Anthropos.

Sanglier, M. and Allen, P. (1989). Evolutionary models of urban systems: an application to the Belgian provinces. *Environment and Planning*, A, 21, 477–498.

Schelling, T. (1974). On the ecology of micro-motives. In *The Corporate Society* (R. Marris, ed.). London: Macmillan.

Shannon, C. E. and weaver, W. (1959/1963). *The Mathematical Theory of Comminication.* Illinois: University of Illinois Perss.

Shilhav, Y. (1983). Communal conflict in Jerusalem – the spread of ultra orthodox neighborhoods. In *Pluralism and Political Geography: People, Territory and State* (N. Kliot and S. Waterman, eds.). London: Croom-Helm.

Simon, H. (1979). *The Science of the Artificial.* Cambridge, MA: MIT Press.

Smith, A. D. (1971). *Theories of Nationalism.* London: Duckworth.

Smith, H. and Williams, M. P. (1986). *Gentrification of the City.* Boston: Allen & Unwin.

Snow, C. P. (1964). *The Two Cultures and a Second Look.* Cambridge: Cambridge University Press.

Soja, E. W. (1989). *Postmodern Geographies.* London, New York: Verso.

Sonis, M. (1991). A territorial socio–ecological approach in innovation diffusion theory: socio–cultural and economic interventions of active environment into territorial diffusion of competitive innovations. *Sistemi urbani* 1-2-3, 29–59.

Stadler, M. and Kurse, P. (1990). The self organization perspective in cognition research: historical remarks and new experimental approaches. In *Synergetics of Cognition* (H. Haken and M. Stadler, eds.), pp. 32–52. Berlin: Springer.

The Concise Oxford Dictionary of Current English (1954). Oxford: Clarendon Press.

Thrift, N. (1983). On the determination of social action in space and time. *Environoment and Planning, D: Socity and Space* 1, 23–57.

Thünen, J. H. von (1826/1966). *von Thünen's Isolated State.* An English translation (P. Hall, ed.). Oxford: Pergamon.

Timmermans, H. (1990). Theoretical aspects of variety-seeking choice behaviour. In *Spatial Choices and Processes* (M. M. Fischer, P. Nijkamp and Y. Y. Papageorgiou, eds.). New York: Elsevier Science.

Tobler, W. R. (1979). Cellular geography. In *Philosophy in Geography* (S. Gale and G. Olsson, eds.). Dordrecht: Reidel.

Tuan, Yi-Fu (1977). *Space and Place: The Perspective at Experience.* London: Edward Arnold.

Toffoli, T. and Margolus, N. (1987). *Cellular Automata Machines: A New Environment for Modeling.* Cambridge, MA: MIT Press.

Tversky. A. (1977). Features of similarity. *Psychological Review* 84(4), 327–352.

Tversky, A. and Gati, I. (1978). Studies of Similarity. In *Cognition and Categorization* (E. Rosch and B. B. Lloyd. eds.), pp. 79–98. Hillsdale, NJ: Lawrence Erlbaum.

Tversky, A. and Kahneman, D. (1971). Belief in the law of small numbers. *Psychological Bulletin* 76(2), 105–110.

Tversky, A. and Kahneman, D. (1972). Availability: a heuristic for judging frequency and probability. *Cognitive Psychology* 4.

Tversky, A. and Kahneman, D. (1974). Judjment under uncertainty: Heuristics and biases. *Science* 185, 1124–1131.

Tversky, A. and Kahneman, D. (1981). The framing of decision and psychology of choice. *Science* 211.

Vygotsky, L. S. (1978). *Mind in Society.* Cambridge, MA: Cambridge University Press.

Wagenaar, W. A. (1993). Heuristics: simple ways for dealing with complex problems. Talk given at a symposium on Natural Science and Human Thought. Villa Vigoni, Italy.

Weber, A. (1929). *Theory of the Location of Industries.* Chicago: Chicago University Press (Translated by C. J. Friedrich).

Weber, M. (1909/1976). *The Agrarian Sociology of Ancient Civilizations.* London: NLB.

Weber, M. (1958). *The City.* New York: The Free Press.

Weidlich, W. (1983). A Non-linear Dynamic Model for the Migration of Human Population.

Weidlich, W. (1987). Synergetics and social science. In *Lasers and Synergetics* (R. Graham and A. Wunderlin, eds.), pp. 238–256. Berlin: Springer-Verlag.

Weidlich, W. (1994). Synergetic modeling concepts for sociodynamics with application to collective political opinion formation. *Journal of Mathematical Sociology* 1(4), 267–291.

Weidlich, W. (1999).From fast to slow processes in the evolution of urban and regional settlement structures: the role of population pressure. In *Population, Environment and Society on the Verge of the 21st Century* (J. Portugali, ed.). A special theme issue, *Discrete Dynamics in Nature and Society.*

Weidlich, W. and Haag, G. (1983). *Concepts and Models of a Quantitative Sociology.* Berlin, New York: Springer.

Wheeler, M. (1956). The first towns? *Antiquity* 30, 132–136.

White, R. and Engelen, G. (1993). Cellular automata and fractal urban form: a cellular modelling approach to the evolution of urban land use patterns. *Environment and Planning*, A, 25, 1175–1199.

White, R., Engelen, G. and Uljee, I. (1997). The use of constrained cellular automata for high-resolution modeling of urban land-use patterns. *Environment and Planning, B: Planning and Design* 24, 323–343.

Wilson, A. G. (1970). *Entropy in Urban and Regional Modelling*. London: Pion.

Wilson, A. G. (1981). *Catastrophe Theory and Bifurcation*. London: Croom Helm.

Wirth, (1938). Urbanism as a Way of Life. *American Journal of Sociology* XLIV, 1–24.

Wischert, W. and Wunderin, A. (1993). On the application of synergetics to social systems. In *Interdisciplinary Approaches to Non-Linear Complex Systems* (H. Haken and A. Mikhailov, eds.). Berlin: Springer.

Wittgenstein, L. (1953). *Philosophical Investigations*. Oxford: Blackwell. (Translated by G. E. M. Anscombe).

Wolch, J. and Dear, M. (1989). *The Power of Geography*. London: Unwin Hyman.

Wolfram, S. (1984). Universality and complexity in cellular automata. *Phisica* D, 10, 1–35.

Wolfram, S. (1985). Some recent results and questions about cellular automata. In *Dynamical Systems and Cellular Automata* (J. Demongeot, E. Goles and M. Tchuente, eds.). London: Academic Press.

Woo, E. S. W. (1994). Urban development. In Y. M Yeung and D. K. Chu. *Guangdong: Survey of Province Undergoing Rapid Change*. Hong Kong: Chinese University Press.

Woods, R. I. (1980). Spatiotemporal models of ethnic segregation and their implications for housing policy. *Environment and Planning*, A , 13, 1415–1433.

Wu, F. (1996). A linguistic cellular automata simulation approach for sustainable land development in a fast growing region. *Compute., Environ. and Urban Systems* 20, 367–387.

Wynn, M. (1985). *Planning Games*. London, New York: E. & F. N. Spon.

Yewlett, C. J. L. (1985). Rationality in Planning: A Professional Perspective. In *Rationality in Planning* (M. Breheny and A. Hooper, eds.), pp. 209–228. London: Pion.

Yovits, M. C. and Cameron, S., eds. (1959). /it Self-Organizing Systems. New York: Pergamon.

Zhang, W. B. (1989). Coexistence and separation of two residential groups a An interactional spatial dynamic approach. *Geographical Analysis* 21(2), 91–102. Zhang, W. B. (1990). Stability versus instability in urban pattern formation. *Socio-Spatial Dynamics* 1(1), 41–56.

Index

Springer Series in Synergetics

Synergetics An Introduction 3rd Edition
By H. Haken

Synergetics A Workshop
Editor: H. Haken

Synergetics Far from Equilibrium
Editors: A. Pacault, C. Vidal

Structural Stability in Physics
Editors: W. Güttinger, H. Eikemeier

Pattern Formation by Dynamic Systems and Pattern Recognition
Editor: H. Haken

Dynamics of Synergetic Systems
Editor: H. Haken

Problems of Biological Physics
By L. A. Blumenfeld

Stochastic Nonlinear Systems
in Physics, Chemistry, and Biology
Editors: L. Arnold, R. Lefever

Numerical Methods in the Study of Critical Phenomena
Editors: J. Della Dora, J. Demongeot, B. Lacolle

The Kinetic Theory of Electromagnetic Processes By Yu. L. Klimontovich

Chaos and Order in Nature
Editor: H. Haken

Nonlinear Phenomena in Chemical Dynamics Editors: C. Vidal, A. Pacault

Handbook of Stochastic Methods
for Physics, Chemistry, and the Natural Sciences 2nd Edition
By C. W. Gardiner

Concepts and Models of a Quantitative Sociology The Dynamics of Interacting Populations By W. Weidlich, G. Haag

Noise-Induced Transitions Theory and Applications in Physics, Chemistry, and Biology By W. Horsthemke, R. Lefever

Physics of Bioenergetic Processes
By L. A. Blumenfeld

Evolution of Order and Chaos
in Physics, Chemistry, and Biology
Editor: H. Haken

The Fokker-Planck Equation
2nd Edition By H. Risken

Chemical Oscillations, Waves, and Turbulence By Y. Kuramoto

Advanced Synergetics
2nd Edition By H. Haken

Stochastic Phenomena and Chaotic Behaviour in Complex Systems
Editor: P. Schuster

Synergetics – From Microscopic to Macroscopic Order Editor: E. Frehland

Synergetics of the Brain
Editors: E. Başar, H. Flohr, H. Haken, A. J. Mandell

Chaos and Statistical Methods
Editor: Y. Kuramoto

Dynamics of Hierarchical Systems
An Evolutionary Approach
By J. S. Nicolis

Self-Organization and Management of Social Systems Editors: H. Ulrich, G. J. B. Probst

Non-Equilibrium Dynamics in Chemical Systems
Editors: C. Vidal, A. Pacault

Self-Organization Autowaves and Structures Far from Equilibrium
Editor: V. I. Krinsky

Temporal Order Editors: L. Rensing, N. I. Jaeger

Dynamical Problems in Soliton Systems
Editor: S. Takeno

Complex Systems – Operational Approaches in Neurobiology, Physics, and Computers Editor: H. Haken

Dimensions and Entropies in Chaotic Systems Quantification of Complex Behavior 2nd Corr. Printing
Editor: G. Mayer-Kress

Selforganization by Nonlinear Irreversible Processes
Editors: W. Ebeling, H. Ulbricht

Instabilities and Chaos in Quantum Optics
Editors: F. T. Arecchi, R. G. Harrison

Nonequilibrium Phase Transitions in Semiconductors Self-Organization Induced by Generation and Recombination Processes By E. Schöll

Temporal Disorder in Human Oscillatory Systems
Editors: L. Rensing, U. an der Heiden, M. C. Mackey

The Physics of Structure Formation
Theory and Simulation
Editors: W. Guttinger, G. Dangelmayr

Computational Systems – Natural and Artificial Editor: H. Haken

From Chemical to Biological Organization Editors: M. Markus, S. C. Müller, G. Nicolis

Information and Self-Organization
A Macroscopic Approach to Complex Systems 2nd Edition By H. Haken

Propagation in Systems Far from Equilibrium Editors: J. E. Wesfreid, H. R. Brand, P. Manneville, G. Albinet, N. Boccara

Neural and Synergetic Computers
Editor: H. Haken

Cooperative Dynamics in Complex Physical Systems Editor: H. Takayama

Optimal Structures in Heterogeneous Reaction Systems Editor: P. J. Plath

Synergetics of Cognition
Editors: H. Haken, M. Stadler

Theories of Immune Networks
Editors: H. Atlan, I. R. Cohen

Relative Information Theories and Applications By G. Jumarie

Dissipative Structures in Transport Processes and Combustion
Editor: D. Meinköhn

Neuronal Cooperativity
Editor: J. Krüger

Synergetic Computers and Cognition
A Top-Down Approach to Neural Nets
By H. Haken

Foundations of Synergetics I
Distributed Active Systems 2nd Edition
By A. S. Mikhailov

Foundations of Synergetics II
Complex Patterns 2nd Edition
By A. S. Mikhailov, A. Yu. Loskutov

Synergetic Economics By W.-B. Zhang

Quantum Signatures of Chaos
By F. Haake

Rhythms in Physiological Systems
Editors: H. Haken, H. P. Koepchen

Quantum Noise 2nd Edition
By C. W. Gardiner, P. Zoller

Nonlinear Nonequilibrium Thermodynamics I Linear and Nonlinear Fluctuation-Dissipation Theorems
By R. Stratonovich

Self-organization and Clinical Psychology Empirical Approaches to Synergetics in Psychology
Editors: W. Tschacher, G. Schiepek, E. J. Brunner

Nonlinear Nonequilibrium Thermodynamics II Advanced Theory
By R. Stratonovich

Limits of Predictability
Editor: Yu. A. Kravtsov

On Self-Organization
An Interdisciplinary Search for a Unifying Principle
Editors: R. K. Mishra, D. Maaß, E. Zwierlein

Interdisciplinary Approaches to Nonlinear Complex Systems
Editors: H. Haken, A. Mikhailov

Inside Versus Outside
Endo- and Exo-Concepts of Observation and Knowledge in Physics, Philosophy and Cognitive Science
Editors: H. Atmanspacher, G. J. Dalenoort

Ambiguity in Mind and Nature
Multistable Cognitive Phenomena
Editors: P. Kruse, M. Stadler

Modelling the Dynamics of Biological Systems
Editors: E. Mosekilde, O. G. Mouritsen

Self-Organization in Optical Systems and Applications in Information Technology 2nd Edition
Editors: M.A. Vorontsov, W. B. Miller

Principles of Brain Functioning
A Synergetic Approach to Brain Activity, Behavior and Cognition
By H. Haken

Synergetics of Measurement, Prediction and Control By I. Grabec, W. Sachse

Predictability of Complex Dynamical Systems
By Yu. A. Kravtsov, J. B. Kadtke

Interfacial Wave Theory of Pattern Formation
Selection of Dentritic Growth and Viscous Fingerings in Hele-Shaw Flow By Jian-Jun Xu

Asymptotic Approaches in Nonlinear Dynamics
New Trends and Applications
By J. Awrejcewicz, I. V. Andrianov, L. I. Manevitch

Brain Function and Oscillations
Volume I: Brain Oscillations.
Principles and Approaches
Volume II: Integrative Brain Function.
Neurophysiology and Cognitive Processes
By E. Başar

Asymptotic Methods for the Fokker–Planck Equation and the Exit Problem in Applications
By J. Grasman, O. A. van Herwaarden

Analysis of Neurophysiological Brain Functioning Editor: Ch. Uhl

Phase Resetting in Medicine and Biology
Stochastic Modelling and Data Analysis
By. P. A. Tass

Self-Organization and the City By J. Portugali

Printing (Computer to Film): Saladruck, Berlin
Binding: Stürtz AG, Würzburg